RENEWABLE ENERGY: RESEARCH, DEVELOPMENT AND POLICIES

RENEWABLE ENERGY AND SUSTAINABLE DEVELOPMENT

RENEWABLE ENERGY: RESEARCH, DEVELOPMENT AND POLICIES

Additional books in this series can be found on Nova's website under the Series tab.

Additional e-books in this series can be found on Nova's website under the e-book tab.

RENEWABLE ENERGY: RESEARCH, DEVELOPMENT AND POLICIES

RENEWABLE ENERGY AND SUSTAINABLE DEVELOPMENT

R. T. DURAI PARBHAKARAN
SANDIP A. KALE
AND
KANDASAMY PRABAKAR
EDITORS

nova publishers
New York

Library of Congress Cataloging-in-Publication Data

Renewable energy and sustainable development (Hauppauge, N.Y.)
Renewable energy and sustainable development / editors, R. T. Durai Prabhakaran, Sandip A. Kale and Kandsamy Prabakar (University of Strathclyde, Glasgow, UK; Trinity College of Engineering and Research, University of Pune, India and Pusan National University, Korea)
pages cm
Includes bibliographical references and index.
ISBN 978-1-63463-429-8 (hardcover)
1. Renewable energy sources. I. Thirumalai, Durai Prabhakaran Raghavalu, editor. Kale, S. A., editor. III. Prabakar, K., editor. IV. Title.
TJ808.R41726 2014
621.042--dc23

2014041532

Published by Nova Science Publishers, Inc. † New York

CONTENTS

PREFACE

Contribution of renewable energy to human life is essential for sustainable life on the earth. The renewable energy industry is growing rapidly to fulfill the energy demand of continuously developing world and has become the focal centre of many researchers across the globe. This development should meet the needs of the present without compromising the ability of future generations to meet their own needs. Sustainable development ties together concern for the carrying capacity of natural systems with the social challenges faced by humanity. This book is intended to highlight various aspects for applications of renewable energy and sustainability.

The solar greenhouse technology used in agriculture brings a new revolution to increase the productivity of various products, employing improved environmental conditions. This aspect is dealt in first chapter. The technology of extension of flat surface is introduced in second chapter for improving the efficiency of flat plate collectors. Third chapter deals with oldest and cleanest source of wind energy. Here a deep thought is extended towards the development of small wind turbine blade materials.

Contribution of Bio energy is marvelous and is available in abundance. There are various methods such as converting biodegradable materials into best quality fertilizer, generating a biogas which can directly contribute to food cooking, this gas also can be used to burn in engines and generate electricity, manufacturing of biodiesels with the use of various bio-products e.g. jatropha leaves, babhool, tobacco etc. Various concepts to this effect are added in subsequent chapters four to eight.

Chapters nine and ten deals with estimating the convertibility of tidal energy which is one of the cleanest and amply available resource in order to create self sufficiency of energy needs. A similar effort is thought for use of geothermal energy in Indian territory, in chapter eleven. A research work is presented in chapter twelve for the performance improvement in hydrogen cells. This source of energy is most suitable for contributing to human life.

In order to make use of these various types of energies for well being of human and other lives on the earth, capacity to store the energy is vital aspect of this process. More the capacity to store the energy in one or either form and retrievable at any time and need, more is the beneficial for sustainability of life. This aspect is explained in the chapter thirteen.

Different considerations are added in chapters fourteen to sixteen, to develop products that contribute for the sustainability of life. Their importance cannot be overemphasized with what is now realized by all the need to practice environmentally friendly practices of product making. Recycling of waste products and converting in to usable products, development of products that needs a minimal amount of energy, or improving the product yield which can contribute for sustainable solutions etc are some subjects which are discussed in these chapters.

We hope that this book will be useful to the researchers, students and technosavi personalities to enrich their concepts and informations. It may also bring inspirational ideas to start new thoughts in the direction of Sustainable developments in their local areas and global aspects too.

R. T. Durai Prabhakaran
Sandip A. Kale
Kandasamy Prabakar

In: Renewable Energy and Sustainable Development
Editors: R. T. Durai Prabhakaran, S. A. Kale et al.
ISBN: 978-1-63463-429-8

Chapter 1

SOLAR GREENHOUSE TECHNOLOGY

A. Ganguly*
Mechanical Engineering Department,
Indian Institute of Engineering Science and Technology (IIEST),
Shibpur, Howrah, West Bengal, India

ABSTRACT

This chapter presents a comprehensive overview of state of the art of solar greenhouse technology with its global and Indian scenario. The different components of the microclimate in a greenhouse are also discussed in details. The various methods of heating, cooling and ventilation technologies used for cultivation of flora in passive and active greenhouses are also highlighted. Subsequently, the various shapes of greenhouse structures and their covering materials have been touched upon in brief. The various modes of heat transfer in a greenhouse structure are also discussed in a separate section. Finally, the concept and the need for establishment of a stand-alone grid independent solar greenhouse has been emphasized upon along with scheme of such a greenhouse powered through solar hydrogen system.

1. INTRODUCTION

Greenhouse technology is an effective method of cultivation of crops, flowers, medicinal plants, etc. under a controlled environment. Solar greenhouses are enclosures covered with suitable transparent materials in which the climatic conditions are artificially regulated for optimum growth of plants. The extent of environmental control required in a greenhouse for cultivation of a plant varies according to the geographical location of the place, type of plant, its stage of growth, prevailing ambient climate, etc. There are several factors that influence the growth of plants such as temperature, humidity, light intensity, carbon dioxide level, etc and the plant growth get retarded or even stopped if one of the factors becomes restrictive. In a greenhouse these factors are taken care of artificially to optimize the growth of plants, which is not possible in open field cultivation.

* Correspondence to: aritra@mech.becs.ac.in.; Ph-+91-33-26684561-63, Fax-+91-33-26682916/4564.

2. History of Solar Greenhouse

The technology of cultivation of plants under shade is not recent. Lemmon recorded the writings of Plato, who in the 4th century B.C. indicated that the plants were grown under cover [1]. In 1385, in France, flowers were cultivated in glass pavilions facing south. Between 1801 and 1805, Czar Alexander-I erected three parallel greenhouses, connecting their ends with two more buildings at Saint Petersburg [1]. The enormous structure was heated using wood burning furnaces. In 19th century the concept of greenhouse cultivation spread throughout England and people constructed greenhouses along with their residential houses which are now called as Solarium. Taft in 1926 provided information on glasshouse constructed in Apothecaries Garden, Chelsea, England, which had tall sidewalls of glass with opaque roofs [1]. By the middle of the 19th century, this technology developed maturity and different varieties of greenhouses were constructed mostly in modular form to permit easy expansion and designed for do it yourself assembly. In 20th century automatic control systems were introduced in greenhouse systems and by 1970, it was possible to build a small greenhouse with automatic environmental controls for less than one third the cost of a small car in western countries of the world [1].

3. Global and Indian Scenario

The technology for solar greenhouse has developed quite rapidly over the years in the different countries of the world to produce vegetables, flowers and medicinal plants under controlled environmental condition. In most of the developed nations the technological development of solar greenhouse systems happened earlier due to their need to save seedlings, plants, vegetables, etc. from extreme cold. At present there are more than fifty countries in the world [1], where cultivation of crops is undertaken on a commercial scale in greenhouses. United States of America has a total area of about 4000 ha under greenhouses, mostly used for floriculture. They have a turnover of more than 2.8 billion US $ per annum [2]. The area under greenhouse cultivation in Spain has been estimated to be around 28350 ha and Italy 26,600 ha mostly used for cultivation of vegetables [1]. In Canada, the greenhouse industry caters both to the flower and off-season vegetable demands in the market. The Netherlands has about 9,600 ha of land under cover and the Dutch greenhouse industry is probably the most advanced in the world [3]. Dutch greenhouse industry however relies heavily on glass framed structures to cope up with cloudy conditions prevailing there for a considerable part of the year. The development of greenhouses in Gulf countries is primarily due to the extremity in climatic conditions. In Saudi Arabia, cucumbers and tomatoes are the crops widely grown in greenhouses mostly equipped with evaporative cooling systems. There they contribute more than 94% of the total production [2]. Egypt has about 1000 ha solar greenhouses mainly made of plastic covered tunnel structures [1]. They use natural ventilation to regulate the inside temperature and humidity for cultivation of crops like tomatoes, cucumbers, melons, etc.

In Asia, China and Japan are the largest users of greenhouses. The technological development in this field is faster in China than in any other country of the world. With a modest beginning in the late seventies, the area under greenhouse cultivation in China has increased to 48000 ha in recent years. Of this 11000 ha is reserved for cultivation of fruits like grapes, cherry, lemon and mango. Japan has more than 40,000 ha under greenhouse cultivation, of which nearly 7500 ha is dedicated only to the cultivation of fruits. Greenhouses in Japan also produce a wide range of vegetables and flowers which meet a considerable share of their total market requirement. Thus, greenhouses are used for cultivation of plants in countries where winter is severe such as in Canada and also permit production in countries where the summer is extreme like that of in the Gulf. Even in countries like the Philippines where there are excessive rains, greenhouses are widely used for cultivation of crops. Thus, in essence greenhouse cultivation is being practiced and possible in all types of climatic conditions.

While greenhouses existed for more than several centuries in various countries of the world, in India this technology started only during 1980s and was mainly used for research purpose. The National Committee on the use of Plastics in Agriculture recommended location specific trials of greenhouses in various regions of the country in 1982 [2]. The establishment of greenhouses for commercial purpose started only in the late eighties. However, in recent years in view of the increased access to international market due to globalization and the increased export potential for agricultural products, there has been a surge in the demand for greenhouse technology in India. But in spite of this, the application of greenhouse technology for commercial purpose in our country still remains insignificant. As available in literature [1], only 1500 ha land is under greenhouse cultivation in India and they are mostly restricted to those parts of the country, where the winter is prolonged and where continuous rain poses threat to cultivate during the monsoons.

In the plains of India, however the climatic and economic conditions are different from those prevailing in the western countries of the world. The solar radiation is abundant and the climate remains hot during most of the months of a year. Comparatively cooler climate is witnessed during the winter season that usually lasts for two months, namely December and January. In addition to high temperature, the coastal areas of the country also witness higher humidity. While excessive heat is detrimental to the growth of crops and flowers, high humidity increases the chance of attack of pests that destroy vegetation in the open field. So, for greenhouses established in the plains of India, cooling and humidity control are the major factors influencing the production of quality crops and flowers. Thus, certain modifications need to be incorporated in traditional greenhouse technology to be used under Indian scenario. The greenhouse designed for a hot and humid climate prevailing in the plains of Indian subcontinent thus requires suitable cooling and dehumidifying arrangement that help to maintain a specified range of temperature and humidity level inside the greenhouse. Cooling can be achieved by applying a measured combination of evaporative cooling and shading mechanisms, while humidity of the greenhouse space can be regulated primarily through ventilation. The ventilation may be fan induced or natural as either can be employed depending upon the requirement.

4. Factors Influencing the Greenhouse Environment

The major environmental factors that influence the plant growth are:

- Temperature
- Light
- Humidity
- Air Composition

4.1. Temperature

Temperature considerably influences the growth of plants that can be cultivated in a particular geographical area. Temperature influences the process of photosynthesis, respiration, ion uptake, transpiration, elongation, etc. Over a limited range at which plant growth will occur, chemical process gets doubled for every 10^0C rise in temperature. Both the rates of photosynthesis and respiration increase with rise in temperature. But the process of photosynthesis also gets limited by the available energy and the concentration of carbon dioxide. An increase in the temperature of the leaf leads to increase its vapour pressure. This increases the rate of plant transpiration assuming the outside vapour pressure remains constant. The effect of temperature on the elongation process varies with the part of the plant. At a given temperature, the rate of elongation of the plant root will not be same as that of the plant stem or that for other parts. When the whole plant is concerned, the maximum growth or elongation will result in some intermediate temperature. Optimum temperature refers to the temperature at which the growth of a plant is the best for a particular climatic condition. This temperature varies with the plant species, its age and the stage of growth. Most plant species also require diurnal fluctuation. The table 1 shows the optimum temperature requirement for few floriculture and horticulture plants grown in greenhouses:

Table 1. Optimum temperature for certain plants cultivated in greenhouse [1]

Plant	Optimum temperature (^0C)	Remarks
Rose	Day temperature: 21, 24, 27, 29 Night temperature: 16-17	Lowest day value for cloudy days, highest value for CO_2 injection.
Cauliflower	27	Germination temperature
Onion	24	Germination temperature
Cabbage	29	Germination temperature
Lettuce	24	Germination temperature
Water Melon	35	Germination temperature

4.2. Light

Light is essential for photosynthesis. Green plants transfer energy from the sun to a usable form of energy for the plant. Light varies in intensity, quality and duration from place to place. Generally, it starts from zero at the beginning of the day, while it reaches a value of

150,000 Lux (Lumen/m^2) during the noontime. The intensity of light available has a direct influence on the process of photosynthesis. Light intensity on cloudy days is quite low and during the night no sunlight is available. So during that period the plant growth gets inhibited. Every plant requires an optimal value of light intensity for its maximum growth. But this varies considerably from plant to plant. Some plants like cucumber need high light intensity, while other plants like spinach, lettuce, tomato, etc. can be grown at lower intensities of light. Since the greenhouse structure does not allow all the incident solar radiation to pass through its coverings, the light intensity in a greenhouse even on a clear day rarely exceeds 50000 Lux. Therefore, more than two glazes are not recommended in a greenhouse as that does not allow sufficient light to reach the inside plantation zone. In areas where the intensity of light is low, the same needs to be increased by using large area of the glazing, selecting a suitable glass or plastics with high transmissivity value for light, orienting the glazing properly and keep them clean (dust free). But it may be noted that additional light beyond a certain limit does not help in the faster growth of the plants.

Supplementary lighting may also be provided using incandescent, fluorescent and High Intensity Discharge (HID) lamps. Incandescent lamps are not preferable for greenhouse applications. Fluorescent lamps with some kind of reflectors are used widely. The HID lamps like High Pressure Mercury Discharge type or High Pressure Sodium Type are widely recommended for lighting in greenhouses as those convert a large fraction of electrical energy into visible light. The solar radiation received on the surface of the earth lies in the wavelength ranging between 0.2 to 3 μ m. In the process of photosynthesis, plants require light mainly in the visible and the infrared spectrum only. It is observed that the rate of photosynthesis is the highest under red and blue colored light. But it is also true that if the plants are exposed to more blue light, their growth gets retarded and they develop a dark color. On the other hand if the plant receives more infrared light, then their internodes become longer which results in taller plants. Ultraviolet radiation is detrimental to plants and may even kill them. So, their entry into the greenhouse structure is restricted by using glazing materials with UV inhibitors.

The photoperiod that is the relative duration of day-night also influences the initiation of flowers and fruits, bulb and tuber formation, etc. There are certain varieties of plants like Spinach and Radish which flower only when the day length is longer than that of night, while there are some other varieties like Chrysanthemum that flower only when the duration of the day is short. Again there are flowers like Rose which does not respond to the variation in day-night duration. It may be noted that the photoperiod can be regulated (increased or decreased) to some extent in greenhouses using artificial light or using shading screens.

4.3. Humidity

Humidity is an important element of the greenhouse microclimate apart from that of temperature and light. Humidity is the amount of moisture present in one kg of air. The relative humidity requirements vary widely between 30 to 80% depending upon the plant species and its stage of growth. There are certain plants like Gerbera, Anthurium, etc. that can grow well even under humidity levels of about 85%. The saplings and the germinating seeds need very high humidity, but as they grow, the relative humidity requirement reduces. High humidity is often related to the development of certain plant diseases. Warm, damp conditions promote the growth of fungi which are highly undesirable from the viewpoint of plant growth.

The relative humidity can be artificially regulated in a greenhouse. In arid climatic regions, it is often necessary to increase the inside relative humidity. This is commonly done using evaporative cooling systems like fan-pad, fogging, misting, etc. in which the relative humidity is increased, while the dry bulb temperature decreases. This is widely used for plants like Carnation, Chrysanthemum, Rose, Gerbera, etc. As the relative humidity drops, the rate of evaporation increases and the plants give up more water through transpiration. Transpiration also increases with increase in the velocity of the air. Unlike that of temperature, accurate humidity control is expensive and difficult in greenhouses. Generally in greenhouses, the rate of ventilation is varied to control the level of humidity. Desiccators may also be employed to regulate humidity in greenhouses. But the desiccators also get saturated with moisture periodically and thus needs to be regenerated. Solar thermal energy may be used for regeneration of the desiccator wheels for regions where the solar radiation is abundant.

4.4. Air Composition

Gaseous composition of the air surrounding a plant are Carbon dioxide (CO_2), Oxygen(O_2), Nitrogen(N_2), Water Vapor and other gases in traces which influence the plant metabolism significantly. Among all the above mentioned gases, CO_2 is the main constituent which influences the rate of photosynthesis directly and ultimately helps in growth, yield, and quality of the flora. Several research workers have shown that the closed greenhouse system offers good opportunity to improve production through the elevation of CO_2 level. Various CO_2 enrichment methods are used like combustion, use of liquid and solid CO_2, etc. The enrichment activity if required is carried out during the daytime when solar radiation is present. Normally, carbon dioxide enrichment begins at or little after sunrise and continue till an hour before the sunset. The enrichment is seldom practiced during ventilation. Plants are influenced simultaneously by all segments of the environment. As solar radiation intensity increases, the greenhouse air temperature increases with a corresponding increase in the rate of photosynthesis. At the same time loss of water also gets affected through the phenomenon of transpiration due to the change in humidity. Thus, it is necessary to consider all the environment factors and need to derive an optimum environmental condition for greenhouse plants.

5. Passive and Active Solar Greenhouses

Based on the working principle, greenhouses may be classified into passive and active greenhouses. Passive greenhouses are those which do not require any energy for their operation. Due to the incident solar radiation, temperature gradients are established which help to maintain natural ventilation, discharging the hot air from the top of the greenhouse and taking in fresh air through side openings. Most of the greenhouses in Egypt are passive type [1]. In active greenhouses, external energy is provided inside the greenhouse structure to maintain a suitable microclimate. The energy may be available by burning a conventional fuel if the greenhouse requires heating or may be supplied in the form of electrical energy to drive the fans, pumps, etc. to maintain the temperature and humidity levels within the permissible range.

5.1. Greenhouse Heating and Cooling System

On the basis of operation, greenhouses may be classified as those that require heating and those which require cooling. Each of them is discussed here.

5.1.1. Heating System

The heating system is required in cold countries where ambient temperature is very low especially during the winter season. Different types of heating arrangements may be used in greenhouses to maintain a suitable inside microclimate:

i) Direct Gain

In this method the greenhouse is designed in such a manner that the maximum solar radiation gets admitted in winter through the roof and walls to heat the internal environment. The glazing surfaces are covered by movable insulation at night to reduce the heat loss from the greenhouse to the ambient.

ii) Indirect Gain

Many times, in direct heat gain method, the fluctuation in greenhouse air temperature is higher than the allowable limits for the optimum growth of the plants. A more effective method for reducing the swings in greenhouse air temperature is to introduce a thermal storage wall on the north side of the greenhouse or constructing a massive wall made up of drums full of water stacked over each other known as water wall. Water wall is more effective in reducing the temperature swings inside a greenhouse.

iii) Rock Bed Storage

In this type of configuration a solar collector is connected to the greenhouse structure through the rock bed. The solar energy entering the greenhouse will heat the air and pass through the rock bed, which stores the energy in the form of heat and releases the same slowly during the off-sunshine hours, thereby reducing the temperature swings inside the greenhouse.

iv) Conventional Heating System

This method of heating is very common in western countries of the world in which conventional fuels like wood, coal coke, oil, etc. are burned inside the greenhouse structure to produce thermal energy. The burning of those fuels also generates carbon dioxide (CO_2) which helps to maintain a suitable level of CO_2 inside the greenhouse for the optimum growth of plants. But one needs to be careful regarding the emission of carbon monoxide which is poisonous. So, it must be ensured that the combustion is complete to eliminate the emission of carbon monoxide gas.

v) Heat Pump

Heat pumps are also used for heating of greenhouses. The schematic of a greenhouse integrated with a heat pump is shown in Fig. 1. The system has two heat exchangers, one is connected to the cold water (CW) insulated tank and the other to the warm water (WW) insulated storage tank. Both the water storage tanks are connected to the heat pump. The water at a temperature of 2-6^{0}C from the cold water storage tank is allowed to pass through a heat exchanger placed inside the greenhouse during the sunshine hours. The water gets heated as the greenhouse acts as a solar collector, Then it is returned to the cold water storage tank.

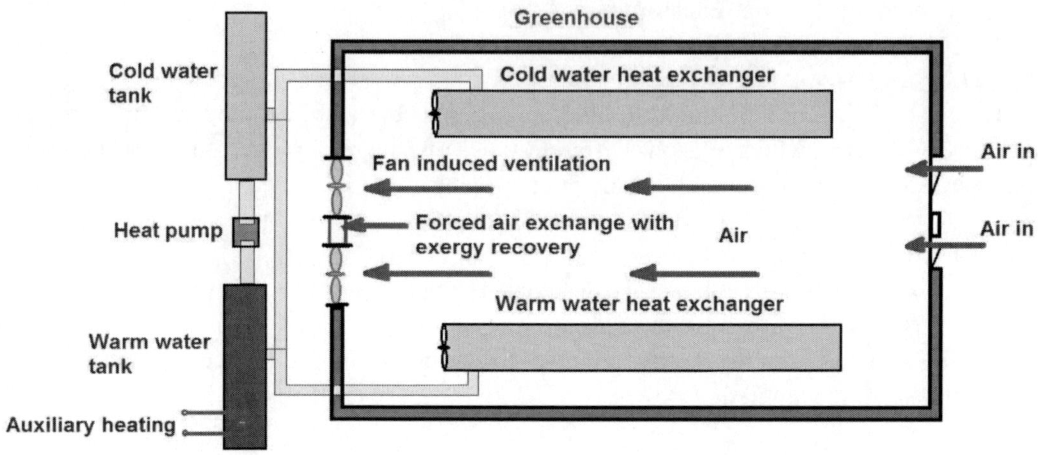

Figure 1. Schematic of a greenhouse integrated with heat pump.

Thus, the water in the cold storage tank gets heated to a temperature of 18-24^{0}C. The water from cold water tank is transferred to warm water tank through the heat pump which is sufficient to heat the greenhouse air during the night through another heat exchanger placed inside the greenhouse. An exhaust fan is also provided for ventilation to control the humidity and the carbon dioxide level. The system was tested in Hannover, Germany over a period of one year [1].

The other possibilities for thermal heating of a greenhouse are as follows:

i) Solar collector integrated into a greenhouse structure under the roof in passive mode of operation.
ii) The use of the greenhouse as a collector to store the heat into storage tank provided below the floor.

5.1.2. Cooling System

These are necessary for solar greenhouses located in equatorial and tropical countries that witness a hot climate for the greater part of a year. During the summer, high intensity of solar radiation increases the air temperature inside the greenhouse to an undesirable level owing to the greenhouse effect. Due to this, the inside environment becomes unfavorable for cultivation of plants unless some method of cooling is applied. The passive and low energy cooling systems applied for buildings can also be used in the greenhouse system as those are simple and inexpensive. However, this can restrict the temperature rise in a greenhouse to a limited extent.

Evaporative cooling systems are widely used in greenhouses to maintain a suitable inside temperature. They are based on conversion of sensible heat to latent heat of evaporated water, where water is supplied mechanically. The temperature of air is reduced by evaporation of water in the air and thus the temperature is reduced at the expense of increase in humidity. The enthalpy of air remains constant in the process. At present the evaporative cooling methods include the fan-pad, fogging/misting and the roof evaporative cooling system. Here all these methods have been discussed with their advantages and limitations. The various active methods of evaporative cooling systems used for greenhouse application are described below:

i) Fan-Pad Evaporative Cooling System

In this type of system induced draught fan(s) are installed at one end of the greenhouse and a cellulose pad on the opposite end of the greenhouse. Water is circulated through the pad to keep it wet. Air is forced to pass through the wet pad openings due to suction from induced draught fan(s). Thus, due to the evaporation of water in the flowing air, the temperature of air leaving the pad is reduced, but at the cost of increase in humidity. In this type of system, thus a temperature and humidity gradient is developed from the pad to the fan end. The temperature just after the pad is minimum and maximum at the fan end, while the humidity is maximum just after the pad and minimum near the fan end. The temperature and humidity gradient thus generated is undesirable as the plants remain in a stressed condition. Also, continuous operation and presence of impurities and salt in water leads to progressive choking of the cooling pad, thereby reducing the effectiveness of the system. Also, this method of cooling requires uninterrupted electric supply to drive the fans and the water pump which becomes a major constraint for its application in rural areas, especially for developing countries like India where a considerable number of villages do not have access to un-interrupted electric supply. To reduce the problem of temperature and humidity gradient along the length of the greenhouse, distributed fan-pad evaporative cooling systems are used where the air is admitted through the side wall openings and is made to flow upwards due to the suction of the induced draught fans placed at the top of the greenhouse.

ii) Fogging/Misting

This method of evaporative cooling uses very small water droplets (2-60 μ m in diameter) to be sprayed into the greenhouse air under high pressure through the nozzles to increase its contact area with air. A fraction of small water droplets evaporate while coming in contact with air and due to high latent heat of vapourization of water, there is reduction in air temperature. Fogging/Misting thus helps to lower the temperature inside a greenhouse and increase humidity. Also, here high temperature gradient is not generated as that obtained in fan-pad ventilated greenhouse. This can also be used with natural ventilation system. The drawbacks of this system are the choking of nozzles with passage of time thereby reducing the effectiveness of cooling and increased chances of attack by microorganisms and pests due to condensation of fine water droplets on the petals of the flowers.

iii) Roof Evaporative Cooling System

Roof evaporative cooling is the process of circulating water on the roof surface resulting in the formation of a water film over the roof surface. This method helps to lower the sensible heat gain of the greenhouse air, thereby reducing its temperature. This is an effective method of reducing the temperature inside a solar greenhouse without altering the inside relative humidity. Figure 2 shows the schematic of a greenhouse with roof evaporative cooling.

iv) Earth to Air Heat Exchanger

Due to constant year round temperature in the range of 26-28^{0}C, the ground potential of earth can be used for cooling of greenhouses in those parts of the year when the ambient temperature is very high. In earth to air heat exchangers, the warm greenhouse air is passed through buried pipes (2-4m depth), where its heat is dissipated to the underground soil and the same can be utilized again. Earth to air heat exchangers integrated with greenhouse can reduce the inside temperature by 3-4^{0}C. But it is observed that such systems face a number of operational problems apart from high installation cost associated with digging of soil. Also, the maintenance cost with those systems is quite high as the buried pipes are subjected to corrosion.

v) Use of Phase Change Materials (PCM)

This is a recent method where the maximum temperature change in a greenhouse can be reduced by storing the energy in a suitable phase change material. It has been observed that the temperature swing between the maximum and minimum values during 24 hours can be reduced by 3-5^{0}C by using a phase change material.

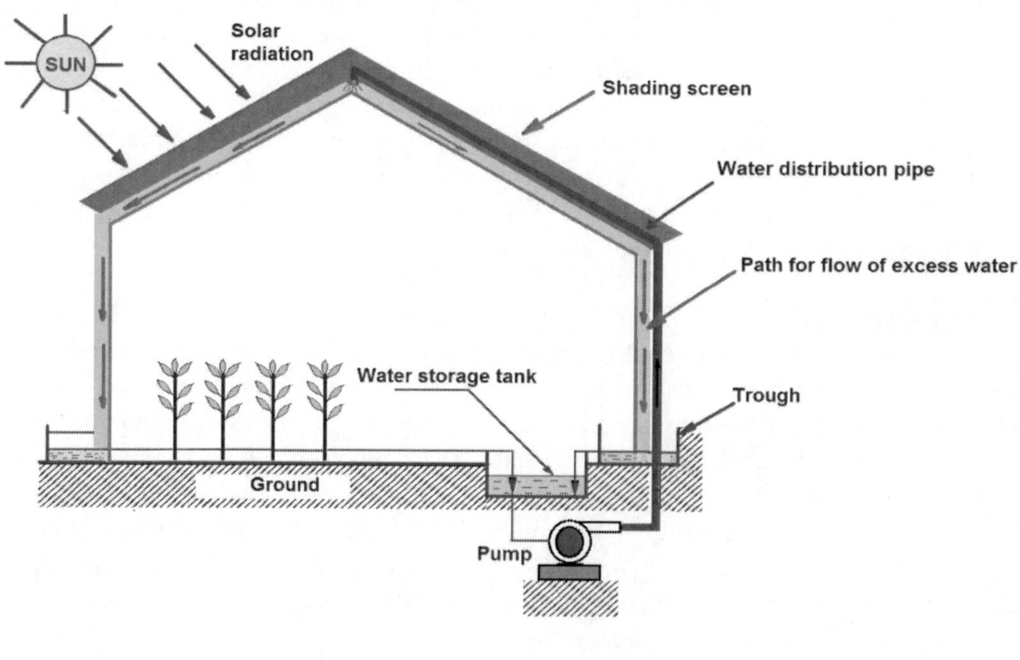

Figure 2. Schematic of a greenhouse with roof evaporative cooling.

vi) Aquifer Coupled Cavity Flow Heat Exchanger System

This is another recent method used for both cooling the greenhouse air during summer days and heating the same in winter nights. The system uses deep aquifer water at the ground surface through an irrigation tube well at almost constant year round temperature of 24^0C. Here the aquifer water flows through a cavity (shallow trench) dug around the perimeter of the greenhouse from one direction and the greenhouse air is made to circulate through an immersed pipe placed horizontally in the trench from the other direction, thus making it a counter flow heat exchanger configuration. The cost of lifting the aquifer water to the ground surface and digging of trench in the agricultural field is negligible as already existing tube wells and trenches meant for irrigation purposes can be used. Figure 3 shows the top view of a greenhouse equipped with Aquifer Coupled Cavity Flow Heat Exchanger System.

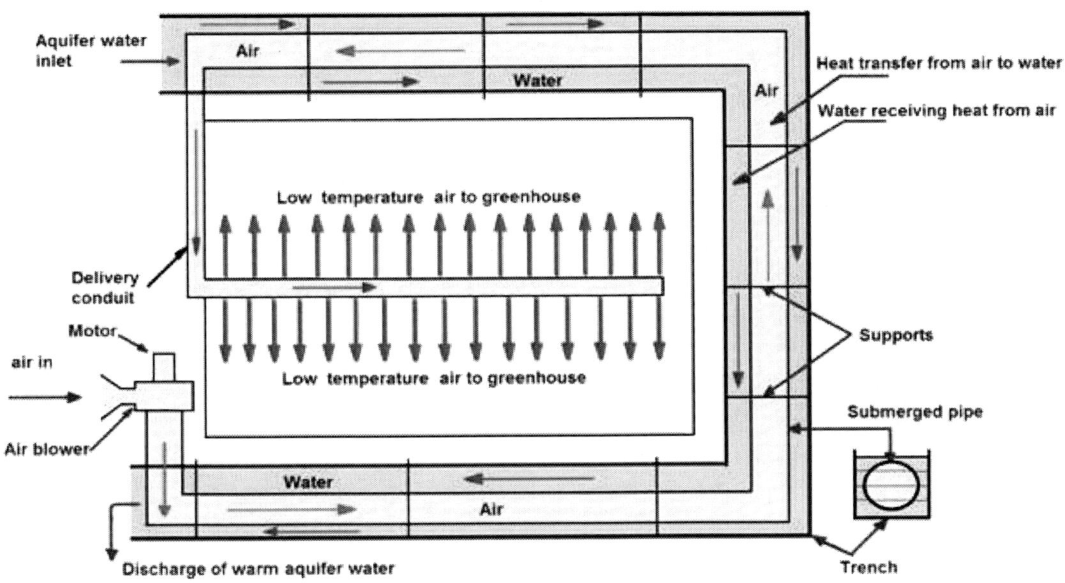

Figure 3. Top view of a greenhouse equipped with Aquifer Coupled Cavity Flow Heat Exchanger.

The various methods of passive cooling used in solar greenhouses are discussed below:

i) Shading and Whitening

The entry of solar radiation through the roof and that of the sidewalls are the primary source of sensible heat addition to the greenhouse air. The entry of excessive solar radiation can be prevented by providing shade nets or thermal screens on the roof and or sidewalls. Shading can also be done using paints, but the problem with the paints is that they get washed away during rains.

ii) Use of Near Infrared Reflecting (NIR) Pigments

The solar radiation spectrum has three important components; Ultraviolet (UV) radiation, Photo-synthetically Active Radiation (PAR) and the Near Infrared Radiation (NIR). The Ultraviolet part is detrimental to the growth of plants as already stated earlier. The PAR component is essential to the plants for photosynthesis. NIR part contains almost half of the total energy and is mainly responsible for the sensible heating of the greenhouse air. Some advanced covering materials have been developed that only allow the PAR component to enter the greenhouse, while the NIR is reflected back. This reduces the sensible heat load on the greenhouse considerably and at the same time does not affect the photosynthesis.

5.2. Ventilation

High ambient temperature coupled with high solar insolation result in the need for constant removal of trapped heat from the solar greenhouses through exchange of air from outside. The circulation of air may be accomplished either in passive mode through natural ventilation or through active mode using fan induced ventilation.

Figure 4. A greenhouse equipped with induced draught and exhaust fans.

5.2.1. Natural Ventilation

This is a passive mode of air circulation that arises due to difference in density between the outside and the inside air. If the ambient temperature is moderately low and sufficient wind velocity exists, then the warm greenhouse air may be exhausted through the vents located in the roof and sidewalls. Thus, the greenhouse temperature can be maintained close to that of the ambient using natural ventilation. This system is effective only during the period when the daytime ambient temperature is not very high.

During the summer season, when both solar insolation and the ambient temperature are high, natural ventilation becomes ineffective and fails to maintain a desirable micro-climate inside the greenhouse.

Under those conditions either natural ventilation along with evaporative cooling or fan induced circulation of air in conjunction with evaporative cooling or other active methods of cooling may be used to maintain a suitable range of temperature inside a greenhouse.

5.2.2. Fan Induced Ventilation

If the rate of heating of the greenhouse air is higher than the rate of heat removal through the vents, then fan induced ventilation becomes necessary. This method of ventilation using induced draught and exhaust fans, blowers, etc helps to maintain a greenhouse temperature close to that of the ambient due to the high rate of air change than that achieved through natural ventilation. In this method the rate of heat removal depends on the capacity of the fan and its rotational speed (RPM). Figure 4 shows a greenhouse equipped with induced draught and exhaust fans.

6. Shapes and Structures of Greenhouses

The greenhouses are available in different shapes and orientations to cater to the requirements of the various climatic zones prevailing in the world. Each zone requires different shapes of greenhouse for providing favorable climatic condition for the growth of plants. Availability of the greatest amount of insolation in winter, coverage of maximum ground area for least cost and a structurally sound facility are some of the criteria for selection of a greenhouse shape. It is observed that worldwide, researchers have used five most commonly used single span shapes of greenhouses for agricultural applications. Those are even-span, uneven-span, vinery, modified arch and Quonset type in both east west and north south orientation. Figure 5 shows the various shapes of the greenhouse used for cultivation of plants. The selection of optimum shape and orientation of a greenhouse can reduce the heating and cooling loads of the installed systems thereby, saving a lot of operating cost. As reported in literature [4] uneven span greenhouse receives the maximum solar radiation during each month of the year at all latitudes, while Quonset shape receives the minimum solar radiation. The pattern and amount of solar radiation available at different latitudes is different for the same greenhouse shape. East west orientation is best suited for year round greenhouse applications at all latitudes as this orientation receives greater radiation in winter and less in summer except near the equator and this effect is more significant at higher latitudes.

There are some other configurations of greenhouses which are exclusively used for floriculture in western countries of the world and also in India. These include the Sawtooth and the Ridge and Furrow Configuration. The Ridge and Furrow Configuration is more common as they are least expensive to built, conserve ground area and are more energy efficient [1]. Such structures are generally used for mass production of single type of flowers like Roses, Carnations, etc. Figures 6 and 7a, b show the schematic of greenhouses having Sawtooth and Ridge and Furrow Configuration.

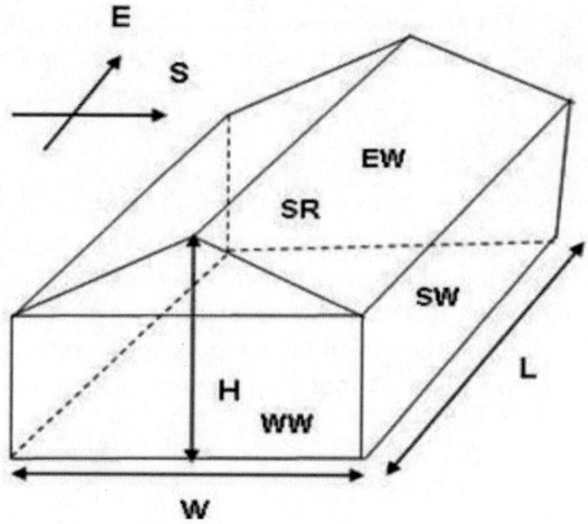

a. Even span greenhouse structure.

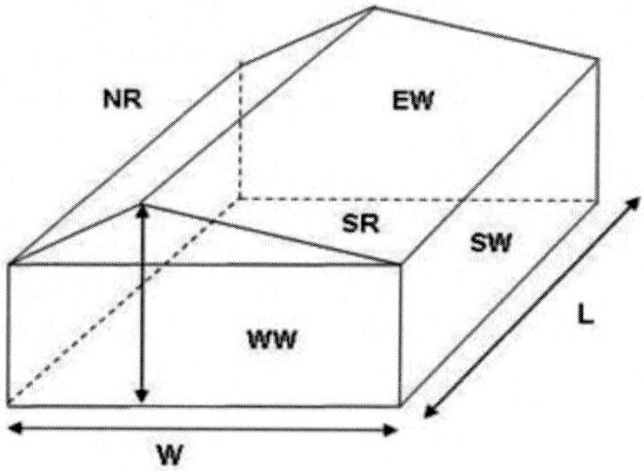

b. Un-even span greenhouse structure.

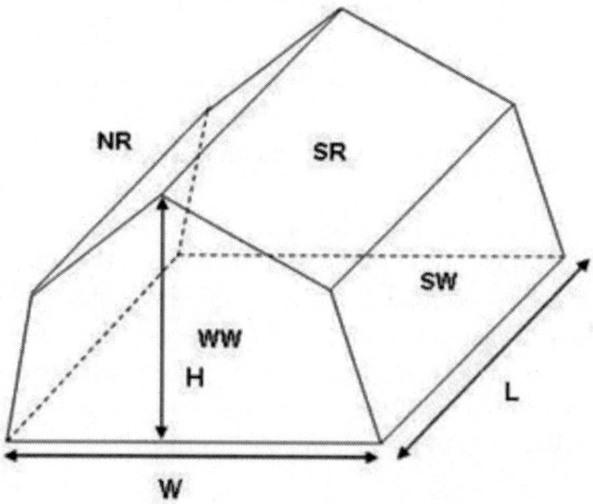

c. Vinery shape greenhouse structure.

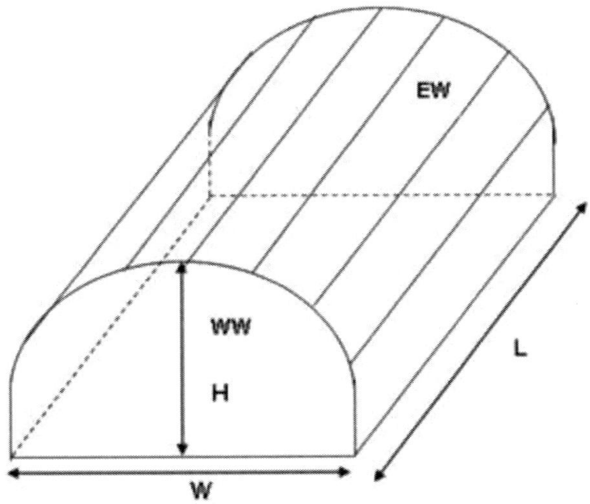

d. Arch shape greenhouse structure.

Figure 5. Various shapes of the greenhouse structure.

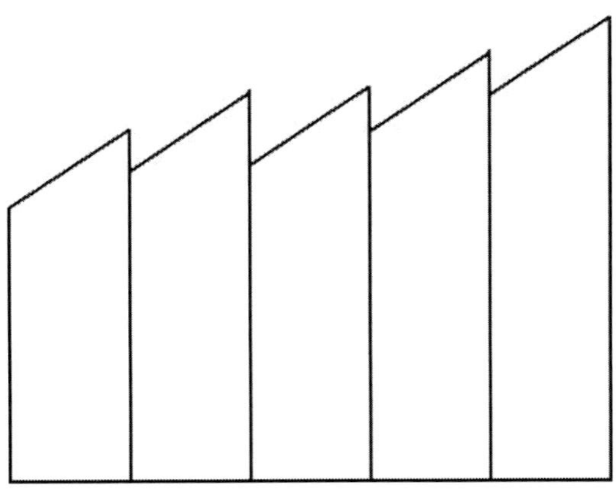

Figure 6. Saw-tooth profile greenhouse.

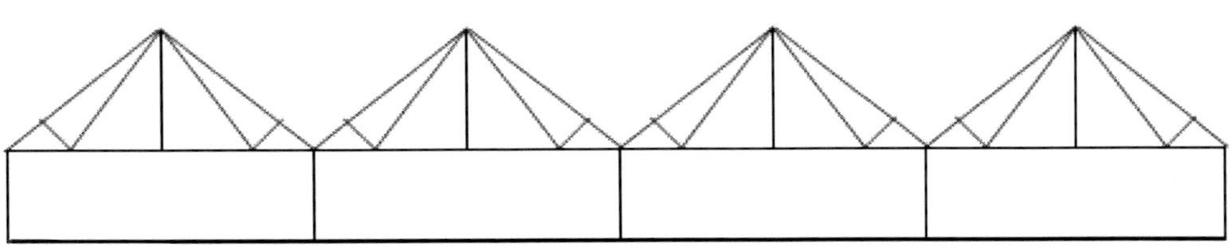

a. Ridge configuration.

Figure 7. (Continued)

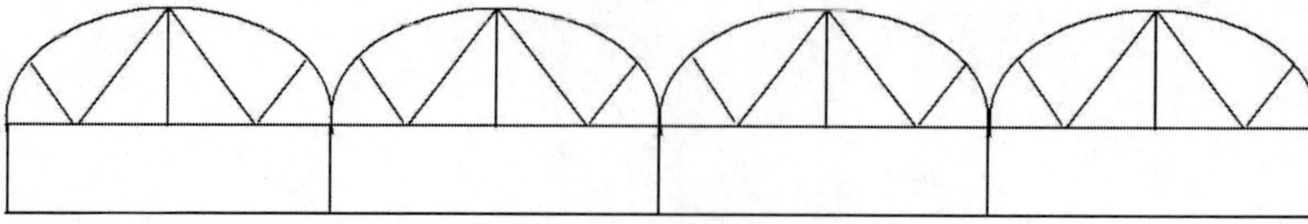

b. Furrow configuration.

Figure 7. Greenhouse Configurations.

7. Solar Greenhouse Covering Material

The solar greenhouses that were constructed before the Second World War used glass as the glazing material as no other material was available. But after World War II, most greenhouse manufacturers and researchers started the use of plastics as a substitute for glass as the covering material. Plastics are considered to be more economical, are easy to install and do provide good light transmission. The covering materials currently available are mostly based on plastics such as thin films or multi layer rigid thermosetting plastic panels. Enhancements include Ultraviolet (UV) radiation degradation inhibitors, Infrared (IR) radiation absorbers and anti condensation drip surfaces. Most of the greenhouses located in southern Europe are covered with plastic film [1]. The majority of greenhouses in China use local materials for the construction of the frame and flexible plastic films for glazing [1]. In this section the advantages and limitations of different glazing materials of greenhouse are discussed in brief.

i. Glass

The maintenance of glass is high, though it provides good light transmission over time. It should be clean to transmit the light. A panel of glass may transmit 84-87% of available visible light, but gets reduced to 70-75% due to the supporting structure that holds it. Unpainted supporting structures and little dust or shading further reduces the transmission to 60-65%.

ii. Acrylic

It is available in thick or thin double wall extruded sheets. It has a long service life of more than 20 years [1], good light transmission characteristics with moderate impact resistance. But its major limitation is that it gets easily scratched. Also, it has high expansion and contraction coefficient and is inflammable and expensive.

iii. Polycarbonate

It is available in thick or thin extruded sheets either of single or double layer. Structurally it may last up to fifteen years, but its colour becomes yellowish over time which reduces the light transmission through it. So growers have to replace the same within 8-10 years. Polycarbonate has a good initial light transmittance of about 77%, excellent impact resistance and a very low flame spread.

iv. Fibre Glass Reinforced Plastic Panels (FRP)

These plastics consist of polyester resins, glass fibers and minor amounts of other resins, stabilizers, etc. Its service life is about 5-12 years [1]; shorter life is for non treated surface. FRP panel has an initial light transmission of about 82% [1] which is comparable to that of glass. Also, the coverings made of FRP have high impact resistance. The corrugated form is stronger structurally, but causes higher heat losses due to larger area of heat transfer.

v. Polyethylene

Polyethylene coverings are the most popular replacement for glass due to high transmissivity of light (about 90%) and due to higher strength. They are also available at low cost. The average service life of polyethylene coverings is 3 years. Low Density Polyethylene (LDPE) sheets are severely affected by the ultra-violet part of sunlight as it breaks down the chemical bonds of the polymer through a process called photo- degradation and ultimately causes cracking, chalking and change in color. Degradation of physical properties such as impact strength, tensile strength, elongation is the other effects. Thus, a basic LDPE film without ultra violet stabilizers lasts only for 4-6 months. By incorporation of proper stabilizers, modern polyethylene films provide a useful life of 3 years or even more.

8. Extent of Environmental Control

The greenhouses are broadly classified as Low Cost or Low Tech Greenhouse, Medium Tech Greenhouse and Hi-Tech Greenhouse depending upon the extent of environmental control [1]. Each of them is discussed in the present section.

a) Low Tech Greenhouse

Low cost or low tech greenhouses are constructed with locally available materials such as bamboo, timber, etc. and may be covered with a polythene sheet of 150-200 micron. They do not have specific control devices for regulating the inside environmental parameters such as temperature, humidity, light intensity, carbon dioxide, etc. These greenhouses are mainly suitable for cold climatic zone where the inside temperature may be maintained 6-10^{0}C higher than that of the ambient [1]. These are mainly used for the cultivation of vegetables in the winter season. These types of greenhouses are also sometimes called shade houses.

b) Medium Tech Greenhouse

They are generally constructed using galvanized iron pipes for permanent framed structure firmly grouted into the ground to withstand the disturbance against wind. The greenhouse structure is covered with ultraviolet stabilized polyethylene sheet of 200 micron thickness [1] and is attached to the frame structure using screws. Exhaust fans and thermostats are used to control the inside temperature. Fan-pad evaporative cooling and misting systems are often used to maintain a suitable temperature and humidity level inside the greenhouse. This type of greenhouse is widely used in floriculture and for vegetable production.

c) Hi-Tech Greenhouse

If more than one environmental factor has to be controlled inside a greenhouse in real time, then automatic computer based controllers are integrated with the greenhouse system. These have an advanced sensor based control which helps to maintain the inside microclimate within the prescribed limit throughout the year. The design of hi-tech greenhouses is similar to that of medium tech, except for the automatic control system. The capital cost of these greenhouses is very high, though the yield and quality of the flowers and other horticultural crops cultivated inside justify the cost.

The study thus reveals that the worldwide greenhouses may be classified in a number of ways depending upon the requirement of the plant and geographical location of the place. No design is perfectly and universally applicable.

9. Heat Transfer in a Solar Greenhouse

There are several modes of heat transfer in a solar greenhouse which is shown schematically in Fig. 8. The major source of the sensible heat generated inside a greenhouse structure results from the incident solar radiation. Also, some amount of sensible heat is generated due to the heat transfer from the various electrical appliances present in the greenhouse and from the various auxiliary sources like that of furnaces (if any present). Some amount of heat infiltrates from outside through the coverings if the greenhouse inside air temperature is lower than that of the ambient. Some amount of latent heat exchange also occurs during the process of transpiration through the stomata of the plant leaves. Some amount of heat gets transferred through the process of conduction to that of the ground. The temperature of the ground at about 3.5 m depth is practically constant. So after certain time owing to the achievement of thermal equilibrium, there is no further heat loss to the ground and the ground temperature gets stabilized to the average greenhouse temperature.

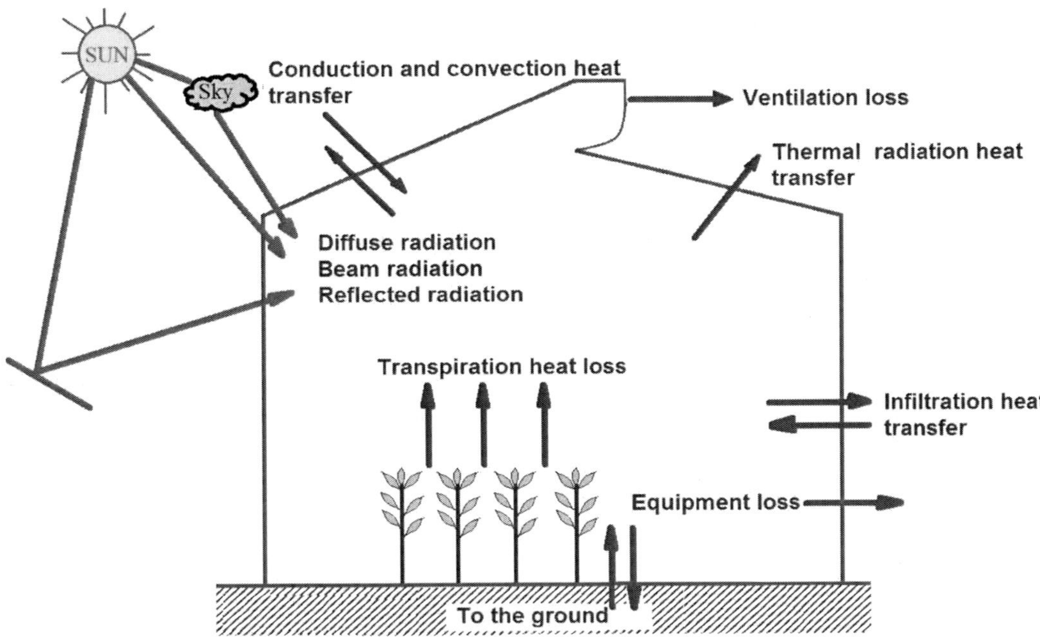

Figure 8. Energy flow in a solar greenhouse.

10. Modeling of Thermal Environment in Greenhouse

There are several variables outside and inside a greenhouse which influence the thermal environment of the greenhouse and finally the crop yield. Several attempts have been made to develop mathematical models based on energy and mass balance to predict the microclimate of a greenhouse in terms of temperature. Some of the approaches like those of Kittas et al. [5], Kittas et al. [6] and that of Ganguly and Ghosh. [7] considered greenhouse as a single component, while there are many other studies where the energy balance of the various components of the greenhouse like the floor, wall, covering material, room air, etc have been done. The modeling is required to optimize the various parameters involved in either heating or cooling of the greenhouse microclimate. The modeling can also be used to optimize the greenhouse air temperature which is one of the most vital constituents of the greenhouse microclimate for maximum crop yield for a given thermal capacity.

11. Standalone Solar Greenhouses

In order to have off seasonal cultivation inside a solar greenhouse, generation of an artificial micro-climate is necessary, which may be maintained using a suitable heating or cooling system with humidity control depending upon the climate of the geographical area. Such greenhouses rely heavily on the grid connectivity for their electricity requirement to maintain the desired inside micro-climate. In developing countries like India, a significant fraction of the population doesn't enjoy the grid connected electricity and significant part of rural areas still remain outside the reach of national or regional grid.

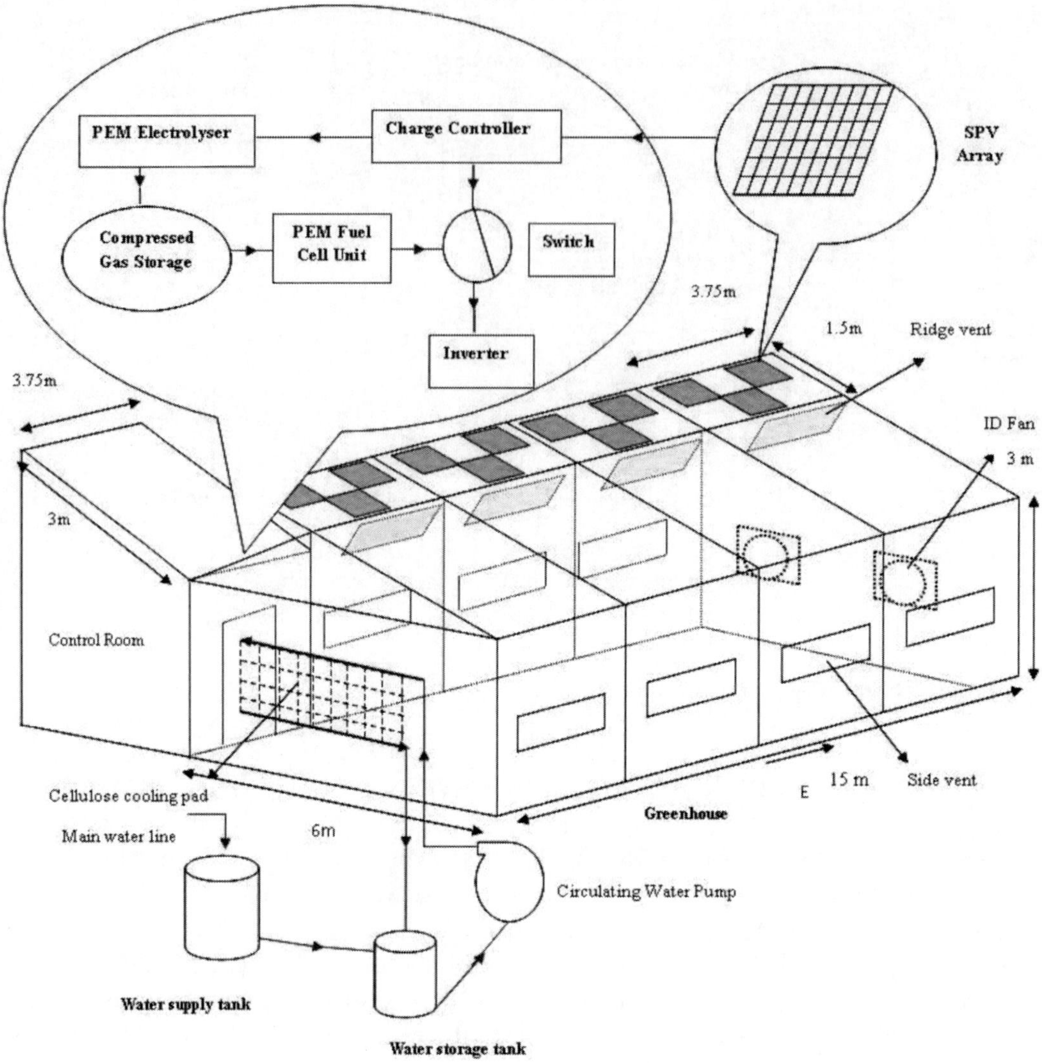

Figure 9. Schematic of a SPV-Fuel Cell-Electrolyser supported greenhouse.

As reported in recent literature [8], there are about 94000 un-electrified villages in India, of which 25000 villages are located in such remote areas that extension of the existing electricity grid is not economically viable there. For greenhouse installations in those areas, the requirement of electricity thus imposes a significant restriction, leading to increased cost and reduced operational reliability. A remote greenhouse connected to a distant grid will also face various operational limitations like low voltage, high power fluctuation and frequent breakdowns during storm, etc. Thus, the development of standalone grid independent solar greenhouses is a genuine need of the hour. For regions where plenty of solar radiation is available, as in the plains of the Indian subcontinent, use of solar energy to power a remote greenhouse provides for a viable option. Solar energy can be harnessed by mounting solar photovoltaic panels on the roof of the greenhouse. Due to non availability of solar energy at night or its intermittent availability during the cloudy days, some form of power back up arrangement becomes necessary. The most common and traditional power back up arrangement uses a regenerative battery that stores the excess energy when the power generation is more than the power consumption and supply the same at the time of deficit. But battery backup has a number of limitations like limited storage capacity, lower charging and discharging rates, fairly high cost, problem of disposal, etc. Also, there are certainly other problems like frequent maintenance of electrolyte levels, removal of sulphur from the electrodes, shorter life span and limited power output as long as the electrolytes are not used up. Solar PV-electrolyser-fuel cell provides an alternative method of energy storage and power back up. The schematic of such a system is shown in Fig. 9 [9]. In this type of power backup, excess electrical energy after meeting the in house requirements is used to produce hydrogen gas in an electrolyser which is compressed using a gas compressor and stored in tanks. The hydrogen gas is consumed by the fuel cell stacks to meet the energy requirement during the off- sunshine hours as well as during the hours of energy deficit. The main advantage of the electrolyser fuel cell assembly is that its energy storage capacity is independent of its size. At the same time hydrogen being a multi-functional fuel, it can be used to meet other energy needs in other applications such as hydrogen combustion engines or chemical plants. That a Solar PV-fuel cell backup system has more benefits than a battery backup has also been suggested by researchers [10]. At present the cost of electrolyser-fuel cell system is more than the battery storage and is less popular than the battery storage. However, extensive research and development works are being carried out in the field of solar hydrogen systems and the costs of these components are also reducing day by day. It is expected that owing to its advantages, the solar hydrogen systems will become more popular in powering remote greenhouse installations in the near future.

12. Summary and Future Direction of Work

This chapter presents a comprehensive overview of state of the art of solar greenhouse technology with its global and Indian scenario. The different components of the microclimate in a greenhouse are also discussed in details. The various methods of heating, cooling and ventilation technologies used for cultivation of flora in passive and active greenhouses are also highlighted. Subsequently, the various shapes of greenhouse structures and their covering materials have been touched upon in brief. The various modes of heat transfer in a greenhouse structure are also discussed in a separate section. Finally, the concept and the need for establishment of a stand-alone grid independent solar greenhouse have been emphasized upon along with a scheme of a greenhouse powered through solar hydrogen system (SPV- fuel cell – electrolyser integrated system).

From this chapter, it is thus evident that a lot of work is being carried out in the field of solar greenhouses. But it is established that though a number of methods of heating, cooling and ventilation technologies are available to regulate the micro-climate in a greenhouse, none of them can be recommended universally for any greenhouse structure. The problem of maintenance of a suitable greenhouse microclimate is more acute for the tropical and sub-tropical countries like India, where cooling and dehumidification are the major objectives. Not many research works are available in literature that describes the modeling of such greenhouse systems accurately. As evident from literature, natural ventilation is the most simple and cost economic method of maintenance of the micro-climate in a greenhouse. But it is unsuitable during the summer season when the ambient temperature values are higher coupled with high humidity. Using shade nets of higher degree lead to blocking of sunlight affecting the growth of the inside flora. The evaporative cooling systems become less effective during hours of high ambient humidity. The operation of artificial ventilation systems demands uninterrupted grid electricity which by itself is a major constraint for rural areas.

So, the future works in this field should be directed towards development of suitable cladding materials for greenhouse structures which should allow the photo-synthetically active radiation to enter the greenhouse, but restrict the entry of Near Infrared radiation into the system. This type of cladding materials would reduce the sensible heat load on the greenhouse without affecting the process of photosynthesis. Also, cost effective methods of humidity control need to be developed. Future works may be focused on the development of solar thermal assisted desiccation system for control of humidity inside greenhouses. Also, not many research and development works have been focused on powering the remote greenhouses through solar photovoltaic systems. As the cost of solar photovoltaic systems is coming down at a very fast rate, it is expected that the same can be used for powering the remote greenhouses in a cost effective manner in the years to come.

REFERENCES

[1] Tiwari G. N. Greenhouse technology for controlled environment. *Narosa Publishing House,* New Delhi (2003).

[2] Tamil Nadu Agricultural University report on Low cost greenhouses for vegetable production at www.agritech.tnau.ac.in/agricultural_engineering (accessed on 25.12.2011).

[3] Jensen M.H; Malter A.J. (1995). Protected agriculture- a global review. *World Bank Technical Paper,* 253, 157.

[4] Sethi V.P. (2009). On the selection of shape and orientation of a greenhouse: thermal modeling and experimental validation. *Solar Energy*, 83, 21-38.

[5] Kittas C; Bartzanas T; Jaffrin A. (2003). Temperature gradients in a partially shaded large greenhouse equipped with evaporative cooling pad. *Biosystems Engineering*, 85(1), 87-94.

[6] Kittas C; Karamanis M; Katsoulas N. (2005). Air temperature regime in a force ventilated greenhouse with rose crop. *Energy and Buildings*, 37(8), 807-812.

[7] Ganguly A; Ghosh S. (2007). Modeling and analysis of a fan-pad ventilated floricultural greenhouse. *Energy and Buildings*, 39(10), 1092-1097.

[8] Mohrail R.M; Kulkarni P.S. (2009). A case study of solar photovoltaic power system at Sagardeep Island, India. *Renewable and Sustainable Energy Reviews*, 13(3), 673-681.

[9] Ganguly A; Ghosh S. (2011). Performance analysis of a floriculture greenhouse powered by integrated solar PV- fuel cell system. *ASME Journal of Solar Energy Engineering*, 133(4), 041001-07.

[10] Ro K; Rahman S. (1998). Battery or fuel cell support for an autonomous photovoltaic power system. *Renewable Energy*, 13(2), 203-213.

In: Renewable Energy and Sustainable Development
Editors: R. T. Durai Prabhakaran, S. A. Kale et al.
ISBN: 978-1-63463-429-8

Chapter 2

Computational and Experimental Assessment of the Role of Extended Surfaces on Performance of Solar Flat Plate Collector

Mahesh Gorawar[1]*, K. Shekar[2], P. P. Revankar[3] and P. G. Tewari[4]
[1,3,4]B.V.Bhoomaraddi College of Engineering and Technology, Hubli, India
[2]Bellari Institute of Technology and Management, Bellary, India

Abstract

The wide ranging applications of solar thermal energy has demanded accelerated research and development in this sector leading to product optimization in terms of cost and functionality. The need for prevention of post harvest losses in the developing countries of the world has further reiterated the impetus of research on solar thermal devices. The proposed work deals with performance studies on three different fin configurations viz horizontal, longitudinal and broken longitudinal fins fitted to a blackened absorber plate of the solar air heater. The three solar air collectors refered in the study had dimensions 1.2×2.0×0.80 m and were fabricated based on solar insolation data of Hubli (15.43° N, 75.03° E) located in Karnataka State of India. The performance study was conducted at different levels of solar insolation, inlet air temperature and mass flow rate of 0.03 to 0.05 kg/s to investigate their influencing parameters in the solar drying process. The experimental results showed that exit air temperature delivered by solar air heater was 80 °C and 52 °C at the controlled mass flow rate of 0.03 kg/s and 0.05 kg/s respectively. The corresponding instantaneous peak efficiency under air mass flow rate of 0.03 kg/s and 0.05 kg/s were respectively at 45 and 60 percent during same part of the day. The computational and experimental studies intended to develop an appropriate solar air heater to deliver process heat that for crop drying application at the location. The solar energy at the location can be effectively harnessed for thermal and electricity generation applications through field studies that provide valuable inputs on its feasibility. The overall study revealed that the location had moderate to good amount of solar energy available for exploitation.

* Corresponding Author. Tel: +91 9986567044, e-mail: mb_gorwar@bvb.edu.

1. INTRODUCTION

In past few decades energy sector has supported world economy by executing an important role in the development of all domains of Agriculture, Industry and Service sectors in the contemporary world. In the light of globalization and the rapid economic growth, energy requirement worldwide has continuously increased leading to faster electric power pool capacity addition. The global concern for sustainable development and climate change has brought focus of engineers and researchers towards development of alternative, clean, inexhaustible source of energy to meet the energy demand of the present and future generation.

In India solar energy is abundantly available source of energy with an estimated nationwide potential of 5,000 trillion kWh per year. The country has 4-7 kWh m^{-2} day^{-1} of average global solar insolation in majority of its the locations. This energy if captured effectively amounts to several thousand times more than total energy needed or consumed nationwide. The reported research studies have identified that nearly 58 % of the country receives in excess to 5 kWh m^{-2} day^{-1} of annual average global solar insolation and are identified as solar hotspots with immense opportunity for techno-economic exploitation of solar energy. Thus solar energy offers succor to poor and energy deprived populace in the country for which every effort needs to be made to exploit the relatively abundant source of energy. It can be used for low and medium temperature range thermal applications.[1]

The Solar energy stands out as brightest long range promise towards meeting constantly increasing energy demand albeit to its main hurdles of low intensity, intermittent nature and non availability during night. Solar energy is the most promising of all the renewable energy resources despite of these limitations. Solar energy is contemplated to have a wide range of applications including water heating, air heating, air-conditioning of buildings, solar refrigeration, photo-voltaic cells, green houses, photo-chemical, power generation, solar furnaces and photo-biological conversions to list a few. Out of these, the utilization of solar energy for power generation and heating/cooling of buildings is the subject of active research these days. The use of electric power for thermal applications constitutes a major part of the domestic energy consumption in India.

The flat-plate collector has a significant place amongst the various solar gadgets for thermal applications commonly used for water heating, space heating and solar drying in the low and medium temperature range domestic thermal applications. The heat extraction rate depends on physical configuration of the absorber plate, mass flow rate and incident solar insolation. The Glazed flat-plate collector is a most common solar collector used in low and medium temperature range solar thermal energy conversion systems. It consists of black selective coated absorber plate with or without heat transfer enhancing devices, used to absorb solar global radiation and convert it into heat, and transfer that heat to working fluid (usually air or water). Based on the working fluid used, flat plate collectors are mainly classified into two types: solar air heater and solar water heater. The flat plate collectors operate in wide range of temperature (30 to 100 °C). Hence these are used in low and medium temperature range thermal applications like domestic water heating, space heating and crop drying. The use of high grade electric energy for thermal applications involves a multistage conversion and hence not recommended owing to poor conversion efficiency.

The performance testing of solar collector constitutes of measuring the useful heat gain by the device as against the input solar radiation it gains from the sun. There are several factors that influence the availability of solar energy at the location including the period during the year, geographical position of the location relative to the equator and the local climatic factors that influence the cloud formation. The useful heat gain by the collector depends upon the choice of the materials of construction, operating parameters of solar collector and the heat transfer characteristics that determine the heat transfer process. The design of the solar thermal conversion device therefore involved optimization with respect to material selection based on the application and the location where the solar conversion device was installed. The parameters influencing performance of solar collector can be analyzed based on computational techniques that are very efficient optimization tools for selection of solar collector parameters. This technique based on mathematical modeling of a physical system termed as simulation mimics real world phenomenon through a set of governing equations. The computational results provide an assessment on the influence of intervening process parameters and the overall output of the system. The computational results minimize the overall lead time of generating equipment based on feasibility and minimizing the number of experimental trial to be performed before arriving at a optimized system. There are several tools available for computational studies on optimization of physical systems that include MATLAB, SCILAB and MATHCAD. These computational tools have capability of performing all mathematical operations which are applicable in prediction of system performance based on the governing equations.

2. Theory and the Design Calculations

This section deals with the overview of the reported research on solar collectors for air heating along with detailed design equations applicable to the performance evaluation of the solar collector.

2.1. The Overview of Solar Thermal Collectors

The use of fins for solar air heating collectors has been a common practice due to poor thermal characteristics of air as working medium as reported by many researchers. The different fin- configurations have been used in crop drying and other thermal applications, based on experimental and theoretical investigations with respect to design, outlet temperature and collector efficiency. The present work includes a literature survey with an objective of assessing relevance of geometry and configuration of flat plate collector.

Jaurker et al. (2006) investigated effect of rib-grooved artificial roughness on heat transfer and friction characteristics of solar heater highlighting that thermo-hydraulic performance of rib-grooved duct was better compared to performance of collector ribbed duct only on account of better heat transfer rates achieved[2]. Tiwade (2009), reported that the heat transfer coefficient of the air collector was maximum for smaller pith of longitudinal fins. The friction factor and pressure drops decrease with increase in Reynolds number, and the thermal efficiency of the air collector increases up to 68 percent [3].

Sharma and Varun (2010) analyzed the thermo-hydraulic performance of air heater in terms of effective efficiency based on the actual thermal energy gain and pumping energy for air heater with roughened duct. The roughness parameters like relative roughness pitch, relative groove position, chamfer angle, relative roughness height and flow Reynolds number were considered to evaluate the combined effect of heat transfer and fluid friction. The air heaters with roughness have better efficiency compared to conventional smooth air heaters and reported that thermal efficiency was maximum at the configuration that has relative roughness pitch: 6, relative groove position 0.4, chamfer angle 18°, relative roughness height 0.04 [4]. It was concluded that one single roughness geometry that gave best exergetic performance for whole range of Reynolds number was not possible due to other intervening parameters of influence. The comparison showed that solar collector with rib-grooved and arc-shaped wire as artificial roughness and solar collector with smooth duct delivered better exergy efficiency for lower and higher range of Reynolds number respectively.

Sharma and Varun (2010) [5]. Sethi et al. (2010), investigated the effect of artificial roughness on fluid pressure and heat transfer indicating artificial roughness lead to increased collector fluid pressure and hence higher pumping power. The heat transfer rate of the collector increased with roughness of the collector that demanded optimization of surface roughness to obtain higher heat transfer rate and low friction [6]. Peng et al. (2010), developed pin-fin integrated absorber and investigated increase the thermal efficiency for varying flow rate on PZ7-11.25 pin-fin array's collector. The correlation equation for a heat transfer coefficient was obtained and the efficiency variations with air flow rate were determined [7]. Chaudhary et al. (2012), compared performance of M shape roughness geometry and smooth geometry. The results showed that the heat transfer rate in artificial roughness of M shape was about 1.7 to 1.8 times than that of smooth duct [8].

Kaushal and varun (2012), conducted an experiment on solar flat plate collector to investigate the effect of Protrusions type roughness with 45° protrusion angle. The pair of protrusion geometry with longitudinal fins was considered during the study. The results indicated an enhancement in heat transfer rate for the following configuration of the collector with aspect ratio of rectangular duct:11, relative roughness pitch : 12-24, relative roughness height:0.03 and ratio of height of protrusion to print diameter :0.3° over the operating Reynolds number in 3600 to18100 range. The heat transfer rate and frictional factor of roughened surface were 2.96 and 2.73 times more than that of smooth duct respectively [9].

Ionut and-Razvan (2012), developed thermal simulation model using Galerkin a finite element method technique. The temperature distribution and velocity profile between PV module and roof wall at different distances were analyzed using fluid flow and energy equations incorporating an implicit convective boundary. The results showed increase in exit temperature with channel width from 50 ^{0}C to 30 ^{0}C and the mean air velocity varied from 0.21 m/s to 0.29 m/s [10].

Nasrin and Alim (2013), compared the thermal performance of solar collector with sinusoidal wave absorber using Water based Ag nano fluid and Water based CuO nano fluid as working fluids. The results showed that Water based Ag nano fluid exhibited better performance compared to other fluid [11]. Chabane et al. (2013), investigated the performance of solar collector with longitudinal fins attached to the back side of the collector and collector without fins. It was observed that at mass flow rate of 0.016 kg/s the efficiency of the collector with and without fins were 51.50 and 43.90 percent respectively [12]

Ramachandra et al., (2011) have reported on the potential techno-economic barriers apart from the climatic uncertainty in availability of solar energy in Indian context. The study has reported on Solar hotspots in the country that receive in excess to 5 kWh/m^2 /day of annual average global solar insolation and are potentially viable locations for installation of either solar PV systems or solar thermal applications. The study also indicates the immense benefits of adopting solar power in terms of employment opportunities at rural levels, leading to the nation emerging as world leader in solar power generation in the near future. [13].

Garg and Adhikari (1999) reported on SPVT collector through computer simulation for selective coating and black paint absorber plates 25- 150 kg/h fluid flow rate and 5 cm duct depth. The thermal performance of selective coated absorber was better than the normal black painted absorber. The exit temperature of air attained 65° C and solar cellelectric efficiency was 9.5 percent (reported as 10 percent at standard conditions) for 25 kg/h and 150 kg/h flow rate respectively [14].

Joshi and Tiwari (2007) modelled the energy efficiency of PVT air-heating collector for the cloudy and cold conditions. It was found that the electrical efficiency of the PVT system increases by 3-4 percent due to lower operating temperature of the solar cell and the energy and exergy efficiencies of the PVT system varied between 55- 65 percent and 12-15 percent respectively [15].

The ensuing discussion on the reported literature on solar flat plate collectors clearly revealed that air as a heat transfer fluid has challenges in terms of heat caring capabilities owing to a lower specific heat and a desirable flow rate that leads to minimum retention with the heated surface. The adoption of enhanced heat transfer area contributes to improvement in the heat transfer characteristics albeit to the associated pressure drops leading to increased pumping power. The solar thermal systems combined with photovoltaic system constitute genera of solar gadgets that are called as Solar-Photovoltaic systems (SPVT). The trend indicates that present SPVT systems serve the dual purpose of supplying thermal energy for process heat requirement along with electric power generated as an additional output to meet small electrical power needs of the integrated system. The technology of SPVT system has evolved towards development of efficient SPV systems through cooling provided on the underneath of solar photovoltaic system and augmenting to enhanced heat of the air medium.

2.2. The Design Equations for the Solar Flat Plate Collector

The present work is based on comparison of the three different configurations of solar heater geometries at the specified location for generation of heated air for thermal application. The detailed calculations used in the present research are summarised in this section. The mathematical model was developed to estimate outlet air temperature, thermal efficiency of collectors and overall efficiency of collectors 1 & 2. The equations were based on the concept of energy balance applied to the collector modelled for determination of various operating parameters like the absorber plate temperature, exit air temperature, useful heat gain and overall efficiency. The developed model was based on energy transfer phenomenon in various components of collector.

$$\text{Collector Efficiency } (\eta) = \frac{Q_u}{AI} = F_r\left((\tau\alpha - U\left(\frac{T_i - T_a}{I}\right)\right) \quad (1)$$

$$h_w = 5.7 + 3.8v \quad (2)$$

$$h_r = \frac{4\sigma T_{pm}^{\ 3}}{\frac{1}{\varepsilon_p} + \frac{1}{\varepsilon_b} - 1} \quad (3)$$

$$\textit{Collector} \text{ overall loss coefficient } (U_L) = U_t + U_b + U_e \quad (4)$$

$$U_t = \left[\frac{N}{\frac{C}{T_p}\left[\frac{T_p - T_a}{N + f}\right]^e} + \frac{1}{h_w}\right]^{-1} + \left[\frac{\sigma(T_p + T_a)(T_p^{\ 2} + T_a^{\ 2})}{(\varepsilon_p + 0.00591Nh_w)^{-1} + \frac{(2N + f + -1 + 0.13\varepsilon_p)}{\varepsilon_c} - 1}\right] \quad (5)$$

$$\text{Bottom loss coefficient } (U_b) = \frac{k_{ins}}{L_{ins}} \quad (6)$$

$$\text{Edge loss coefficient } (U_e) = \frac{k_e}{x_e}\left[\frac{2(L+W)H}{LW}\right] \quad (7)$$

$$\text{Collector efficiency factor } (F') = \frac{\frac{1}{U_L}}{\left[w\left(\frac{1}{U_L\left[D + (W - D)F\right]} + \frac{1}{C_b} + \frac{1}{\pi D h_{fi}}\right)\right]} = \frac{U_o}{U_L} \quad (8)$$

$$\textit{Fin} \text{ efficiency factor (F)} = \frac{\tanh\left[m\left(\frac{W - D}{2}\right)\right]}{m\left(\frac{W - D}{2}\right)} \quad (9)$$

$$\text{Collector heat removal factor } (F_r) = \frac{m c_p}{A_c U_L F'}\left[1 - Exp\left(-\frac{A_c U_L F'}{m c_p}\right)\right] \tag{10}$$

$$\text{Useful heat energy gain by collector } (Q_u) = F_r \times A_c\left(I_0 \tau\alpha - U_L\left(T_{pm} - T_a\right)\right) \tag{11}$$

$$\text{Electrical efficiency } (\eta_{el}) = \frac{V_m \times I_m}{I \times A_p} \tag{12}$$

$$\text{Overall efficiency } (\eta_o) = \frac{[Q_u + (V_m \times I_m)]}{I \times A_{total}} \tag{13}$$

3. Experimental Setup and Methodology

This section deals with the detailed account of experimental set-up and the methodology adopted for the performance analysis of the solar flat plate collector for generation of heated air using the forced convection type of solar collector. The various measuring devices used for the performance analysis of solar air heater are dealt in this section detailing out the specification of each system that includes the thermocouples for temperature measurement, orifice-meter for air flow measurement, sunshine recorder and pyranometer for solar insolation measurement.

3.1. Experimental Set-up Used for Experimental Investigations

The experimental work intended to investigate the influence of different operating parameters on the overall performance of solar flat plate air heating collector located at the Hubli location. Three different configurations of the solar flate plate collector represented in figure 3.1 that were investigated included:

Collector 1: had rectangular container box of dimensions 2.0 m length × 0.97 m width ×0.08 m height fabricated using 12 mm thick waterproof plywood that had very low thermal conductivity to function as side insulation to the collector. The lower part of the rectangular collector box had glass wool insulation of 70 mm thick to prevent loss of heat from the underneath of the collector. The absorber plate consisted of 1.9m x 0.96m galvanized iron sheet of 0.3 mm thickness smeared with the 300 microns layer of selective coating (black chrome). The inner side walls of the duct collector had aluminium foil to reflect back the thermal radiations on to the absorber plate thereby enhancing the nett heat gain by the collector as it acted as a glazing surface. The L shaped equally spaced transverse fins of 600×25×25mm and 0.5mm thickness are placed on absorber surface, The fins placed on absorber plate were bent to an angle of 45^0 facing towards the air inlet, this will serve the purpose of providing retention time to incoming cold air to mix with hot air as the air movement take a zigzag path. The collector box was covered with a semi transparent glass to glass photovoltaic laminate with dimensions 1.245m × 0.96m × 7 mm and 0.755m × 0. 960m

×7 mm of plain glass. The PV panel has a rating of 100 Wp power (not shown in figure 3.1). The collector 1 was provided with a 37.5mm diameter PVC pipe at inlet and outlet. The electric blower was used to circulate the controlled volume of air. The air mass flow rate was measured using orifices with the help of differential water column manometer. After installation, the collector was left operating several days under normal weather conditions for weathering processes.

Collector 2 and Collector 3: consisted of SPV hybrid air collector of size 2.0 m × 1.0m × 80 mm were constructed based on local needs. The collector 2 was covered with PV laminated glass of 4 mm thick and collector 3 was covered with plane transparent glass of 4 mm. The absorber plate of collector 2 and collector 3 were made of galvanized iron sheet of 1.99 m ×0.96 m with a 0.3mm thickness smeared with 300 microns black chrome) of thickness. To increase the heat transfer rate the absorber plate of collector 2 and collector 3 were augmented with broken transverse ribs and longitudinal fins. A schematic view of the constructed single flow solar air collectors are shown in figure 3.1.

3.2. Experimental Methodology

The experiments were conducted in two phases. The first phase of experiments were conducted using hybrid solar air collector (collector 1) constructed based on local needs. In second phase two solar air collectors mounted with K- type thermocouples positioned evenly on the top surface of the absorber plates, at identical positions along the direction of flow. The inlet and outlet air temperature were measured by two well insulated thermocouples while the top and bottom surface temperatures were measured using surface thermocouples. The thermocouples readings are recorded using ESD make temperature recorder. Solar insolation was measured using pyranometer. The experimental observations were made for the three collectors over a wide range of operating conditions that included the time of the day, ambient temperature, wind speed, mass flow rate of air through the collector, temperature at inlet and outlet of the collector, glass surface temperature and absorber plate temperature. The standard prescribed methodology was adopted for the measurement of all the above parameters based on averaging experimental data obtained for three trials to facilitate for random and systemic errors in experimentation. The important objectives of the proposed research included: Assessment of the solar energy potential at the location and its variation, Investigation of operating parameters on the performance of solar flat plat collectors of three specified configurations and development of an efficient solar flat plate configuration that was suitable for the given location.

3.3. Computational Methodology

The computational work used the governing equations for the operation of solar flat plate collector and simulated their performance to aid in development of a suitable design for the solar air heating system. The computational analysis was executed on MATLAB version 12 based on the governing equations for heat transfer analysis of the solar flat plate collector and the results obtained are compared with experimental values obtained from field trials.

Figure 3.1. Schematic view of experimental setup (collector 1, 2 and 3).

4. Results and Discussions

This section presents the detailed observation made as part of investigative studies on three different configurations of solar flat collector for air heating applications based on experimental observations and computational studies. The results of the study have been categorized into investigations on operating parameters and performance parameters. The experiments were conducted to investigate performance of three designs of solar collectors namely: collector 1, collector 2 and collector 3.

4.1. Investigations on Operating Parameters

The diurnal variations in the operating parameters over a period spread from 9.00 hours 17.00 hours over the day included solar radiation at the location, exit air temperature and collector instantaneous thermal efficiency. The variations in solar insolation at the horizontal location close to solar collector represented in fig 4.1 was compiled based on the data collected using the Pyranometer with provision to record instantaneous solar insolation in W/m^2 using electronic printer. The trend of variation of solar insolation indicated a peak value during the noon session with a magnitude close to 650 W/m^2 that decreased in the afternoon session to a magnitude of 275 W/m^2 at 15.00 hours time. The magnitude of solar insolation during the early part of the day was lower at 500 W/m^2 recorded at 10.00 hours. The variation in the solar insolation indicated fluctuation in available energy thereby leading to non-uniformity in the output energy delivered by the system. This necessitates the development of an intermediate system for storage of the thermal energy to match the point of generation with point of utilization.

4.2. Investigations on Performance Parameters

The performance parameters here referred to the characteristics that contributed to the overall operation of the solar air heater to deliver the thermal energy for end utilization. The important parameters investigated during the study included Absorber plate temperature, exit air temperature, overall efficiency of collector and temperature of integrated solar PV system. The variation of all these parameters with time for different combination of mass flow rate of air as working fluids The forced convection type solar air heaters were investigated for 0.03 kg/s, 0.04kg/s and 0.05kg/s flow rates and experiemntal results and computatioanl results are presented in the figures 4.2 to 4.12 indicated below.

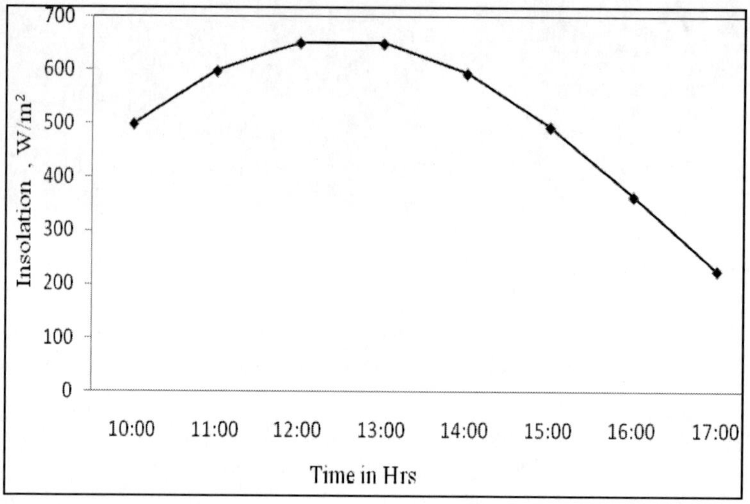

Figure 4.1. Diurnal variation of solar insolation at Hubli.

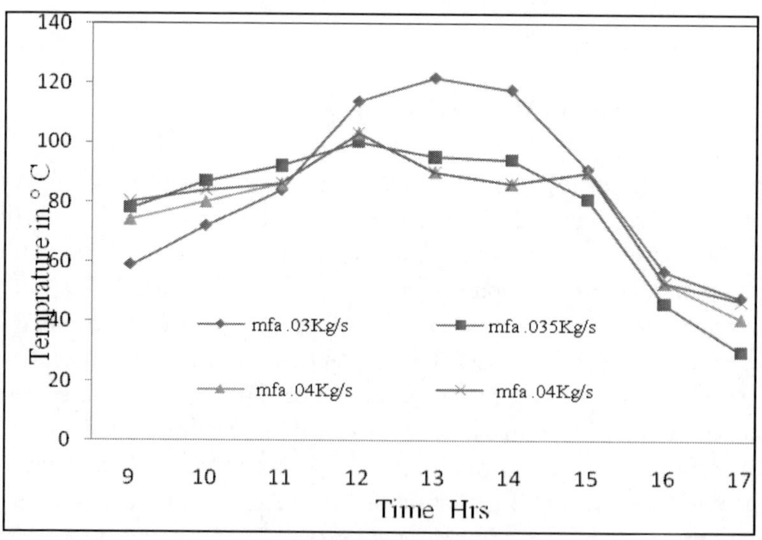

Figure 4.2. Variation of absorber plate temperature of collector 1.

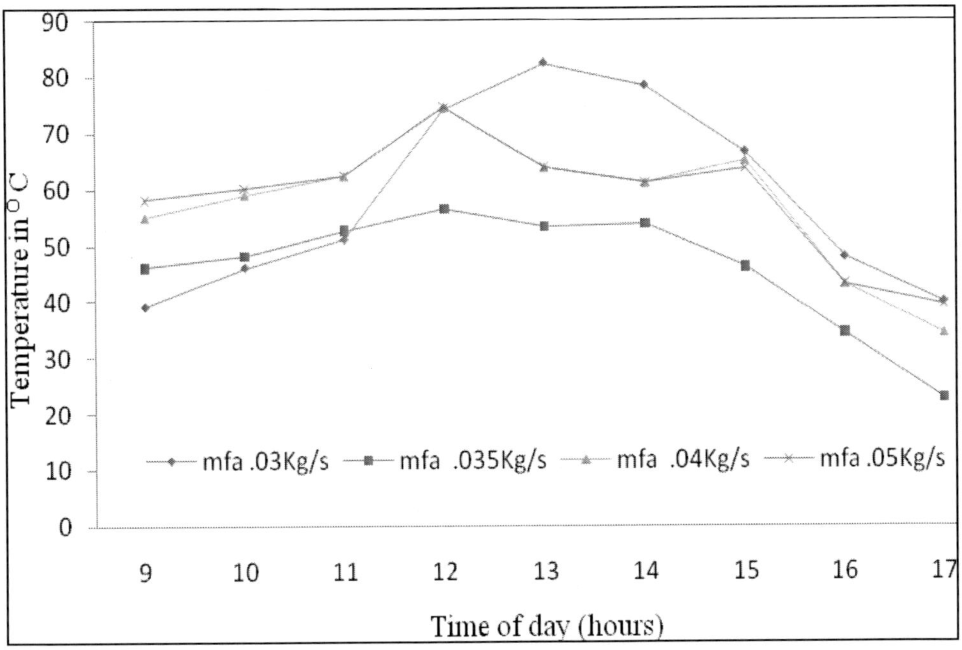

Figure 4.3. Variation of exit air temperature of collector 1.

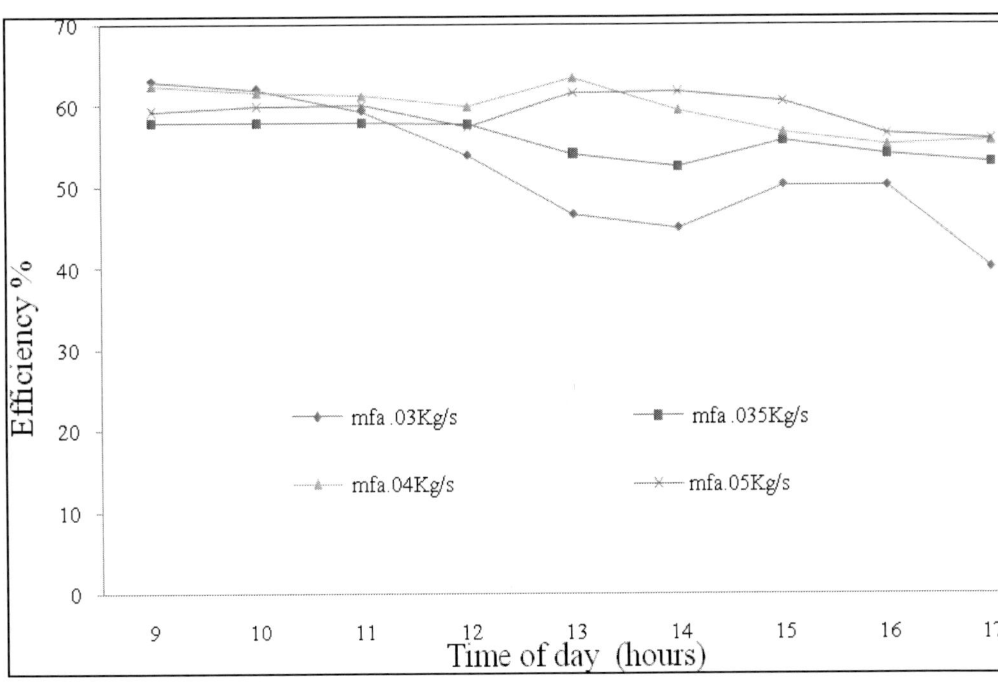

Figure 4.4. Variation of instantaneous efficiency for Collector 1 with time.

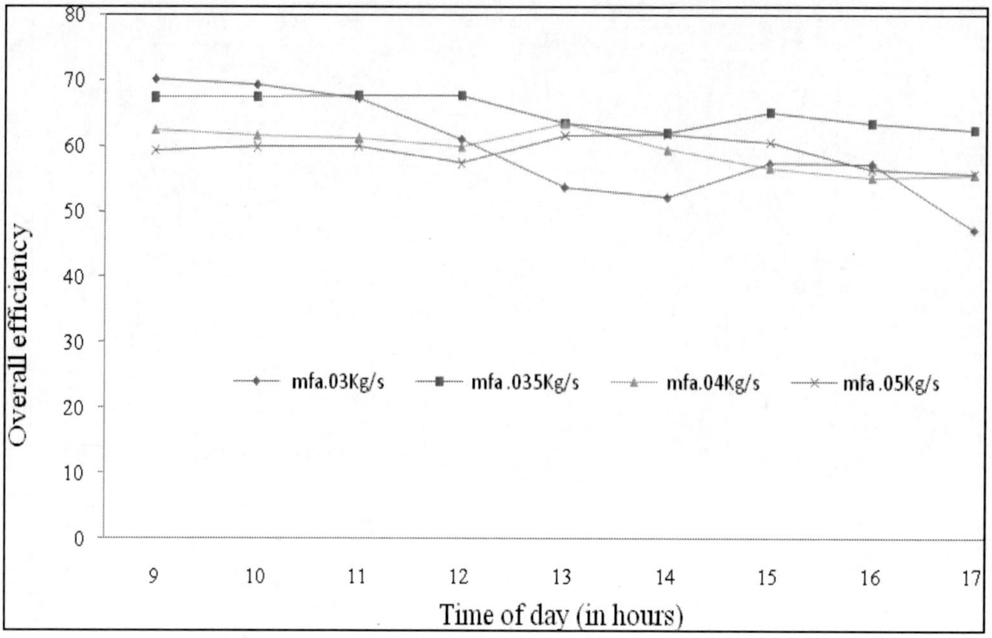

Figure 4.5. Variation of overall efficiency for collector 1 with time.

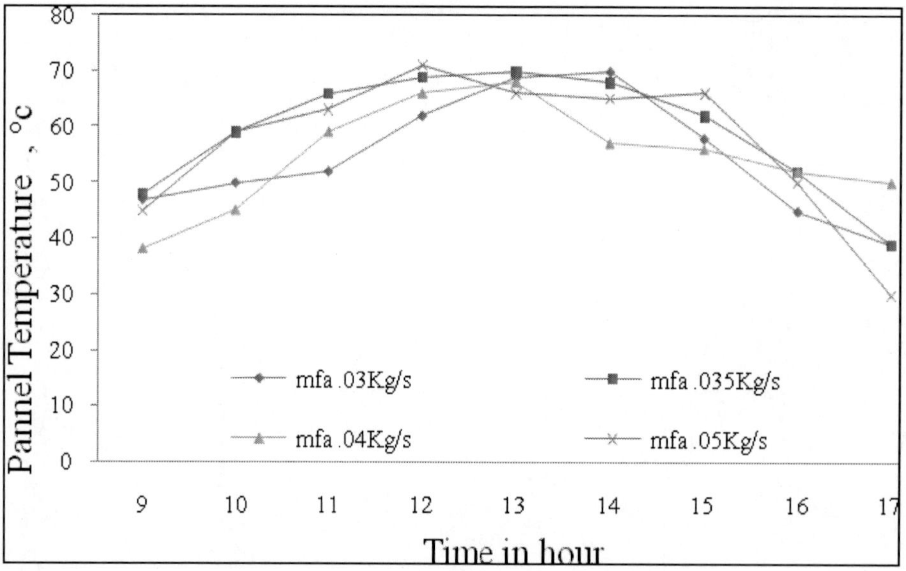

Figure 4.6. Variation of PV panel temperature of collector 1with time.

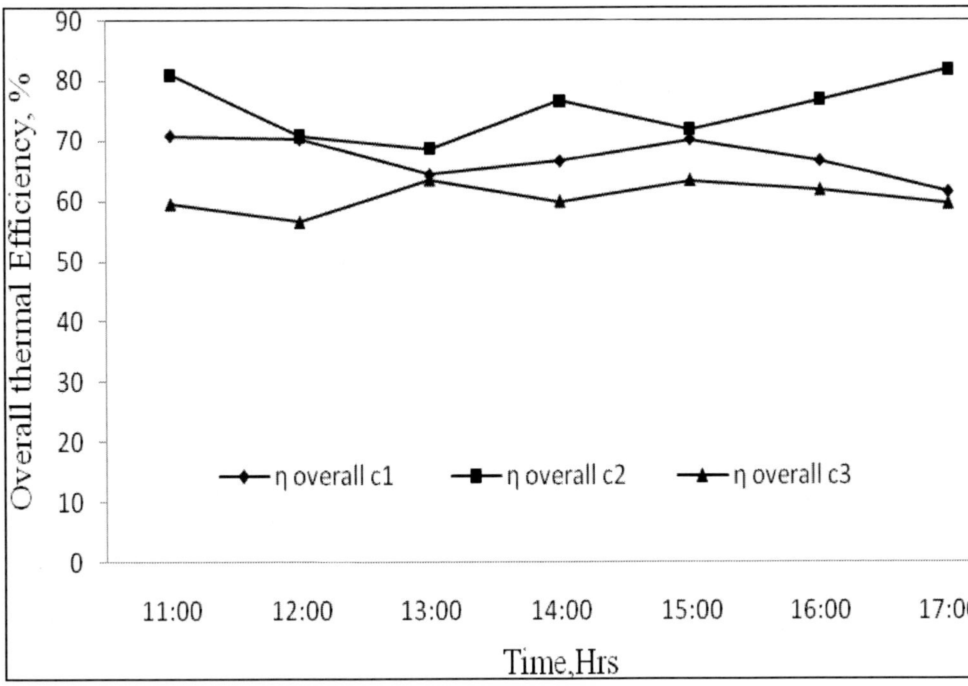

Figure 4.7. Variation of overall efficiency of collector 1, 2, &3 with time.

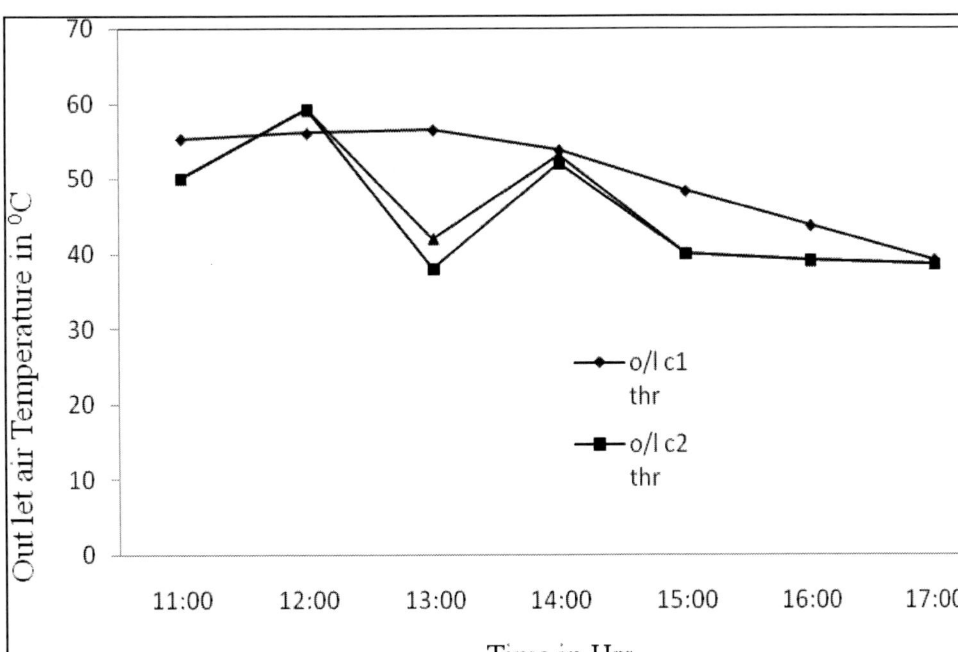

Figure 4.8. Predicted outlet temperatures for collector 1,2, &3.

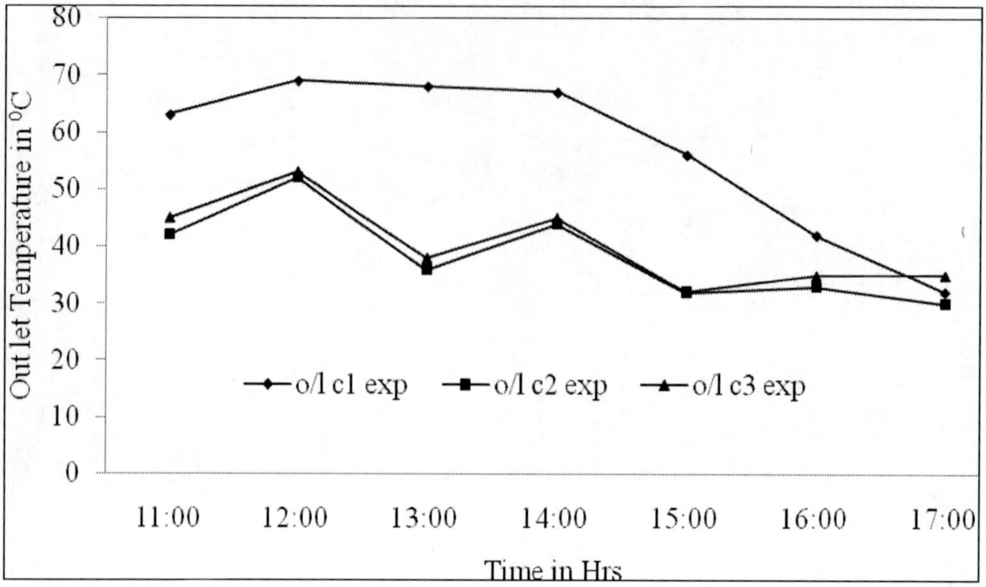

Figure 4.9. Variation of outlet temperature of collectors.

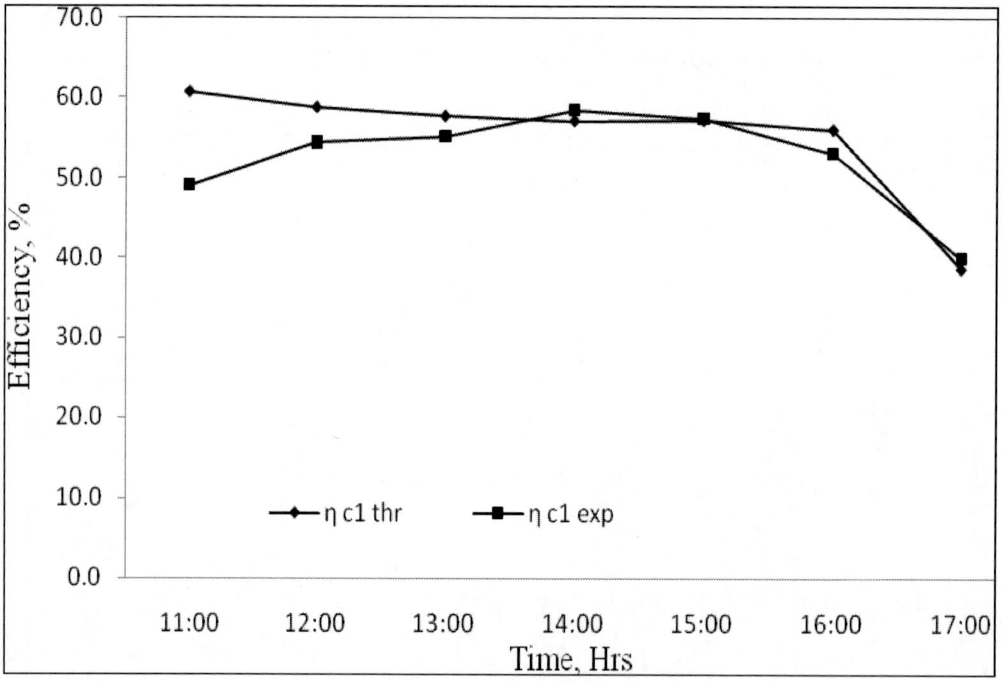

Figure 4.10. Experimental and theoretical η inst of collectors.

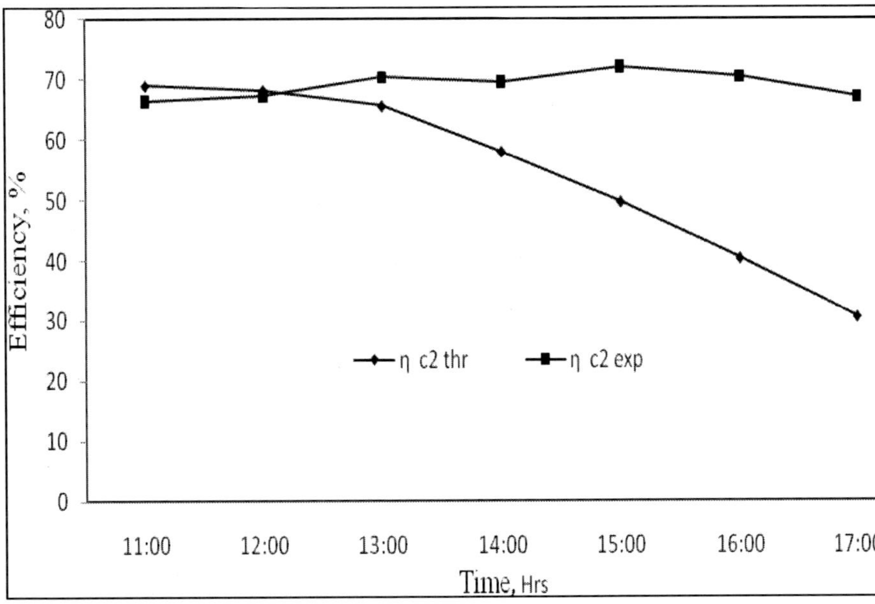

Figure 4.11. Comparision of Experimental and theoretical H inst of collector 2.

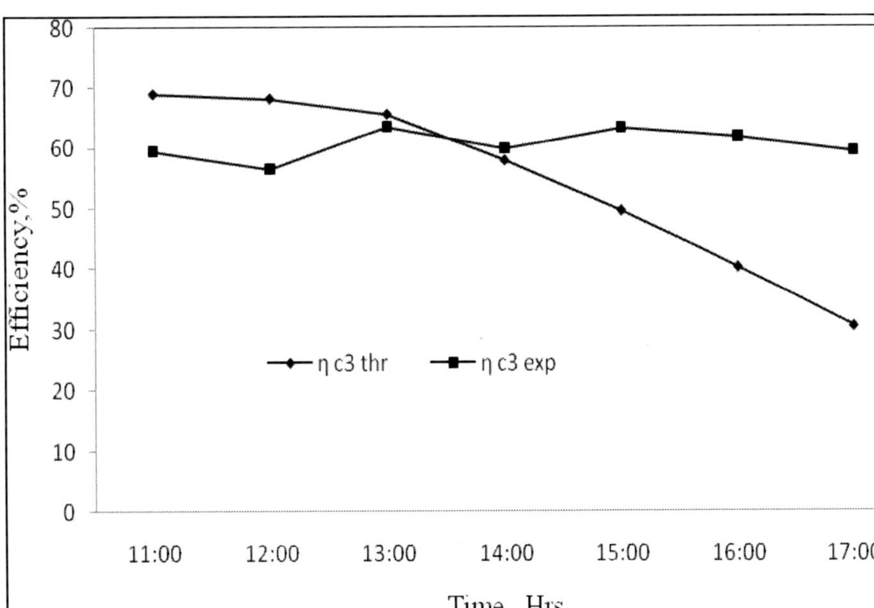

Figure 4.12. Comparision of Experimental and theoretical η inst of collector 3.

4.2.1. Investigations on Performance of Collector 1

The Figure. 4.2 depicts variations of absorber plate temperature with time. It was indicated that the average plate temperature was reported during peak sunshine hours, notable difference was observed in curves for the mass flow rate form 0.03 kg/s to 0.05 kg/s. The dependence on mass flow rate of air clearly indicated that for higher mass flow rate the temperature was lower owing to greater heat removal from the plate. The maximum tempearture was 120 °C for 0.03 kg/s flow rate during 12:00 to 14:00 hours of the day.

The Fig. 4.3 depicts variations in exit temperature with time that indicated a similar trend as the absorber plate temperature. The higher mass flow rate of air leads to decreased exit air temperature which was higher during the duration of 12:00 to 14:00 hours owing to greater input solar insolation for all flow rates of air to the collector. The highest recorded air temperature was 80°C for a mass flow of 0.03 kg/s while the lowest occurred at 0.05 kg/s that was obtained during the period close to solar noon.

The Fig. 4.4 shows variations in instantaneous thermal efficency of solar air collector with time that indicated lower values during period close to solar noon owing to higher inlet air temperatures that contributed to enhanced heat losses. The earlier and later part of the day exhibited a better instantaneous efficiency due to lower thermal losses. The dependance of losses indicated that instantaneous efficiency was higher for mass flow rate of 0.05 kg/s at 60 percent as compared to 0.03 kg/s that indicated 45 percent instantaneous efficiency that was supported by the fact that at lower mass flow rate operating temperatures were higher that contributed to more thermal losses and lower efficiency.

The Fig. 4.5 indicated overall efficiency of the hybrid system, showed similar trend of variations as the instantaneous efficiency except for relatively higher values. This was due to additional amount of electric energy contributed to the useful energy term of the efficiency factor. The dependance of losses indicated that overall efficiency was higher for mass flow rate of 0.035 kg/s at 68 percent as compared to 0.03 kg/s that indicated 55 percent overall efficiency due to improvement in electircal efficiency of the SPV system.

The Fig. 4.6 dipicts variations in SPV panel temperature with time that indicated that panel temeperature was higher for the duration when solar insolation was higher that occurred during the period closer to solar noon. In relation to the exit air and absorber plate temperture the panel temperature was relatively lower but depicted the same trend in variations exhibited by the former parameters. The maximum value of SPV panel temperature was obtained for the lower mass flow rate of air that indicated poor heat extraction from the collector for lower mass flow rate of air. On the contrary higher mass flow rate of air ensured better heat transfer rate to the working fluid maintaining the SPV panel temperature lower, which also leads to its better performance.

4.3. Comparative Performance of Collector 1, 2 and 3

Figure 4.7 shows the comparison of overall thermal efficiency of three collectors collector-1,2 and 3 with time. The overall efficincy includes the thermal and electrical efficiencies of the collectors. Collector 1and 2 were hybrid solar PV air collectors and collector 3 was plane solar air collector. Hence collector 1 and 2 exhibited higher overall efficeincy as compared to collector 3. The highest overall efficiency was achived in collector 2 up to 75 percent for mass flow rate of rate of .03kg/s.

Figure 4.8 represents comparison of predicted outlet air temperature for three collectors. It was indicated that theroritical outlet air temperature of collector 2 and 3 were higher at solar noon. The trend then exhibits an eventual drop in temperature. The maximum tempearture was reported as 59 °C for flow rate of 0.03 kg/s during peak sun shine hours of the day.

Figure 4.9 represents comparison of experimental outlet air temperature for 3 collectors with different geometry and fin confugurations. It indicated that experimental outlet air temperature of collector 1 was higher because of higher insolation, and higher ambient temperature during the conduction of test in the month of April, in comparition to the later two collectors which were subjected to test during the month of June. The trend then exibits an eventual drop in temperature.

Figure 4.10 dipicts comparison of theoretical and experimental instantaneous thermal efficiency of collector 1 with time. The results show larger varaition of theoretical values of instaneous effieciency during day. It varies between 35 to 70 percent. However experimental results shows a small variation in value of instanatneous efficiency during a day. It varies from 65 to 73 percent. Lower values of instaneous efficiency during the period close to solar noon, owing to higher inlet air temperatures that contributed to enhanced heat losses. The earlier and later part of the day exhibited a better instantaneous efficiency due to lower thermal losses. The variation in theoretical and experimental results were less than 4 percent.

The Figure 4.11 dipicts comparison of theoretical and experimental instantaneous thermal efficiency of collector 1 with time. The experiment indicated lower values during the period close to solar noon owing to higher inlet air temperatures that contributed to enhanced heat losses. The later part of the day exhibited a higher instantaneous efficiency due to lower thermal losses.

The Figure 4.12 shows the comparison of theoretical and experimental results of instantaneous thermal efficiency of collector 3 with time. It was observed that the experimental results indicated lower values during period till 14.00 hr owing to higher inlet air temperatures that contributed to enhanced heat losses. The earlier and later part of the day exhibited a better instantaneous efficiency due to lower thermal losses. The highest experimental instantaneous efficiency was achieved up to 65 percent for the mass flow rate of 0.03kg/s.

CONCLUSION

The salient conclusions drawn from the proposed study to review designs and analyze a thermal and overall efficiencies of three different types of solar collectors with L type transverse (collector 1), broken transverse (collector 2) and longitudinal (collector 3), operating at varying solar insolation and mass flow rate (0.03 kg/s to 0.05 kg/s) are summarized below.

- The experimental investgations on the solar air heating system clearly highlighted the role of fins(extended surfaces) in the enhancement of heat transfer characteristics of the system through increased turbulence in flow over the absorber surface thereby contributing to a higher efficiency. There was a need for long term studies on the solar collector that would reveal the serviceability aspect of solar collector while the present investigation was limited to only the heat transfer characteristics
- The Solar air collectors with finned surfaces attained plate temperature up to 120 °C with a decrease in its magnitude with increase in mass flow rate of working fluid. The lower flow rate of air lead to higher air temperatutre in the order of 80°C at 0.03

kg/s flow while the 0.05 kg/s air flow delivered air at 52 °C during the period close to solar noon.

- The instantaneous efficiency was directly proportional to air flow rate, with 60 percent instantaneous efficiency at 0.05 kg/s as compared to 45 percent at 0.03 kg/s owing to lower flow rate producing relatively higher operating temperatures.
- The overall efficiency that accounted instantaneous thermal efficiency and electrical efficiency of SPV system was higher at higher mass flow rate (68 percent at 0.035 kg/s) compared to lower mass flow rate of air (55 percent overall efficiency at 0.03 kg/s) owing to the fact that both components of overall efficiency were directly proportional to mass flow rate of air.
- The average exit air temperature that was used for thermal applications like crop drying depended again on the mass flow rate of air and was recorded to vary in the range of 45 to 65 °C for the experiments conducted. The temperature range obtained was congenial for solar cabinet drying of fruits and vegetable.
- The use of SPVT system enhances the overall useful energy extracted from the incident solar energy as indicated in the higher value of overall conversion effiicency as compared to individual thermal and elecrical efficiencies. The comparision between the three configuration of solar collectors based on overall conversion efficiency clearly revealed that collector with broken transverse rib fins performed better than the other two configuration (longitudnal and transverse fins) for the entire part of the day.
- The experiemental investigation obtained from the present study are very useful in designing of solar flat plate collector that can be emlpoyed for crop drying applications that was very relevent in the context of alarming rise in post harvest losses in the country.

ACKNOWLEDGMENTS

The authors express gratitude for the financial assistance provided by Visvesvaraya Technological University, Belgaum under the Research Promotion scheme and Management, B.V. Bhoomaraddi college of Engineering and Technology, Hubli, for the execution of the project.

REFERENCES

[1] http://www.mnre.gov.in

[2] A.R. Jaurker, J.S Saini and B.K Gandhi, “Heat transfer and friction characteristics of rectangular solar air heater duct using rig-grooved artificial roughness”. *Solar Energy* Vol. 80, 895–907, 2006.

[3] P.C.Tiwade., “Experimentation and performance analysis on continuous longitudinal fins for solar air heater” *International Conference on Engineering trends in Engineering and Technology,* 2009.

[4] Mridul Sharma, Varun "Performance estimation of artificially roughened solar air heater duct provided with continuous ribs" *J. Energy and Environment.* Vol. 1: 897-910, 2010.

[5] Manish Kumar Chauhan, Varun, Sachin Chaudhary "Performance evaluation of roughened solar air heater having M-shaped as roughness geometry on the absorber plate" *J. Energy and Environment* Vol 3, Issue 6, 881-894, 2012.

[6] Muneesh Sethi, Mridul Sharma, Varun "Effective efficiency prediction for discrete type of ribs used in solar air heaters" *J. Energy and Environment* Vol 1, Issue 2, 333-342, 2010.

[7] D. Peng, X. Zhang, H. Dong, and K. Lv, "Performance study of a novel solar air collector," *Applied Thermal Engineering,* vol. 30, pp. 2594-2601, 2010.

[8] Sachin Chaudhary., Varun., Manish Kumar Chauhan., "Heat Transfer And Friction Factor Characteristics Using Continuous M Shape Ribs Turbulator at Different Orientation on Absorber Plate Solar Air Heater" *Energy and Environmental* Vol3, Issue1, 2012, pp.33-48.

[9] Maneesh Kaushal., Varun., "Heat Transfer And Friction Factor Characteristics rectangular channel solar air heater duct having protrusions as roughness element" *Energy and Environmental* Vol3, Issue-6, 967-976, 2012

[10] Ionut, Razvan Caluianu, Florin Baltaretu, "Thermal modelling of a photovoltaic module under variable free convection conditions", *Applied Thermal Engineering* 86-91, 33-34, 2012.

[11] Rehena Nasrin and M. A. Alim, "Performance of nanofluids on heat transfer in a wavy solar collector", *International Journal of Engineering, Science and Technology*, Vol. 5, No. 3, 58-77,2013

[12] Foued Chabane., Noureddine Moummia., Said Benramachec., "Experimental analysis on thermal performance of a solar air collector with longitudinal fins in a region of Biskra, Algeria". *J. Power Technologies* vol 93, Issue 1, pp52–58, 2013.

[13] T.V. Ramachandra, Rishabh Jain, Gautham Krishnadas, Hotspots of solar potential in India, *Renewable and Sustainable Energy Reviews* 15 (2011) 3178–3186

[14] H.P.Garg, R.S.Adhikari, System Performance Studies on a Photovoltaiclthermal (PV/T) Air Heating Collector, Renewable Energy 16 (1999) 725-730.

[15] A.S. Joshi, A Tiwari, Energy and exergy efficiencies of a hybrid photovoltaic–thermal (PV/T) air collector, *Renewable Energy* 32 (2007) 2223–2241.

In: Renewable Energy and Sustainable Development
Editors: R. T. Durai Prabhakaran, S. A. Kale et al.
ISBN: 978-1-63463-429-8

Chapter 3

MATERIALS FOR SMALL WIND TURBINE BLADES

Sandip A. Kale**, Swanand R. Kulkarni, Suraj D. Shravagi and Ganesh P. Bharambe
Trinity College of Engineering and Research, Pune, India

ABSTRACT

Renewable energy sources play a significant role in meeting the increasing power demand across the globe in various aspects. Solar and wind energy are the major renewable energy sources available in abundance. Wind energy has become more popular than solar energy because of its compatible cost with conventional energy sources. Large wind turbine technology has reached a peak in the market. Small wind turbine market is growing slowly and demands reliable and durable technology. Blade and generator are the key elements of small wind turbines. Many researchers are working to enhance reliability and performance of blades and generators. Performance of wind turbine blade is dependent on profile and blade material. Selection of blade material is important in designing the wind turbine blade. Wood, metals, glass fiber reinforced polymer, carbon fiber reinforced polymers, natural fiber reinforced polymer, and nanocomposites are the various materials available in the market for small wind turbine blade. These materials have different properties. This chapter presents detailed Strengths, Weaknesses, Opportunities and Threat (SWOT) analysis of the materials available in the market.

1. INTRODUCTION

Energy demand across the world is continuously increasing due to population growth, per capita energy consumption and large industrialization. Conventional fossil fuels are available in limited quantity and may be not able to meet energy demand in the future. Use of renewable energy sources is spreading at a fast rate. Solar, wind, biogas, ocean, and tidal energy are various renewable energy sources. Among these, solar and wind energy are the most promising [1-3]. Cost and payback period of solar energy is comparatively more than

* Corresponding Author. Tel:+91 8888821160, Email:sakale2050@gmail.com.

conventional energy sources. The wind energy cost is compatible with conventional energy sources [4]. Today large wind turbines have grabbed significant positions with worldwide cumulative installation of 336 GW in first half of 2014 [5]. Small wind turbine market is also growing and forecast to reach 3700 MW by the 2020 [6]. Small wind turbines are capable of generating power in low or medium wind regions. A reliable small wind turbine at a compatible price is a need of market [4]. Many researchers and industries across the globe are working to fulfill this market demand.

Blade is a key element of a wind turbine. The performance of the wind turbine blade mainly depends on the aerodynamic profile and angle of the blades as well as the materials [7]. Wind turbine blades are exposed to diverse working conditions, such as dust and shiny environment, heavy rainfall and snowfall, high wind velocities in the rugged topography, drastic changes in the temperature, moisture, and ultraviolet radiation. In addition, wind turbine blades are subjected to high fatigue. It results in crack formation and growth. The load induced in large and small wind turbine blades is different because of rotational speeds. Large wind turbine blades rotate slowly and small wind turbine blades rotate several times faster. As a result, centrifugal forces in small wind turbines are major as compared to that in large wind turbines. In addition, small wind turbine blades experience more fatigue cycles as a function of rotational speed. Expected life of blade is 20 years and number of fatigue load cycles may exceed 10^8 in small wind turbines [8]. Appropriate blade design and material selection is required to ensure this lifelong reliability. A focus is required on material selection and associated manufacturing process of small wind turbine blade. The blade material should have high strength and stiffness, high strength to weight ratio, resistance to fatigue damage, resistance to environmental loading and low-cost. The wind turbine manufacturing industries are focused on lightweight, cost-effectiveness, easy availability and environmental friendly material for manufacturing of blades.

A wide range of materials is available for small wind turbine blades. This range includes wood, metals, glass fiber reinforced polymer, carbon fiber reinforced polymer, natural fiber reinforced polymer and nanocomposites. Each of these materials has its optimistic and off-putting characteristics. It is necessary to compare these characteristics. This chapter presents characteristics of these materials through SWOT analysis with its Strengths, Weaknesses, Opportunities in the future and any possible Threats.

2. Materials for Small Wind Turbine Blades

Wood, metals, Glass Fiber Reinforced Polymer (GFRP), Carbon Fiber Reinforced Polymer (CFRP), Natural Fiber Reinforced Polymer (NFRP) and nanocomposites are the major materials used in the manufacturing of small wind turbine blades. This section presents different properties of these materials. In strength design of wind turbine blade, it is important to consider the specific strength and specific stiffness of materials. Specific strength is strength-to-weight ratio (σ_f / ρ) and specific stiffness is the stiffness-to-weight ratio (E/ρ). These are dependent on material type and its density. The advantage of composite materials is that they offer highest values of these ratios, which meet specific needs required for wind turbine blades. Figure 1 shows a comparison of specific strength and specific stiffness for some conventional, composite and nanocomposite materials [7]. Within the composite family, it is easy to discern the positions of GFRP, CFRP and Carbon Nanotubes (CNTs).

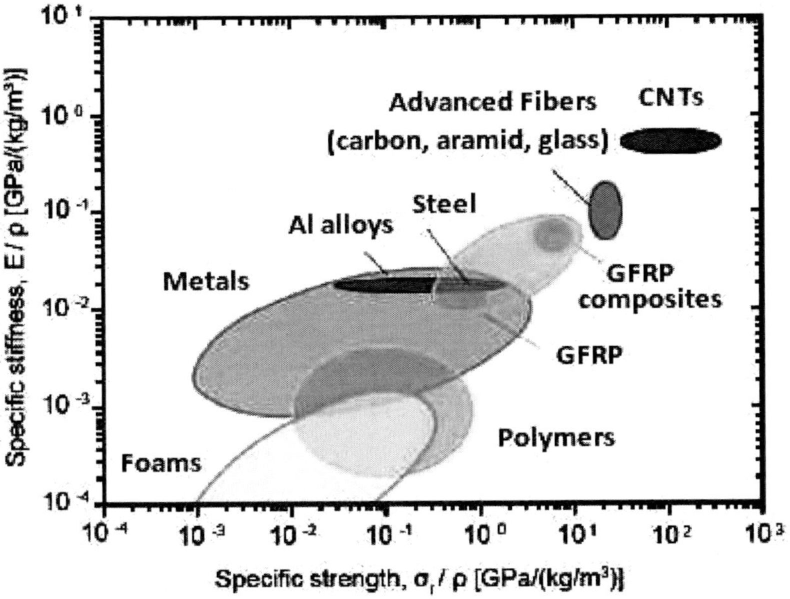

Figure 1. Specific stiffness versus specific strength for materials.

2.1. Wood and Canvas

The various types of wood suitable for small wind turbine blades are Nepali Timber, Radiata Pine, Australian Hoop Pine, Sitka Spruce Timber and Douglas Fir. Wood is the natural composite made up of cellulose and lignin. Structural damping of the wind turbine blade improves its life and performance. It improves the aero-elastic stability of the wind turbine blade. Wood is one of the best damping material. It has good strength to weight ratio. Anisotropic property of the wood favors its use in the manufacturing of the blade. It shows well fatigue characteristics and tensile strength along the length of the fiber growth. Further, it is easily available and economical. Use of wood in the manufacturing of small blade lowers the cost [1, 23].

Use of wood in the manufacturing of blade is limited because of its lesser resistance to environmental degradation and drastic impact on the environment. To overcome these problems, wood blades are laminated with weather resistant coatings. In addition, special tapes made up of polyurethane or aluminum are used to keep the tip of the blade safe against wind erosion. However, the cost of laminating blades is high. As blade length increases, it becomes more difficult to get knot and defect free logs. Fabrication of the wind turbine blade from the two different logs of the wood affects the strength and increases the manufacturing cost [10, 23].

2.2. Metals

As compared to wooden material, conventional metals have high stiffness to weight ratio and better resistance to environmental degradation. In the 19th century, galvanized steel was used instead of wood. Subsequently aluminum was used for manufacturing of wind turbine blades because of low density. Density of the steel is 8000 kg/m^3 and that of aluminum is 2700 kg/m^3 [10, 24]. Use of low density material such as aluminum as a wind blade material considerably reduces the weight of the blade.

Steel is a common, relatively inexpensive material and used extensively in industries. However, it is difficult to manufacture a complex twisted shape from steel. Besides having low density and good cast ability of the aluminum, its use in the manufacturing of wind blade is limited due to low fatigue life and high material cost [10].

2.3. Composites

A composite material is a combination of two or more chemically dissimilar constituents. It enhances desirable characteristics of the used constituents. This recompenses weaknesses of original materials. The constituents of a composite are immiscible, distinct and identifiable. To get desirable properties of wind turbine blades, different composites such as GFRP, CFRP, NFRP, nanocomposites, etc are developed by researchers. These composites consist of long fibers to give longitudinal strength & stiffness and resin matrix improves fracture toughness and de-lamination strength [9]. The use of such developed materials for manufacturing of wind turbine blades has increased exponentially.

Strength to weight ratio and stiffness to weight ratio of composites are many times greater than steel and aluminum. In addition, fatigue properties are generally better than common engineering metals. These are also lighter than the metals and this reduces the inertial and fatigue load on the rotor shaft and blade.

2.3.1. Glass Fiber Reinforced Plastics

The glass fibers are manufactured in the form of woven matrices or knitted cloth. They are formed into continuous strand, chopped strand mat or chopped fibers (whiskers). These mats are fabricated with variable density and thickness. Whiskers, fibers, and matrices of the glass fibers are manufactured by various methods. For polymer matrix composites, glass fibers are the most common of all reinforcing fibers, which are amorphous solids. Chemically, glass is primarily composed of silica (SiO_2) that acts as a backbone in the form of -SiO_4 tetrahedral. GFRP with woven matrices has high strength and Young's modulus. They can withstand the working temperature to the order of 650ºC. Low thermal expansion of these composites provides conformability to the dimension. The tensile strength of the composite is considerably higher than that of the constituents. Properties of the fibers also depend upon the orientation and stacking sequence. Figure 2 shows the glass fibers with four different orientations as uniaxial, biaxial, triaxial and random.

As shown in Table 1 and Table 2, the glass fibers are lightweight and their resistance to corrosion is very high. The process of manufacturing of the wind turbine blade with glass fiber as a constituent is well established.

Table 1. Properties of wind turbine blade materials [10-15, 18, 24]

Sr. No.	Material	Density kg/m^3	Young's modulus GPa	Breaking strength MPa	Breaking Strain	Cost per kg in $
1.	Wood	400-625	0.5 – 2.5	40 -90	0.5 – 1.5	0.3-5
2.	Steel	8000	200	200-1000	3-5	0.6-2.5
3.	Aluminum	2500	70	50-300	4-12	2.5-5
4.	Glass fiber	2660	30-80	2000-3500	2.5	2-3
5.	Carbon fiber	1600	100-1000	380-6200	1.6	7-90
6.	Plant fiber	1350-1550	73	400-1500	1.4-3.2	1-3
7.	Nanocomposite	Depends upon type of constituents				

Table 2. Qualitative comparison of wind turbine blade materials [10-15, 18, 24]

Sr. No.	Material	Renewable source	Recyclable	Biodegradable	Eco-friendly Manufacturing process	Corrosion resistance	Availability
1.	Wood	Yes	Yes	Yes	Yes	Least	High
2.	Steel	No	Yes	Yes	No	Medium	Medium
3.	Aluminum	No	Yes	Yes	No	Medium	Medium
4.	Glass fiber	No	No	No	No	High	Least
5.	Carbon fiber	No	No	No	No	High	Least
6.	Plant fiber	Yes	Yes	Yes	Yes	Low	Least
7.	Nano-composite	No	No	No	No	High	Least

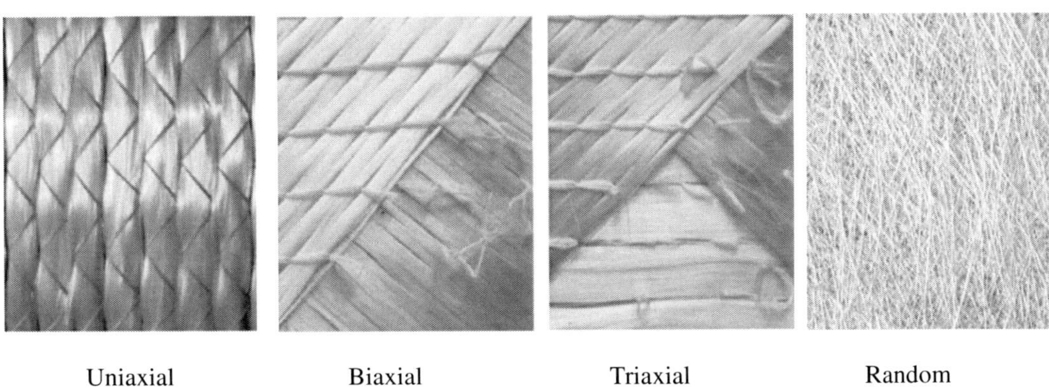

Figure 2. Orientation of glass fibers [21].

The properties of the composite depend upon the permeability of the mats. The permeability depends on several factors such as the nature of the reinforcement, the direction and arrangement of fibers, the stacking sequence of plies, the temperature of the resin, the position of injection-vent ports, and so forth. Mats formed with the glass fiber have better permeability than that formed with any other fiber. These fibers are not hygroscopic and they are less prone to ultraviolet degradation.

The fatigue strength of the fiberglass is much more than conventional metals. GFRP has fatigue strength of 140 MPa, many times greater than that of weld-able aluminium as 17 MPa at 10^7 cycles. Its use in the area with high snowfall and rainfall is highly beneficial. High stiffness to weight ratio and dimensional stability of the glass fiber reinforced composite makes it a suitable material for wind turbine blades. [10, 13, 14]

Manufacturing of blade using GFRP is a time-consuming process. Manufacturing of glass fiber is not an eco-friendly process because it involves harmful chemical emissions of the silica material.

2.3.2. Carbon Fiber Reinforced Polymers

CFRP have high specific strength and specific stiffness. They are reinforced with resin-matrix to formulate composites. Carbon fiber has a carbon content of 80.95%. The carbon content is varying depending upon the type of heat treatment. The properties and the cost of these fibers are directly proportional to the carbon content. Properties of carbon fibers are based on graphene (hexagon type) layer networks present in carbon. Figure 3 shows the cross-section of a cross-plied carbon/epoxy laminate. [15]

CFRP are more expensive than any other fiber. The cost of CFRP is 7-20 times that of glass fiber Reinforced Composites [13]. Manufacturing of CFRP is not an eco-friendly process. The fiber misalignment reduces compressive and fatigue strength considerably.

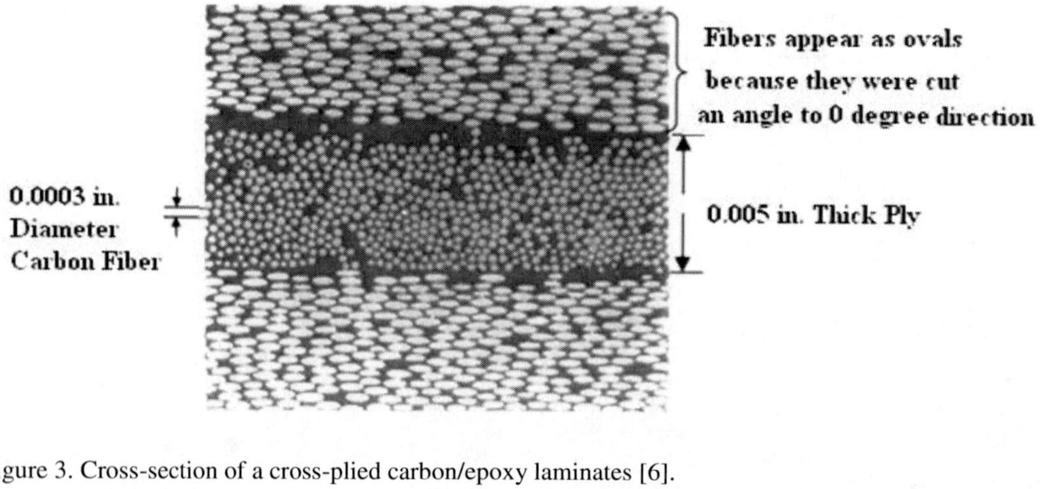

Figure 3. Cross-section of a cross-plied carbon/epoxy laminates [6].

2.3.3. Nanocomposites

Nanocomposites are manufactured with randomly reinforced nano-particles of metals or ceramics as reinforcement. Carbon Nanotubes (CNTs) are the nano materials based on the graphite, which is the stable form of carbon at ordinary temperature and pressure. These tubes are used as continuous fiber in various structural applications. These tubes have a very high aspect ratio of diameter to length. The diameter range is 1-80 nm and the length can reach several micrometers. Such unique characteristics contribute to improve durability of structures and structural components. CNTs are manufactured with several processes. The Chemical Vapor Deposition (CVD) process is a commonly used process. This process is not eco-friendly.

These materials have high specific stiffness, specific strength, thermal and electrical conductivity. For a same cross-section, the strength of nanocomposites increases many times that of the CFRP. These improved properties increase their use in the manufacturing of the blade.

However, from an ecological point of view, these materials are likely to have adverse impacts on the environment [10].

2.3.4. Natural Fiber Reinforced Polymer

Though the GFRP, CFRP and CNTs have better properties than the conventional metals, their manufacturing processes are not an eco-friendly. These materials are highly dependent on petroleum-based resources, which are depleting rapidly. The carbon content is emitted during manufacturing processes. Hence, researchers are working optimistically to develop ecofriendly composite materials for wind turbine blades. NFRP can satisfy these requirements.

These are the plant fibers reinforced with either thermoset or thermoplastic epoxy material. Natural fiber itself is a composite material, where hollow cellulose fibrils are held together by lignin and pectin in hemicellulose matrix [11, 12]. Researchers are focusing on development of the NFRP, which has high resistance to moisture absorption along with the above mentioned properties. NFRP may prove an alternative to conventional and composite materials for manufacturing of the wind turbine blades [17, 21-23]. These fibers have low density, low-cost, non-abrasive properties, and are renewable. Natural fibers such as sisal, flax, hemp, kenaf, bagasse, banana, jute, abaca and bamboo are easily available and need low processing cost. Other example of natural fiber composite is the Mongols bows, which are made by combining cattle tendons, horn, bamboo and silk where natural pine resin are used for binding.

Table 3 shows the world production and origin of the various types of fibers. NFRP are classified based on their origin, types of plant, animal and mineral types. Depending upon the source, natural fibers are broadly divided into three categories: (i) plant fibers, (ii) animal fibers and (iii) mineral fibers. The Figure 4 shows the Banana fiber and a cross-section of banana fiber cell [10- 12]. It can be recycled to optimize the use of material.

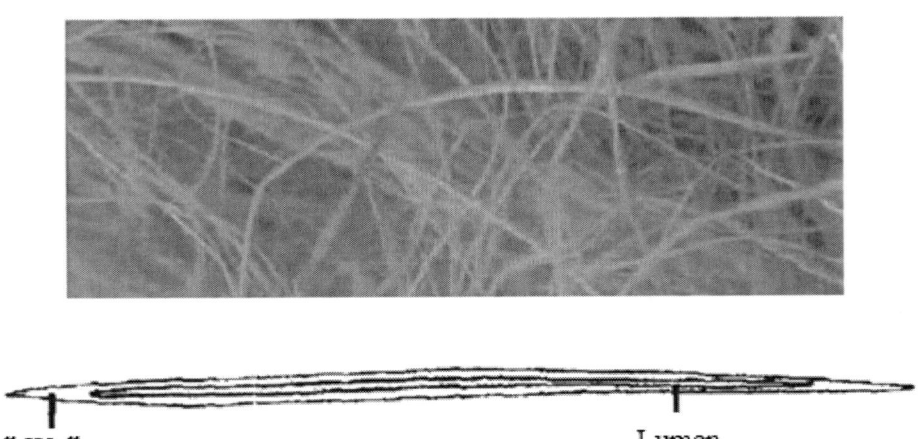

Figure 4. Banana fiber and cross section of banana fiber cell.

3. SWOT Analysis of Materials for Wind Turbine Blades

Material	Strengths	Weaknesses	Opportunities	Threats
Wood & canvas	• Lightweight • Good damping property • Good strength to weight ratio in the direction of fiber growth • It is a natural composite and can form composite with plastic • Low cost	• Hygroscopic in nature • Length of blade is limited by wood size and type • Requires large land and fast growing plants to compensate demand • More manufacturing time	• Life and performance can be improved after lamination • Can also used as a reinforcement in the NFRP • Recyclable	• Wood cutting has adverse effect on the environment • Change in length due to temperature and humidity varies the stress concentration and load conditions on the rotor shaft • High crack growth rate causes sudden failure
Conventional metals	• Specific strength and specific stiffness are greater than wood • Does not absorb moisture • High Young's modulus, hardness and breaking strength • Readily available	• Difficult to manufacture into complex shapes • Low fatigue strength • Heavier than wood • High inertial and gravitational forces on the rotor shaft	• Fibers can be used as reinforcement • Can be used along with composite patching from outside to optimize cost, material, and properties	• May attract the thunder during rainy season and cause damage. • Good conductors of electricity
GFRP	• Specific strength and stiffness is greater than wood and metals • High permeability than any other fibers • Cheaper among all composites • Chemically inert • Can be formed into intricate shape • Good dimensional stability	• Time consuming manufacturing process • Not ecofriendly	• Sandwich can be used along with carbon fiber & NFRP • Specific automatic and time saving manufacturing process can be developed for blades	• Manufacturing of raw materials depends upon the petroleum products which are rapidly depleting • Increases the carbon content in the environment
CFRP	• Specific strength and specific stiffness are greater than GFRP • Chemically inert • Can be formed into intricate shape • Good dimensional stability	• High cost • It is not easily available • Time consuming manufacturing process • Not ecofriendly • Permeability is lower than the glass fiber • Sensitive to fiber misalignment and waviness	• Sandwich can be used along with GFRP & NFRP • Specific automatic and time saving manufacturing process can be developed for blades	• Manufacturing of raw materials depends upon the petroleum products which are rapidly depleting • Increases the carbon content in the environment • Non uniform interface bonding results in defects

Material	Strengths	Weaknesses	Opportunities	Threats
NFRP	• Environmental friendly • Compatible specific strength and specific stiffness • Can be formed into intricate shape • Sustainable manufacturing process • Low cost of raw material and manufacturing	• Hygroscopic in nature • Requires coating material • Low range of working temperature • Less developed manufacturing process of raw material	• Life and performance can be improved after lamination • Can also used as a reinforcement in the NFRP • Recyclable • Increased interest and awareness towards clean energy & environmental issues	• High crack growth rate will cause sudden failure • Change in length due to temperature, humidity and load, varies the stress concentration and load distribution at the interface.
Nanocomposites	• Variability in the properties as per type of material choice • Good dimensional stability • Can be formed into intricate shape • High range of working temperature • High specific stiffness and specific strength	• High cost of raw material • It is not readily available • Time consuming manufacturing process • Not ecofriendly • Higher energy consumption in the fabrication of material • Skilled workers required • Prone to coagulation and segregation of the nanopowder during fabrication of the nanocomposite	• Can be used as a coating material to reduce the effect of the impact of the dust particles • Can be used with NFRP to reduce the effect of the temperature, humidity and radiation	• Recycling and sustainability is difficult • Repair cost after damage will be high • Failure analysis is difficult • Stress distribution is non-uniform • Life prediction is difficult • High thermal and electrical conductivity

Table 3. Plant Fibers [11, 12]

Sr. No	Fiber Source	Species	World production (10^3 tonns)/year	Origin of fiber
1	Jute	Corchors	2,300	Stem
2	Bamboo	1250 species	10,000	Stem
3	Cotton lint	Gossypium sp.	18,450	Fruit
4	Sunhemp	Crorolaria juncea	70	Stem
5	Kenaf	Hibiscus cannabius	970	Stem
6	Wood	1000 species	1,750,000	Stem
7	Sisal	Avage sisilana	378	Leaves
8	Roselle	Hibiscus sabdariffa	250	Stem
9	Hemp	Cannabis sativa	214	Stem
10	Coir	Cocos nucifera	100	Fruit
11	Ramie	Boehmeria nivea	100	Stem
12	Abca	Musa textiles	70	Leaves
13	Flax	Linum usitatissimum	830	Stem

Conclusion

A concise description of various properties of the materials used for small wind turbine blades is presented in this chapter. Strengths, Weaknesses, Opportunities and Threats analysis for different materials such as wood, metals, GFRP, CFRP, NFRP and nanocomposites is carried out specifically over the properties, manufacturing process, cost and availability. Wood is lighter, eco-friendly, easily available and cheaper material. Fatigue and strength properties of wood are poorer than that of the metals and composites. Lamination is required on the wooden blade due to its hygroscopic nature and manufacturing of such blades is a lengthy process and needs a skilled worker. Still, many developing countries are using wood as a blade material for small wind turbine. Metals have moderate strength and fatigue properties. They are not preferred for blades because of limitations associated with the manufacturing process and fatigue strength. Composites are widely used materials for small wind turbine blades. GFRP is commonly used composite because of its many desirable properties such as high specific strength and specific stiffness, easy availability, low-cost and ease in manufacturing. Nowadays, use of CFRP is also increasing because of its higher strengths. Its use is limited due to high cost and less availability. Nanocomposites and NFRP are the emergent materials for blades. Nanocomposites have many desirable properties. Manufacturing of raw materials for GFRP, CFRP and nanocomposites are not eco-friendly processes. NFRP can become a good option along with other composites. All composites should have easy, automatic and time-saving manufacturing processes.

Acknowledgments

We are thankful to Prof. Kulkarni Hema, Trinity College of Engineering and Research, Pune, India, for her support during completion of this chapter.

References

[1] Wood, D., (2011). *Small Wind Turbines: Analysis, Design and Applications.* Springer-Verlag London Limited.

[2] Kale, S. A., Varma, R. N. (2014). Aerodynamic Design of a Horizontal Axis Micro Wind Turbine Blade Using NACA 4412 Profile. *International Journal of RenewableEnergy Research.* Vol. 4, No. 1. 69-72.

[3] Wang, F., Bai, L., Fletcher, J., Whiteford, J., Cullen, D. (2008). The methodology for aerodynamic study on a small domestic wind turbine with scoop. *Journal of Wind Engineering and Industrial Aerodynamics.* 96. 1–24.

[4] Kale, S. A., Sapali, S. N. (2013). A review of multi-rotor wind turbine systems. *Journal of Sustainable Manufacturing and Renewable Energy*. Vol. 2, No. 1-2.

[5] 2014 Half-Year Report. (2014). World Wind Energy Association.

[6] Zhang, P. *Small Wind energy Report.* (2012). World Wind Energy Association.

[7] Brahim, A. (2013). Designing Composite Wind Turbine Blades from Cradle to Cradle. Recent Advances in Composite Materials for Wind Turbine Blades. *World Academic Publishing - Advances in Materials Science and Applications.*

[8] Lion Mishnaevsky Jr. (2013). Composite materials in wind energy technology. *Thermal to mechanical Energy conversions: Engines and requirements.*

[9] Kishore, D., Singh, I., Dvivedi, A., Kumar, P. (2013). Natural Fibre Reinforced Polymer Composites for Wind Turbine Blades: Challenges and Opportunities. Recent Advances in Composite Materials for Wind Turbine Blades. *World Academic Publishing - Advances in Materials Science and Applications.*

[10] Composite, Volume 21, ASM handbook, 2001.

[11] Laly, A. P., Zachariah, O., Sabu, T. (2003). Dynamic mechanical analysis of banana fiber Reinforced polyester composites. *Composites Science and Technology.* 63 283–293.

[12] Mukhopadhyay, S., Fangueiro, R., Arpaç, Y., Şentür, Ü. (2008). Banana Fibers – Variability and Fracture Behaviour. *Journal of Engineered Fibers and Fabrics.* Vol. 2, Issue. 3.

[13] Schubel, P. J., Shah, D. U., Clifford, M. J. (2013). Can flax replace E-glass in structural composites? A small wind turbine Blade case study. *Composites: Part* B 52. 172–181.

[14] Yu-Chung TSENG and Chih-Yu KUO. (2011). *Engineering and Construction Torsional* Responses of Glass- Fiber/Epoxy Composite Blade Shaft for A small Wind Turbine. *Procedia Engineering.* 14. 1996–2002.

[15] Xiaosong Huan. (2009). *Fabrication and Properties of Carbon Fibers. Materials.* doi:10.3390/ma2042369. 2369-2403.

[16] Samir Ahmad and Izhar-ul-Haq. (2012) Wind Blade Material Optimization. *Advances in Mechanical Engineering.* Doi.10.5729/ame.vol2.issue4.48. 2 (4).

[17] Corchur O Bradiagh (2004). *First recyclable Wind turbine Blades.* Reinforced Plastics.

[18] Phillipe Christou. (2007). Advanced material for wind blade manufacture. Reinforced plastics.

[19] Norlin, P. (2013). The role of sand, which cores in wind turbine blades. *DIAB technologies.*

[20] UV degradation prevention of fiber reinforced composite blade. (2009). Wachita Sate University, *Department of Mechanical Engineering.*

[21] K. Suresh Babu, N. V. Subba Raju, M. Srinivasa Reddy, D. Nageswara Rao. (2006). The material selection for typical wind turbine blades using a MADM approach & analysis of blades. MCDM, Chania, Greece, June 19-23.

[22] Ancona D. and Jim Mcveigh, J., (2001) "Wind Turbine – Materials and Manufacturing Fact Sheet", Princeton Energy Resources International, LLC.

[23] Leon Mishnaevsky Jr., Peter Freere, Rakesh Sinha, Parash Acharya, Rakesh Shrestha, Pushkar Manandhar. (2011). Small wind turbines with timber blades for developing countries: Materials Choice, development, installation and experiences. *Renewable Energy.* 36.

[24] Joseph R. Davis. (1998). *Metals Handbook* Desk Edition. Second Edition. ASM.

In: Renewable Energy and Sustainable Development
Editors: R. T. Durai Prabhakaran, S. A. Kale et al.
ISBN: 978-1-63463-429-8

Chapter 4

ANAEROBIC DIGESTION SYSTEMS

***B. Deepanraj*[1,*], *V. Sivasubramanian*[2] *and S. Jayaraj*[1]**
[1]Department of Mechanical Engineering, [2]Department of Chemical Engineering,
National Institute of Technology, Calicut, India

ABSTRACT

Commercial energy has become very essential for the modern human living. There is shortage of commercial energy forms due to the fast depletion of fossil fuels and increase in the demand for energy due to increase in population coupled with industrialization. To overcome the above problem, renewable energy sources can be invoked and made ready to serve the mankind on demand. Out of the many renewable options, wind, solar and biomass energies are considered as the major sources. This chapter focuses on the biogas technology, particularly with respect to anaerobic digestion. It has become very attractive way to utilize biomass sources for fulfilling partially the commercial energy requirements. Biogas can be produced through anaerobic digestion process, which provides multiple benefits to the users and aid to protect the environment. This chapter provides an overview of the anaerobic digestion system including various stages, chemical reactions and microorganisms involved in the biogas production. The various parameters that affect biogas production, pretreatment techniques to enhance the digestion process, the feedstock suitable for biogas, classification of anaerobic digestion systems and the types of digester available were also described in this chapter.

1. INTRODUCTION

Rapid depletion of fossil fuel resources, ever increasing petroleum price and the increasing concern for environmental issues have challenged researchers to develop new techniques to obtain clean and sustainable energy through the utilization of all available forms of renewable energy sources [1,2].

* Corresponding author email: babudeepan@gmail.com.

Among the various renewable energy sources available, exponentially accumulating waste biomass is potentially creating a problem as well as a solution in itself to the emerging crisis with respect to the available forms of energy source crisis. Biomass is the third largest primary energy in the world, after coal and oil [3]. It remains the primary source of energy for more than half of the world's population and provides about 1250 million tons oil equivalent (Mtoe) of primary energy which is about 14% of the world's annual energy consumption [4]. Correctly managed biomass utilization is a renewable and sustainable activity that can deliver significant reduction in the net carbon emissions when compared with fossil fuels.

Biomass sources are suitable for different applications such as cooking, steam generation, electricity generation, process heating, shaft power applications and also for producing various chemicals as by-products. Each and every application of the biomass source will primarily based on either physical, chemical or biological conversion. Figure 1 shows the various biomass conversion techniques with corresponding products and applications. Among these techniques, biogas technology appears to be an assuring mode for utilizing biomass sources for fulfilling local energy requirements. This system generates biogas, as a mixture of methane and carbon dioxide with traces of other gases through a biochemical process called anaerobic digestion.

Anaerobic digestion process involves the decomposition of organic matter by microorganisms in an oxygen free environment [6]. It is a natural process occurring in an anaerobic environment such as the ocean/lake sediments and digestive tracts of human beings/animals.

This process comprises of a series of complex reactions that convert a wide array of polymeric organic substances such as carbohydrates, proteins and lipids, having carbon atoms at various oxidation and/or reduction states, to one-carbon molecule. The products of anaerobic digestion are biogas and the digestate. The biogas produced from anaerobic digestion typically contains 60 to 75 percent methane, 20 to 40 percent of carbon dioxide and trace amounts of other gases [7, 8].

Biogas can be combusted to generate heat and electricity. After removing carbon dioxide and hydrogen sulfide using scrubbers, the biogas (methane) can be used in automobiles and fuel cells. The typical composition of biogas is given in Table 1. The digestate from the anaerobic digestion is a good soil improving material which can be used as fertilizer.

Table 1. Composition of biogas [9]

Constituents	Amounts (%)
Methane (CH_4)	50-75
Carbon dioxide (CO_2)	20-45
Nitrogen (N_2)	0-10
Hydrogen (H_2)	0-1
Hydrogen sulfide (H_2S)	0-3
Oxygen (O_2)	0-2

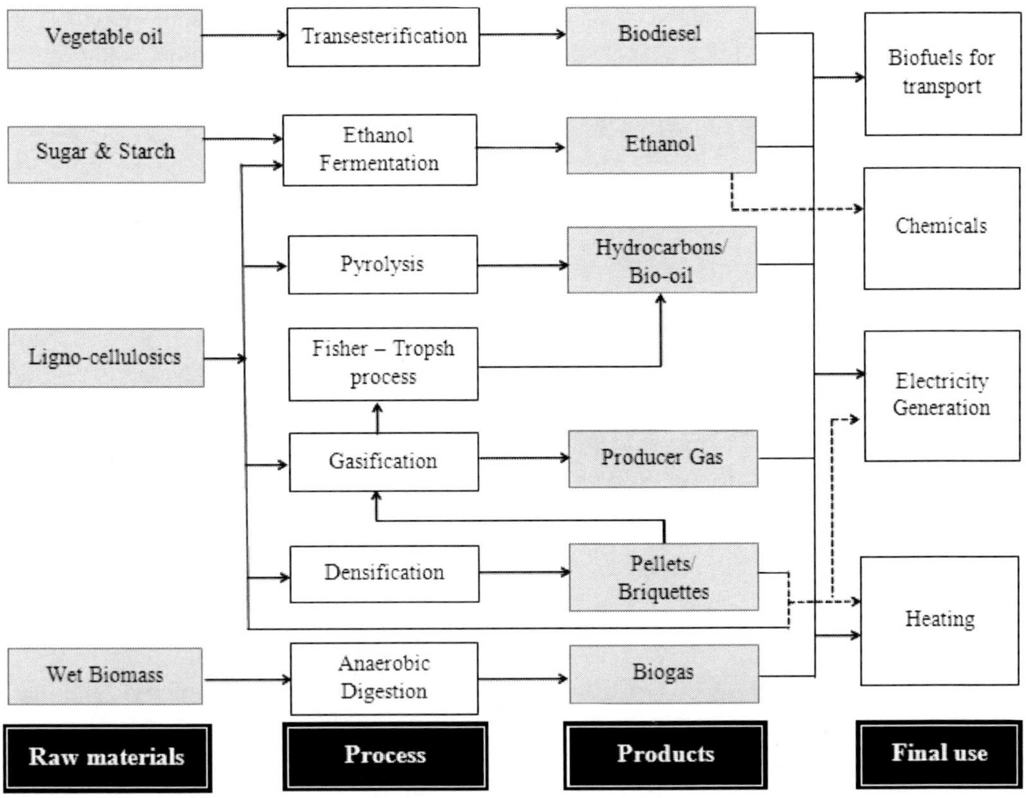

Figure 1. Biomass conversion techniques [5].

2. STAGES OF ANAEROBIC DIGESTION

Anaerobic digestion process comprises of a series of bio-chemical reactions which occurs in four different stages, namely hydrolysis, acidogenesis, acetogenesis and methanogenesis [10, 11]. Each stage is facilitated by distinct group of microorganisms. In first stage, all complex organic compounds are broken down to simpler compounds by hydrolysis. The next two stages involve acid formation and methane is obtained in the last stage. The reactions and the bacteria involved in various stages are enumerated in Table 2 and the process flow chart is shown in Figure 2.

2.1. Hydrolysis

In general, hydrolysis is a chemical reaction which splits water into H^+ cations and OH^- anions. This process converts the larger (complex) organic polymers into simple smaller soluble molecules. During hydrolysis, the large polymers namely carbohydrates, proteins and lipids are broken down into smaller molecules such as simple sugars, amino acids and fatty acids respectively. Some of the hydrolytic products including hydrogen and acetate might be used by the methanogens in the later stages. Hydrolysis stage is carried out by groups of facultative or obligate fermentative bacteria by the excretion of extracellular enzymes [14]. Lipases convert lipids to fatty acids, proteases converts proteins to amino acids and the polysaccharides such as cellulose, starch and pectin are hydrolyzed to monosaccharaides by cellulases, amylases and pectinases [15].

2.2. Acidogenesis

Acidogenesis is the second stage in anaerobic digestion, where the soluble compounds produced through hydrolysis are converted into volatile fatty acids (C1-C5), hydrogen, carbon dioxide, ethanol and some organic nitrogen and sulfur compounds [16]. The acids produced in this stage are acetic acid (CH_3COOH), propionic acid (CH_3CH_2COOH), butyric acid ($CH_3CH_2CH_2COOH$) and valeric acid ($CH_3CH_2CH_2CH_2COOH$). The acetic acid formed in this stage is directly taken to last stage and the other products are taken to third stage for further degradation by acetogens.

2.3. Acetogenesis

Acetogenesis is the third stage in anaerobic digestion process, where the creation of acetate, a derivative of acetic acid takes place. In this stage, the volatile fatty acids having more than two carbon atoms (from acidogenesis stage) are converted into acetic acids, hydrogen and carbon dioxide with the help of acetogens [10].

2.4. Methanogenesis

Methanogenesis is final stage of anaerobic digestion in which the methanogenic bacteria (methanogens) produce methane by consuming acetic acid, hydrogen and some carbon dioxide. Around 66% of methane is formed from acetic acids by means of acetate decarboxylation and remaining 34% of methane is formed from carbon dioxide reduction [12].

Table 2. Chemical reaction and bacteria involved in the anaerobic digestion

Stage	Type of conversion	Bacteria involved
Stage-I Hydrolysis $(C_6H_{10}O_5)n + nH_2O = n(C_6H_{12}O_6)$	Proteins to soluble peptides and amino acids	Clostridium, Proteus vulgaris, Vibrio, Bacillus, Peptococcus, Bacteriodes,
	Carbohydrates to soluble sugars	Clostridium, Acetovibrio celluliticus, Staphylococcus, Bacteriodes
	Lipids to fatty acids or alcohols	Clostridium, Micrococcus, Staphylococcus

Stage	Type of conversion	Bacteria involved
Stage II Acidogenesis $C_6H_{12}O_6 + 2H_2O \rightarrow 2CH_3COOH + 4H_2 + CO_2$ $C_6H_{12}O_6 + 2H_2 \rightarrow 2CH_3CH_2COOH + 2H_2O$ $C_6H_{12}O_6 \rightarrow CH_3CH_2CH_2COOH + 2H_2 + 2CO_2$ $C_6H_{12}O_6 \rightarrow 2CH_3CH_2OH + 2CO_2$ $C_6H_{12}O_6 \rightarrow 2CH_3CHOHCOOH$	Amino acids to fatty acids, acetate and NH_3	Lactobacillus, Escherichia, Bacillus, Staphylococcus, Pseudomonas, Sarcina, Desulfovibrio, Selenomonas, Streptococcus, Veollonella, Desulfobacter, Desulforomonas.
	Sugars to intermediary fermentation products	Clostridium, Eubacterium limosum, Streptococcus
Stage III Acetogenesis $CH_3CH_2OH + H_2O \rightarrow CH_3COOH + 2H_2$ $2CH_3CH_2OH + 2CO_2 \rightarrow CH_4 + 2CH_3COOH$ $CH_3CH_2COOH + 2H_2O \rightarrow CH_3COOH + 3H_2 + CO_2$ $CH_3CH_2CH_2COOH + 2H_2O \rightarrow 2CH_3COOH + 2H_2$ $CH_3CHOHCOOH + H_2O \rightarrow CH_3COOH + CO_2 + 2H_2$	Higher fatty acids or alcohols to hydrogen and acetate	Clostridium, Syntrophomonas wolfeii
	Volatile fatty acids and alcohols to acetate or hydrogen	Syntrophomonas wolfei, Syntrophomonas wolinii
Stage IV Methanogenesis $CH_3COOH \rightarrow CH_4 + CO_2$ $CO_2 + 4\,H_2 \rightarrow CH_4 + 2H_2O$	Acetate to methane and carbondioxide	Methanosaeta, Methanosarcina
	Hydrogen and carbondioxide to methane	Methanobacterium formicicum, Methanobrevibacterium, Methanoplanus, Methanospirilum

Source: [12, 13].

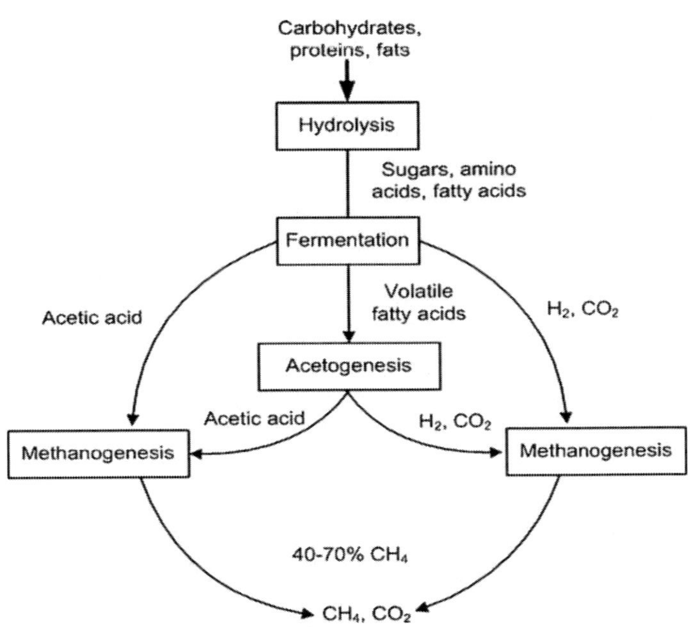

Figure 2. Stages of anaerobic digestion process [14].

3. Factors Affecting Anaerobic Digestion Process

Many factors affect the performance of anaerobic digestion processes. These factors includes solid to water content, pH level, temperature, retention time, organic loading rate, C/N ratio, mixing/agitation, pretreatments, etc.

3.1. Solid to Water Content

Water and raw material are added simultaneously to generate slurry with sufficient consistency. Production of biogas becomes inefficient, if the slurry is too dilute or too dense. The optimum solid concentration may vary from 7-25% depending on the raw material composition [13]. Sewage waste contains very low solid content and so the optimum level can be maintained by adding solid matters like crop residues, weeds plants, etc. Deepanraj et al. [17] experimented the effect of total solid contents (2.5, 5, 7.5, 10 and 12.5%. of TS) on biogas production using food waste in a 2 liter batch digester and found that 7.5% of total solids achieved better performance on biogas production than other solid concentrations.

3.2. pH Level

The optimum pH level of an anaerobic digester is 6.7 to 7.5 [18]. The pH level will vary during the digestion process. The volatile fatty acids production rate is much higher than the methane production rate, and this will reduce pH level below the optimum range thereby inhibiting methanogens which is very sensitive to acidic conditions. [7, 10]. The reduction of pH can be controlled by adding chemicals like sodium carbonate, sodium bicarbonate, gaseous ammonia, ammonium hydroxide, lime, potassium and sodium hydroxide. Sivakumar et al. [19] studied the effect pH (5-8) using spoiled milk as feed and reported that pH between 6.5 to 7.5 gives maximum biogas yield which is shown in Figure 3.

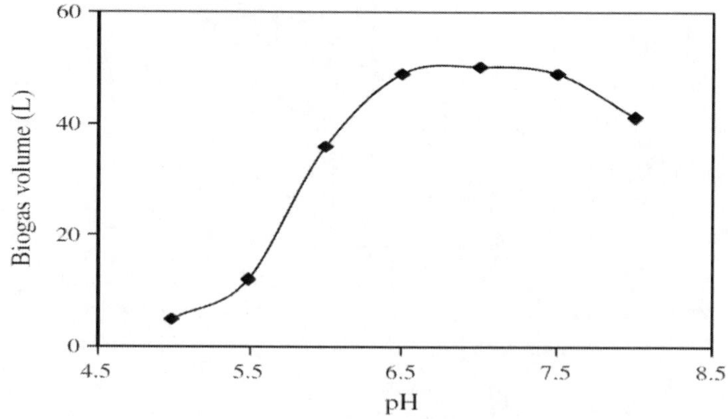

Source: [19].

Figure 3. pH influence on anaerobic digestion of spoiled milk.

3.3. Temperature

Biogas production from organic substrates is strongly affected by the temperature during anaerobic digestion. The digestion of organic material during biogas production takes place in three main temperature ranges [7]: 45–60°C: thermophilic, 20–45°C: mesophilic and below 20°C: psychrophilic. The rate of biogas production increases with an increase in temperature. In anaerobic digestion process, only mesophilic and thermophilic temperature ranges are considered important. The main reason for avoiding the psychrophilic temperature is that at such low temperatures the rate of organic matter conversion into biogas is minimized, since the activity of microorganisms is limited. The anaerobic bacteria used for digestion process are sensitive to temperature fluctuation; hence, it is essential to maintain and control the process at a constant temperature. Thermophilic bacteria are highly efficient in terms of process parameters like, retention time, loading rate and gas yield; however, they need higher heat input, also sensitive to temperature fluctuations and environmental variables compared with mesophilic bacteria. The influence of temperature on the rate of anaerobic digestion process is better explained by a graph shown in Figure 4. Ji-shi et al. [21] evaluated the performance of anaerobic batch reactors under different temperatures (25°C to 55°C) and reported that the reactor with temperature 55°C produced more biogas which is shown in Figure 5.

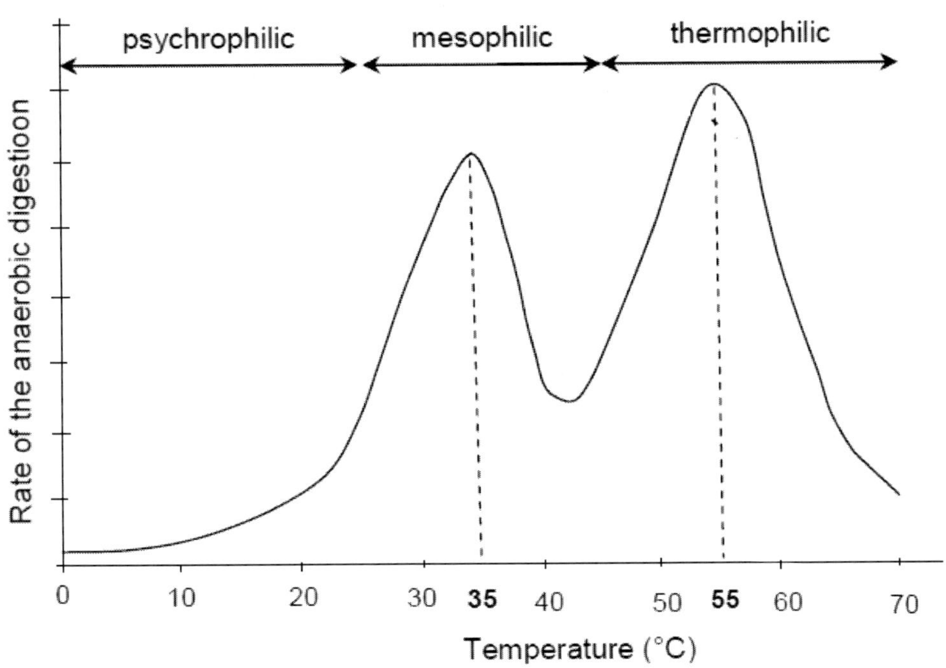

Source: [20].

Figure 4. Influence of temperature on the rate of anaerobic digestion.

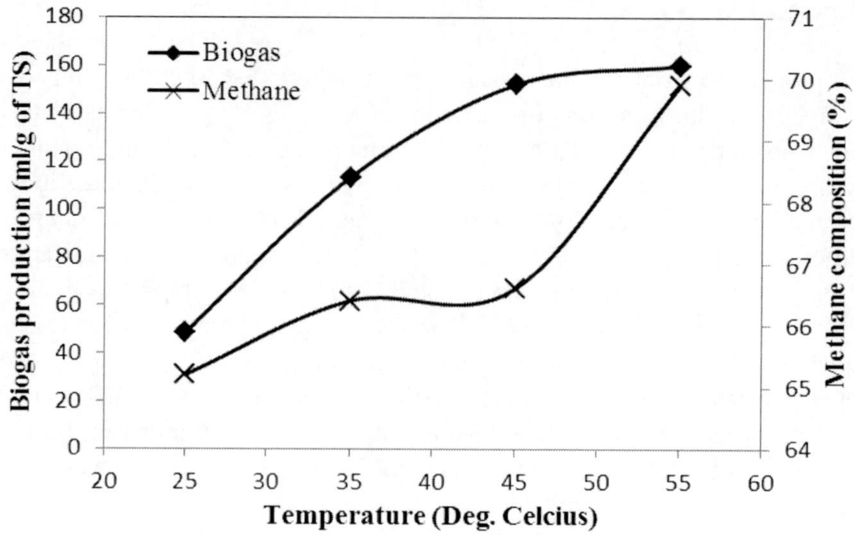

Figure 5. Effect of temperature on anaerobic digestion of municipal solid waste [21].

3.4. Retention Period

The time period for which the organic material remains inside the digester for biogas generation is known as retention period. The retention period varies depending on the type of feedstock and the temperature range used [22]. Solids retention time (SRT) and hydraulic retention time (HRT) are the two significant retention times in anaerobic digestion process. SRT refers the time that bacteria (solids) remain inside the digester. HRT is commonly used to denote substrate retention time. It is the time spent by the input slurry, inside the digester from the instant of its entry to its exit [16, 13]. San Thy et al. [23] investigated the effect of retention time (10, 20 and 30 days) on biogas production using pig manure and found that biogas yield was higher during the hydraulic retention time of 20 to 30 days.

3.5. C/N Ratio

The ratio between the amount of carbon and nitrogen present in the raw materials is represented by the C/N ratio. The carbon-nitrogen (C/N) ratio is one of the important factors in the production of biogas. The elements of carbon (in the carbohydrates) and nitrogen (in the form of proteins, ammonia nitrates) are the major food for anaerobic bacteria [22]. The consumption of carbon by bacteria is 30 times faster than its nitrogen consumption. Therefore, the availability of carbon in the substrate should be 20-30 times higher than nitrogen (i.e., C/N ratio must be between 20 and 30) [24-26]. If the C/N ratio is more, then rapid consumption of nitrogen by methanogens takes place and results in lower bio-gas production. Lower C/N ratio leads to accumulation of ammonia cause the pH value to exceed 8.5, which is again toxic to the methanogens. To maintain the optimum C/N ratio in the digester, substrates of high C/N ratio can be co-digested with low C/N ratio substrates. The typical C/N ratios for different organic materials are shown in Table 3.

Table 3. Typical C/N ratios of different materials

S.No	Material	N (%)	C/N ratio
1	Cow, buffalo, sheep, pig and horse manure	1.4-3.8	15-40
2	Poultry manure	6.3	5.2
3	Slaughter house waste	7-14	2-4
4	Night soil	6	6-10
5	Fish scarps	6.5	5.1
6	Kitchen vegetable scraps	3.3	16
7	Grass clipping and hay	2-4	10-20
8	Sawdust	0.1-0.25	200-511
9	Corn stalks	0.8	60
10	Bagasse, Wheat and rice straw	0.3-0.5	120-150

Source: [7, 16].

3.6. Organic Loading Rate

Organic loading rate (OLR) is an important parameter which affects the biogas production in anaerobic digestion, specifically when the digestion process is carried out in continuous flow mode. OLR is a measure of biological conversion capacity of the system [13]. It is expressed as the amount of raw material (kg of volatile solids) fed to the digester per unit volume per day [7, 16]. Overloading affects the digestion process by the accumulation of acids. The optimum loading rate is between 0.5 kg and 2 kg of total volatile solids per unit volume of the digester per day, which depends on type of raw material, retention time and the process temperature [16]. Babaee and Shayegan [27] examined the effect of organic loading rate (1.4, 2 and 2.75 kg/ m^3-d) on methane production using vegetable waste as feed and suggested that OLR of 1.4 kg VS/m^3-d achieved better methane yield compared to other two.

3.7. Inoculum

For a balanced microbial population, the anaerobic digestion process required long time. To overcome this, inoculum is added to the reactor to speed up the stabilization of digestion process. Effective reactor start-up depends on the quantity and quality of the inoculum. Dhevagi et al. [28] evaluated various inoculum sources for anaerobic digestion process and reported that goat rumen fluid at the rate of 8 per cent (v/v) was more efficient in biogas production. Also, compared to other inocula tested, the digesters inoculated with goat rumen fluid showed higher population of cellulolytic anaerobic bacteria.

3.8. Mixing/Agitation

Mixing or agitation in the digester is done to maintain consistency in slurry, to maintain uniform temperature, process stability and to combine the substrate with microorganisms. This prevents thermal stratification and scum formation in the digester and also prevents solid deposition at the bottom and corners of the digester as dead zones [29]. The mixing process can be done either by using mechanical stirrers or by recirculation of the digester slurry using centrifugal pumps [12]. Even without mixing device, there is a certain agitation through the raising gas, through the movement of substrates with different temperatures and by the inflow of fresh substrate. Keanoi et al. [30] studied the effect of with/without agitation of agricultural waste on biogas production and found that agitation gave an increase of 7.56% on biogas yield over without agitation. Some of the results for operating parameters from available literature are presented in Table 4.

Table 4. Effect of different operating parameters

Parameter experimented	Feedstock	Parameter range	Result	Reference
Solid to water content	Cattle manure	Substrate with seven different solid concentrations 2.6, 4.6, 6.2, 7.4, 9.2, 12.3 and 18.4%. of TS	Maximum biogas yield was obtained at 7.4% of TS followed by 9.2% and others	[31]
Solid to water content	Jatropha seed cake	Four different solid concentrations of 10, 15, 20 and 25% of total solids were investigated	Maximum biogas yield obtained at 20% TS followed by 15, 25 and 10% TS	[32]
pH	Cow dung	Three different pH 4, 7, 9 were investigated	Better biogas yield was obtained at pH 7, followed by 9 and 4	[33]
pH	Food waste	Five different pH 5, 6, 7, 8 and 9 were investigated	Better biogas yield was at pH 7, followed by 8, 6, 9 and 5	[34]
pH	Spoiled milk	Seven different pH (5, 5.5, 6, 6.5, 7, 7.5, 8)	pH 7 produced better yield than other pH	[19]
Temperature	Food waste	Temperature range from 30 to 55°C	Highest gas production and COD removal rate was achieved at 50°C	[35]
C/N ratio	Beef cattle feces and water hyacinth	Three different C/N ratios (20, 25 and 30)	Highest CH_4 content was provided by C/N ratio of 30.	[36]
C/N ratio	Waste water and bagasse	Seven different C/N ratios (6.62, 9.27, 13.19, 19.56, 24.53, 31.76 and 64.58) were investigated.	C/N ratio of 25.53 gave better yield than others.	[37]

4. PRETREATMENTS

Improving the hydrolysis by pretreatments can increase the biogas yield by increasing the performance of anaerobic digestion. Pretreatment breaks down the complex structure of organic compounds into simpler molecules and makes them more susceptible to microbial degradation [26]. Various pre-treatments can be used to improve the biogas yield such as mechanical, thermal, chemical, biological and combined pretreatments [38, 39]. These treatments help to solubilize and/or to reduce the size of organic compounds, in order to make them more easily biodegradable.

The mechanical pretreatment of biomass is one of the important methods used for intensifying the bio-methanation process. The most commonly used mechanical pretreatment methods include milling/grinding, high pressure homogenizer, ultrasonication, mechanical jet, etc [40]. Particle size of the substrate plays a major role in the bio-methanation process. Smaller particle size will increase the substrate utilization because smaller particle size can provide increased microbial activity. By breaking the large structures into shorter chains, the speed and the efficiency of the hydrolysis increase which makes the biodegradability easy [40]. Milling/grinding is used to reduce the particle size of the substrate. For materials like hardwoods, instead of milling/grinding, chipping can be applied [41]. The size of the substrate will be 0.2-2mm after milling/grinding and 10-30mm after chipping. Sharma et al. [42] investigated the effect of particle size on agricultural and forest residues with five particle sizes (0.088, 0.40, 1.0, 6.0 and 30.0 mm) and reported that the maximum quantity of biogas was produced from 0.088 and 0.40 mm particle size.

Ultrasonication is one of the useful and effective mechanical pretreatment methods to enhance biodegradability of the material. Application of high intensity ultrasound modifies the structure of the material which makes the biodegradability easier compared to untreated sample [43]. Available literature reported that the application of ultrasonic pretreatment in anaerobic digestion of waste activated sludge increases the biogas production and reduces the volatile solids. Wang et al. [44] investigated the effect of ultrasonication on waste activated sludge and reported that the methane generation was increased by 64% due to ultrasonic pretreatment.

Thermal pretreatment includes conventional thermal hydrolysis, microwave thermal treatment and steam explosion. Biogas production can be increased when the organic and inorganic compounds in the feedstock are partially solubilized during thermal pretreatment. During the thermal pretreatment, the organic and inorganic compounds in the feedstock are partially solubilized before hydrolysis which reduces digester volume and increases the biogas production [45]. From literature, it is found that the optimum temperature is 160 to 200°C and the treatment time is 30 to 60 minutes. Temperatures above 250°C should be avoided during pretreatment, because above 250°C unwanted pyrolysis reactions start to take place. Thermal treatment can be done either by conventional, steam explosion or by autoclave/microwave. Carrere et al. [46] investigated 6 different waste activated sludge samples using autoclave type pretreatment technique with a temperature range of 60-210°C and reported that the biodegradability of the substrate and the COD solubilization increases with increase in temperature due to pretreatment.

Chemical pretreatments are used to solubilize the main components of lignocellulosic biomass such as cellulose, hemi-cellulose and lignin. Alkaline pretreatment, acid pretreatment, ozonolysis pretreatment and oxidation pretreatment are the techniques available in the chemical pretreatment. During alkaline pretreatment, various reactions like solvation, saponification, peeling and hydrolytic reactions takes place in order. These reactions result in swollen state of the biomass, which enhances the accessibility of bacteria and enzymes [47]. Various alkali agents like KOH, NaOH, $Ca(OH)_2$ and $Mg(OH)_2$ are used in the alkaline pretreatments. The alkaline pretreatment causes swelling of organic particles, making them more susceptible to enzymatic attack by improving the biodegradability [39]. Ofoefule et al. [48] studied the effect of alkaline pretreatment on water hyacinth using KOH and reported that the methane composition of biogas increased from 60% to 71%.

In acid pretreatment, dilute or strong acids like acetic acid, nitric acid, etc are used as agents to treat the biomass chemically. The purpose of acid treatment is to make the better accessibility of cellulose for bacteria/enzymes by solubilizing the hemicellulose and removing the lignin [47]. During strong acid pretreatment, the solubilization of hemicellulose and precipitation of solubilized lignin are more pronounced compared to dilute acid pretreatment. Moset et al. [38] evaluated the methane yield of pig slurry with acid pretreatment (H_2SO_4) and found that 10% acid treated pig slurry generates 20% higher methane yield than untreated slurry.

Ozonolysis pretreatment uses ozone, which can effectively remove lignin and the part of hemicellulose. Ozone pretreatment is normally carried out at room temperature, and main parameters involved in the pretreatment are moisture content of the sample, particle size, and ozone concentration in the gas flow. Among these, moisture content has the most significant effect on the solubilization. This method is very best in chemical pretreatments, because it does not leave acidic, basic, or toxic residues [49]. Yeom et al. [50] studied the effect of ozone treatment on municipal waste water sludge in an anaerobic batch reactor and obtained increased biodegradation from 36% to 77%.

Biological pretreatment is carried out by using biological agent (microorganisms) with the biomass materials [51]. The microorganisms usually degrade most of the lignin and hemicellulose but very small part of cellulose. This is because cellulose is more resistance to the biological attack. Various micro-organisms such as brown-rot fungi, white-rot fungi, and soft-rot fungi, are commonly used for the degradation of lignin and hemicellulose in the substrate. Brown-rot fungi are mainly used to attack cellulose and the white and soft rot-fungi are used to attack both cellulose and lignin [52]. Some of the results for pretreatment from available literature are presented in Table 5.

5. FEEDSTOCK

All biodegradable biomass materials are well suitable for anaerobic digestion. Feedstock may be solid, slurries, and both dilute and concentrated liquids. Common feedstock includes agricultural wastes and crop residues, animal wastes, aquatic waste, forest residues and municipal solid wastes [16, 60]. Table 6 gives the detailed list of feedstock under each category.

Table 5. Effect of different pretreatment techniques in biogas production

Pretreatment type	Feedstock used	Digester condition	Results obtained	Reference
Ultrasonication (50kHz, 120W, 50 minutes)	Aquaculture effluent	Psychrophilic, semi-continuous reactor	Chemical oxygen demand (COD) was increased by 10% and methane production increased from 0.06 m^3 /kg of COD removed to 0.08 m^3 /kg of COD removed.	[53]
Ultrasonic (20 kHz)	Palm oil mill effluent	Batch reactor (32-37°C)	Increase in methane production by 16%.	[54]
Thermal (microwave, 175°C)	Waste activated sludge	Semi-continuous mesophilic reactor	Increase in biogas production from 636 l/kg of VS to 839.6 l/kg of VS.	[55]
Chemical (Alkali treated-NaOH)	Pulp and paper sludge	Batch reactor	0.32 m^3 CH_4/kg $VS_{removal}$, was produced which is 183.5% of the untreated sludge. Soluble chemical oxygen demand increased upto 83%.	[56]
Chemical (Acid treated- H_2SO_4)	Sunflower de oiled cake	Mesophilic reactor	Increase in methane production from 195 mL CH_4/g VS to 302 mL CH_4/g VS.	[57]
Chemical (Ozone treated)	Textile waste water	Continuous reactor	Chemical oxygen demand removal rate increased upto 65%.	[58]
Biological (white-rot fungi)	Olive mill wastewaters	Continuous plug flow reactor	Chemical oxygen demand removal increased from 34 to 65 %	[59]

Table 6. Feedstock for anaerobic digestion

Category	Types of waste
Agricultural wastes and crop residues	Crop Stubbles, straw, spoiled fodder, sugar cane trash and bagasse, weeds, tobacco waste, rice and coffee husks, fruit and vegetable processing wastes, oil cakes, etc.
Animal wastes	Cattle dung, goat and sheep manure, pig manure, elephant dung, fishery waste, slaughter house wastes
Aquatic plants	Algae, sea weeds, water hyacinth, etc.
Forest residues	Dead trees, plants, twigs, barks, roots, leaves, etc.

Source: [16, 22].

5.1. Agricultural Wastes and Crop Residues

Many agricultural crops, byproducts and processes yield residues can be used as feedstock for biogas generation. Among the world countries, India stands second in producing wheat, rice, cotton, tea, cashew nuts and some other crops. In India, crops are not cultivated specifically for the purpose of anaerobic digestion to produce biogas. However, there is abundant biomass source available in the form of agricultural waste and crop residue which can be used as feedstock for anaerobic digestion [61]. Crop stubbles, spoiled fodder, straw, sugar cane trash and bagasse, tobacco waste, weeds, rice and coffee husks, fruit and vegetable processing wastes, oil cakes, etc. are some of the agricultural wastes and crop residues that can used in anaerobic digestion [16, 22]. Biogas yield potential from various agricultural wastes and crop residues are shown in Table 7.

5.2. Animal Wastes

Animals produce a large amount of wastes, as animal breeding activity has highly increased. The world's cattle population is estimated as 1300 million. India stands first rank with the population of 281.7 million. Cattle dung is used as one of the solid waste feedstock for biogas production in rural India. For the year 2010, the estimated cattle and buffalo populations are 224 and 97 million respectively. It has been calculated that from a livestock population of about 458 million, 354 million tons of dung as solid waste is available per year. China, Korea, Taiwan and Philippines use pig manure as feedstock [7]. Widely used animal wastes worldwide are namely cow dung, pig waste, poultry manure, horse dung, camel dung, elephant dung, fishery waste, slaughter house wastes, etc. Out of these some wastes like poultry excreta are rich in organic nitrogen and relatively lower carbon source [68]. This leads the process to slow down and less productive, this problem can be overcome by co-digestion of different feedstock having low and high nitrogen values. Quantities of fresh dung available from various animal sources and their biogas yield are presented in Table 8.

Table 7. Biogas yield from Agricultural and crop residues

Agricultural and crop residues			Vegetable and fruit wastes		
Residues	Methane yield (m^3 per kg of volatile solids)	Ref.	Wastes	Methane yield (m^3 per kg of volatile solids)	Ref.
Sugar beat leaves	0.21	[62]	Cauliflower leaves	0.190	[67]
Sunflower (whole crop silage)	0.3	[62]	Cabbage leaves	0.309	[67]
Maize (whole crop silage)	0.39	[62]	Carrot (leaves)	0.241	[67]
Tops of sugar beets	0.32-0.34	[63]	Carrot (petiole)	0.309	[67]
Maize straw	0.34	[15]	Radish shoots	0.304	[67]
Sunflower oil cake	0.227	[64]	Onion outer peel	0.400	[8]
Corn cob mix	0.35-0.36	[8]	Banana peel	0.24-0.32	[8]
Wheat grain	0.37-0.4	[8]	Pine apple peel	0.357	[67]
Wheat straw	0.3	[15]	Mango peel	0.37-0.52	[8]
Oat straw	0.32	[63]	Grape pressings	0.28	[8]
Rice straw	0.3-0.38	[65]	Potato peel	0.267	[67]
Oat husk	0.242	[66]	Pomegranate peel	0.312	[67]
Rice husk	0.13	[15]	Lemon pressings	0.473	[67]
Coffee husk	0.20	[15]	Coriander leaves	0.325	[67]
Grass cuttings	0.3	[63]	Coriander stem	0.309	[67]

Table 8. Fresh dung availability from different animals and their biogas yield [16, 69 and 70]

Animal	Fresh dung (kg/animal/day)	Volatile matter (%)	Biogas yield (liter per kg of volatile solids)	Biogas yield (liter per kg of dry solids)	Average biogas yield (liter per animal per day)	Methane composition in biogas (%)
Cow	10	71.3	72	60	400	50
Buffalo	15	81-84	79	65	600	50
Camel	20	81.1	119	100	700	53
Horse	15	43.4	111	91	600	55
Elephant	40	91	272	-	800	44
Pig	1.2	54.3	149	107	100	56
Sheep/Goat	2.0	-	113-135	90-110	100	54
Poultry	0.10	63.3	191	126	6	56

5.3. Aquatic Plants

Aquatic plants such as water hyacinth, various species of micro and macro algae, sea weeds, etc are the best feedstock for biogas production, as they have easily hydrolysable sugars and low lignin content. Also, do not compete with land resources used in arable food crop cultivation. Water hyacinth and micro algae are reportedly have higher gas yields [7]. It is calculated that the efficiency of biomass production per hectare of microalgae are about 5–30 times than that of crop plants [71]. The relatively high lipid, protein and starch contents and the absence of lignin makes microalgae as an excellent feed stock for methane gas production. Biogas yield from different aquatic plants are given in Table 9.

Table 9. Biogas yield from different aquatic plants

Species	Biogas yield (liter per kg of volatile solids)	Methane composition in biogas (%)
Water hyacinth	398	77
Euglena gracilis	485	67
Scenedesmus obliquus (Green algae)	181.81	74.3
Scenedesmus obliquus (Green algae)	241.28	77.1
Phaeodactylum tricornutum	421	75.1
Chlorella kessleri	335	65
Dunaliella salina	505	64
Arthrospira platensis	481	61

Source: [72, 73].

6. Classification of Anaerobic Digestion System

Anaerobic digester systems can be classified based on feeding mode (Batch and continuous system), moisture content of the substrate (Wet and dry), process temperature (mesophilic and thermophilic), etc. These categories are described below.

6.1. Batch and Continuous Systems

In the batch process, the substrate is fed into the digester only once (with or without inocula) and sealed completely for the entire retention time. After the retention period, the digested substrate is emptied from the digester and the new substrate is filled to start the process again. In this type, the production of gas is non-continuous because at the beginning and end of the process, the gas production is low and at the middle of the process, the gas production is high. So, in order to get constant gas supply from batch systems, many batch type digesters can be operated in parallel [18]. In continuous process, the digester is fed continuously with the substrate and the equal amount of digested substrate is removed, resulting in constant and continuous production of biogas. This type of system uses mechanical agitator or biogas recirculation to mix the digester contents continuously [68].

6.2. Wet And Dry Systems

Based on the total solid concentration of the feed substrate the process can be a wet or dry system. Dry digestion system processes the substrates with solid concentration between 20-40%. Wet system processes the substrates with solid concentration less than 15% [11, 12]. Dry system processes thick substrates and slurry and hence needs more energy input to process and move. In wet systems, the substrates can be transported easily through standard pumps with less energy input. Wet systems provides increased gas production rate because the circulation of material is easy and the more contact between the bacteria and their food [12].

6.3. Thermophilic and Mesophilic Systems

The temperature of the mesophilic systems digester is maintained at less than 45°C, whereas in thermophilic system, the temperature of the digester is maintained in between 45°C to 60°C [12]. Mesophilic systems are more stable and tolerant to changes in environmental conditions than thermophilic systems. In thermophilic system, the higher process temperature enhances the reaction faster and decreases the substrate retention time compared to mesophilic system [74]. To achieve higher temperature in case of thermophilic system, it requires more energy input.

6.4. Single Stage and Multistage Systems

In a single stage system, all the stages of the anaerobic digestion takes place in a single, sealed reactor [75]. The major problem in this system is during the acid formation stage, the acidogenic bacteria reduce the pH of the digester which affects the methane formation due to difference in the growth of acidogenic and methanogenic bacterial communities [11]. In two stage or multi stage systems, separate digester vessels are employed for both acid and methane formation to bring maximum control over microorganisms during acid and methane formation. In the first digester vessel, hydrolysis, acidogenesis and acetogenesis takes place and the product of first vessel is taken to the second vessel for methane formation. The temperature of the second digester vessel can be maintained either mesophilic or thermophilic based on the need. For dry batch systems or wet continuous systems, single stage systems are preferred, whereas in two-stage system, continuous and wet processes are preferred [75].

7. Types of Anaerobic Digesters/Reactors

7.1. Floating Drum Type Digester

Floating drum shaped digester has a cylindrical or dome-shaped digester, metallic floating drum (gas holder), inlet pipe, outlet pipe and the partition wall. The floating drum holds the gas and maintains constant pressure by moving up and down with the help of central guide pipe [16]. Common models of floating drum digesters are KVIC (Khadi and Village Industries Commission) model, Ganesh model, Pragathi model, Nepal model, ASTRA model etc. Figures 6, 7 and 8 shows the schematic of KVIC, Pragathi and Nepal model floating drum plant. Floating drum type digesters have many advantages like maintenance of biogas at constant pressure, integrated arrangement for scum breaking, volume of gas can be observed by sensing the drum position, etc.

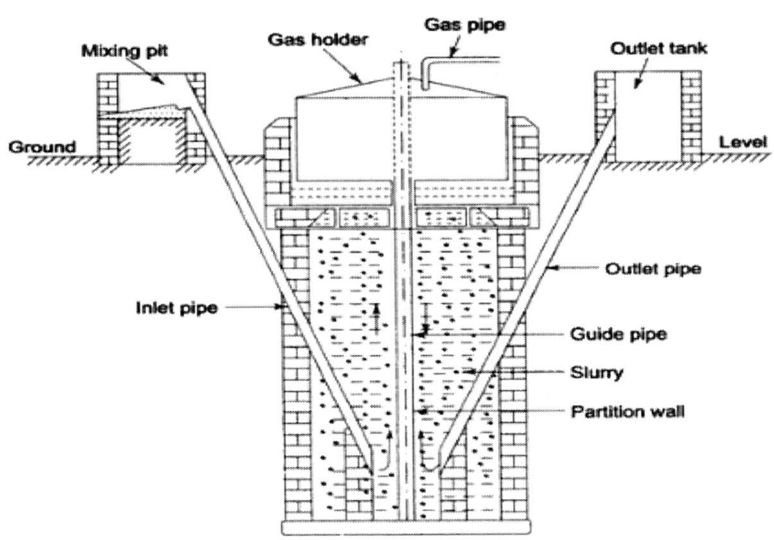

Figure 6. Schematic of KVIC model digester [7].

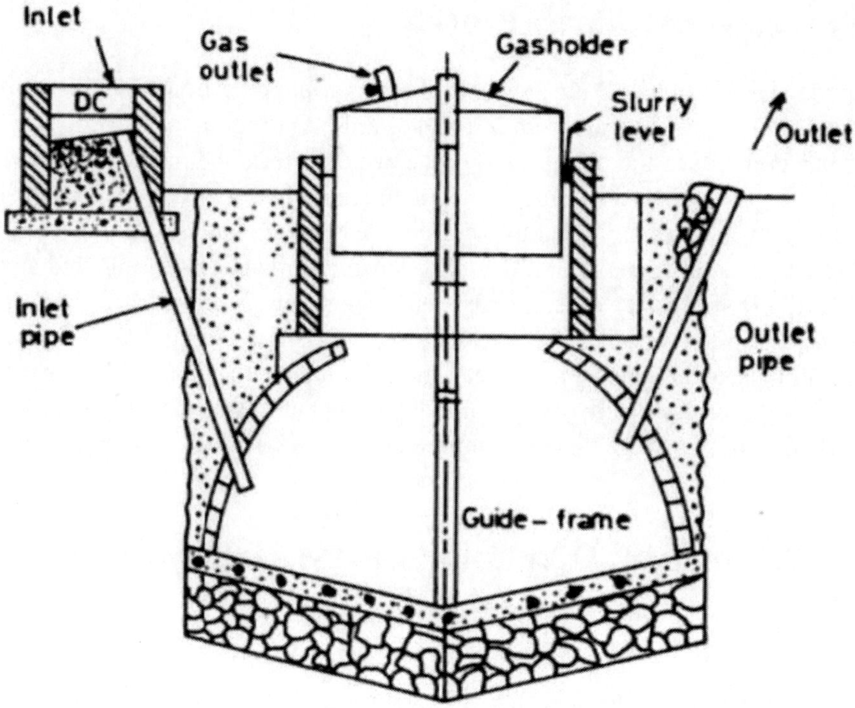

Figure 7. Schematic of Pragathi model digester [7].

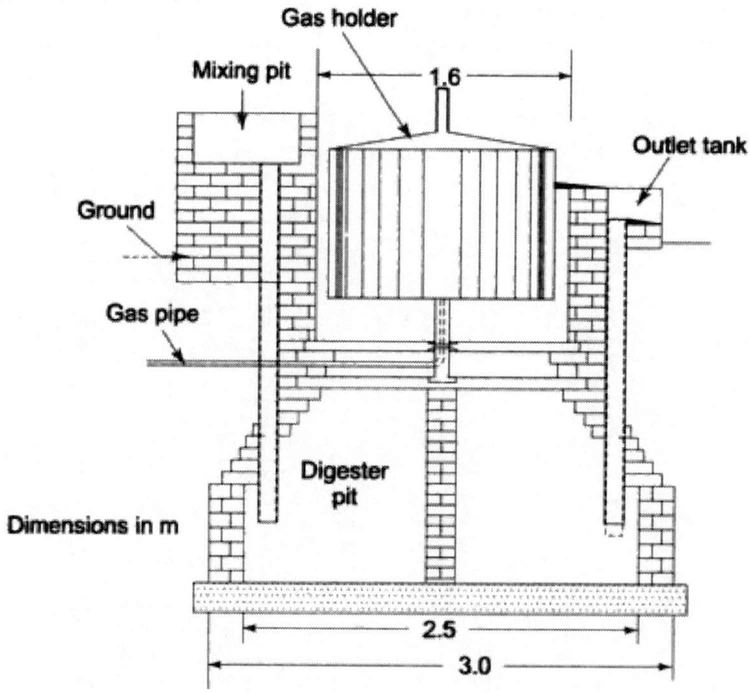

Figure 8. Schematic of Nepal model digester [7].

7.2. Fixed Dome Digester

The fixed dome plants were first developed by Chinese [16]. This type of plant consists of an immovable closed, dome shaped digester as shown in Figure 9. As like floating drum digesters, the gas pressure inside the plant is not constant, because the gas produced is stored between the slurry and fixed roof of the plant [76]. Common models of fixed dome plants are Janata, Chinese fixed dome plant, Camartec model, Deenbandhu model, etc. The cost of fixed dome plants are less compared to floating dome plants. The schematic of Chinese fixed dome plant, Camaratec, Deenbhandhu and Nicarao design are shown in Figure 9, 10, 11 and 12 respectively.

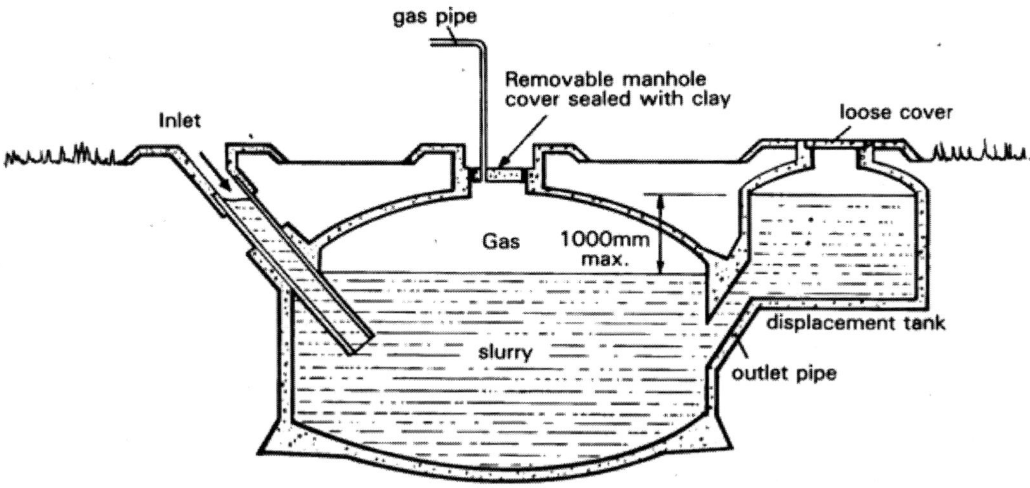

Figure 9. Chinese fixed dome plant [77].

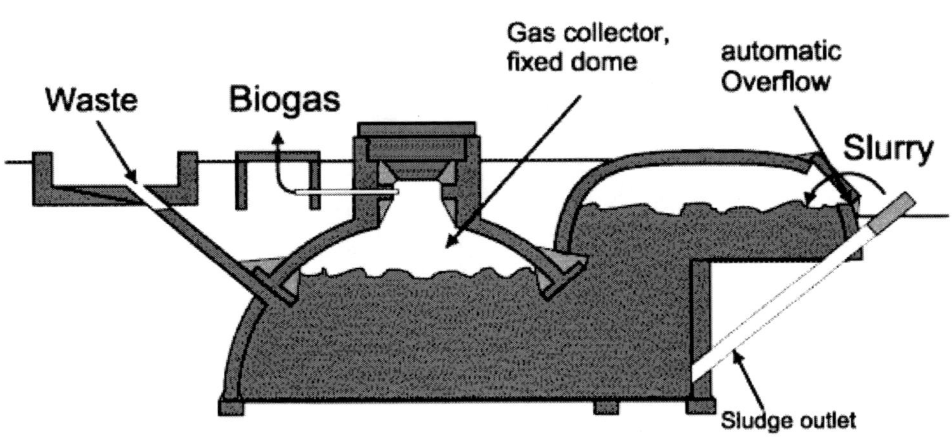

Figure 10. Camaratec model fixed dome plant [78].

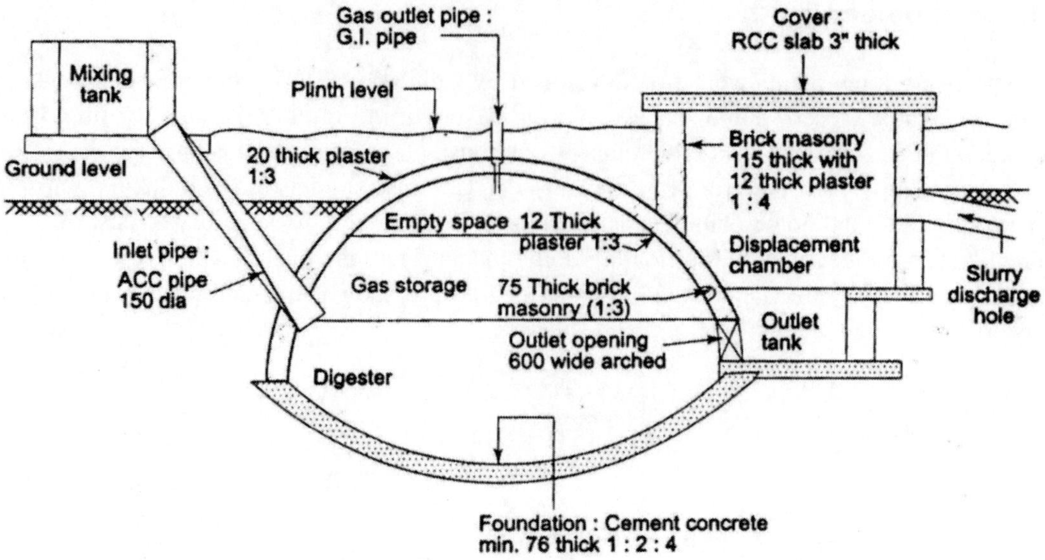

Figure 11. Deenbhandhu model fixed dome plant [7].

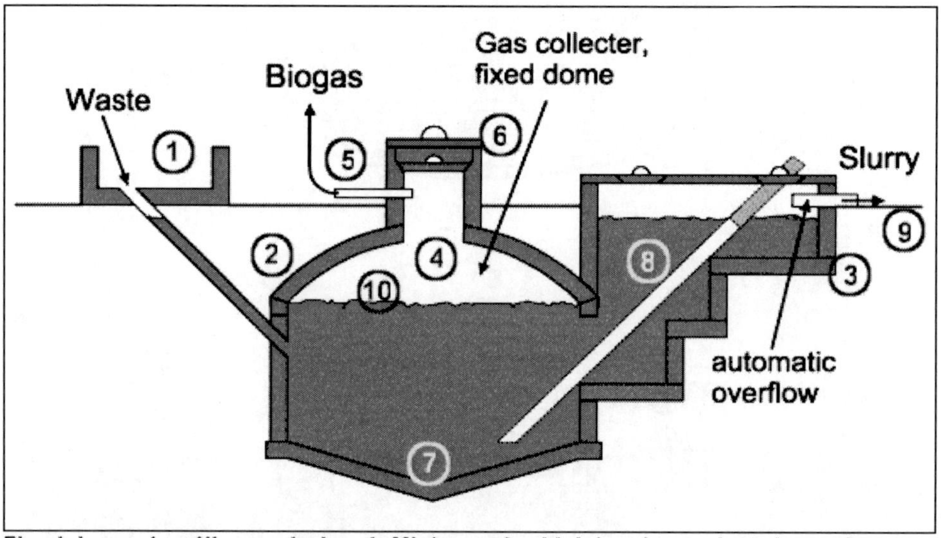

Fixed dome plant Nicarao design: 1. Mixing tank with inlet pipe and sand trap. 2. Digester. 3. Compensation and removal tank. 4. Gasholder. 5. Gaspipe. 6. Entry hatch, with gastight seal. 7. Accumulation of thick sludge. 8. Outlet pipe. 9. Reference level. 10. Supernatant scum, broken up by varying level.

Figure 12. Nicarao design fixed dome plant [78].

7.3. Plug Flow Digester

A plug flow digester has a long, narrow concrete tank (typically a 5:1 ratio; 5 times as long as the width) with a rigid or flexible cover as shown in Figure 13. The inlet and outlet to the digester are at opposite ends, and feeding is carried out semi-continuously, with the feed displacing an equal amount of effluent at the other end. The tank is built partially or fully below grade to limit the demand for supplemental heat. Plug flow digesters are used only at dairy operations that collect manure by scraping [79]. Plug flow digesters do not require mechanical mixing as like other digesters. The shape of the floor and walls of the digester facilitate the movement of all material through the digester to minimize short circuiting flow.

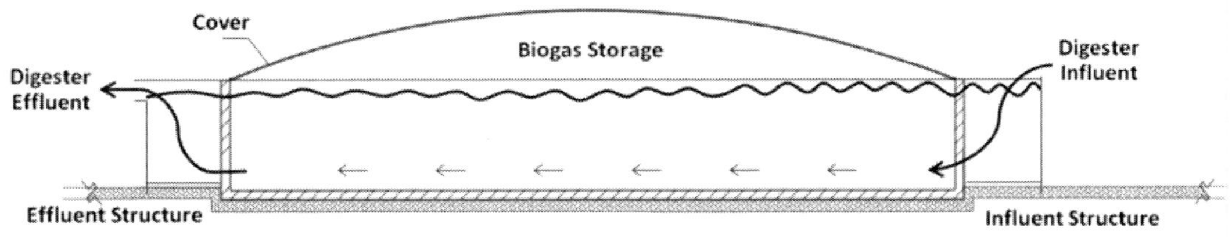

Figure 13. Schematic of plug flow digester [79].

7.4. Anaerobic Baffled Reactor

An anaerobic baffled reactor is a simple rectangular tank, with physical structure similar to a septic tank, and is divided into various compartments, by means of partitions (baffles) from the roof and bottom of the tank [80]. The schematic view of anaerobic baffled reactor is show in Figure 14. Because of the series of baffles under which the substrate is forced to flow, the contact time with the active biomass is more which results in improved biological treatment. Most of the settleable solids present in the feed are removed in the sedimentation chamber at the beginning of the reactor, which typically represents 50 % of the total volume. They are simple to build and simple to operate, as well as very robust to hydraulic and organic shock loading.

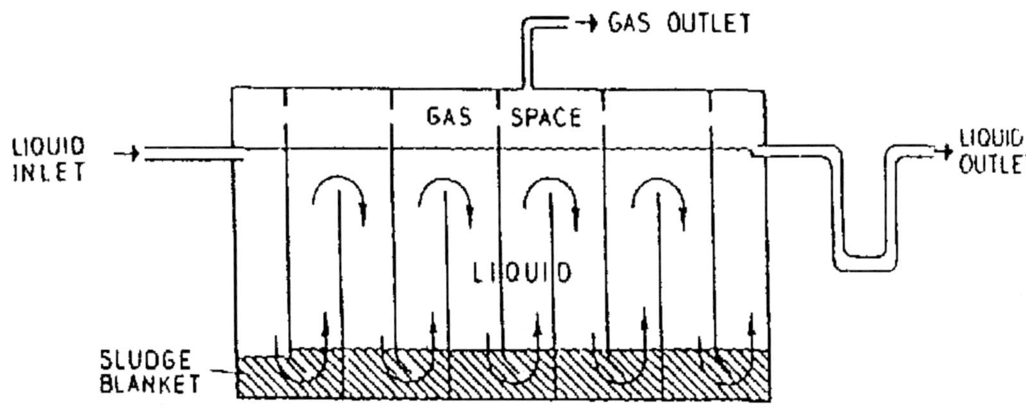

Figure 14. Anaerobic baffled reactor [80].

8. Products of Anaerobic Digestion

8.1. Biogas

The primary product of anaerobic digestion is biogas which contains methane (55-65%) and carbon dioxide (30-40%) as major composition with water vapour, hydrogen sulfide and hydrogen as trace elements. The physical and chemical properties of biogas obtained through anaerobic digestion are given in table 10. The biogas can be directly used for cooking, lighting and in combined heat and power (CHP) units. The biogas can be upgraded by removing carbon dioxide and hydrogen sulfide using scrubbers and the upgraded biogas (methane) can be used in automobiles and fuel cells. Biogas burns up with a clear blue flame without releasing smoke. The calorific value of biogas is about 5650 kcal/m^3 of gas. The energy equivalent of one cubic meter of biogas with 60% methane content is equal to 0.7 cubic meter of natural gas, 0.714 liter of petrol, 0.62 liter of kerosene, 0.417 liter of furnace oil, 3.5 kg of firewood, 0.433 kg of LPG, 1.5 kg of coal, 12 kg of manure and 4,698 kWh electricity [80]. Table 11 shows the comparison of constituents in biogas and natural gas.

Table 10. Physical and chemical properties of biogas

Property	Values
Calorific value (MJ/kg)	21.5
Ignition limit (vol. % in air)	6-12
Density (kg/l)	1.2
Ignition temperature (°C)	650-750
Wobbe Index	6.9

Table 11. Comparison of constituents in biogas and natural gas [9]

Constituents	Biogas	Natural gas
Methane (%)	50-75	91
Ethane (%)	0	5.1
Propane (%)	0	1.8
Butane (%)	0	0.9
Pentane (%)	0	0.3
Carbon dioxide (%)	20-45	0.61
Nitrogen (%)	0-10	0.32
Hydrogen (%)	0-1	0
Hydrogen sulfide (%)	0-3	Approx. 1
Ammonia (ppm)	Approx. 100	0

8.2. Digestate

In anaerobic digestion process, digestate is the by-product. The digestate can be fibrous, liquor, or a sludge-based combination of the both, which is an extra material that cannot be used by the microbes [10]. In multi-stage digestion systems, different forms of digestate come from different digester tanks. But in single-stage digestion systems, the two fractions will be combined, and the separation can be done by further processing. The application of digestate in agriculture serves the double benefit, because it can be used as fertilizer and soil improving material. Due to the removal of carbon during digestion process, the digestate is rich in nitrogen and phosphate than the substrate and hence acts as a superior fertilizer than the compost. Essential plant nutrients like nitrogen, potassium and phosphorous are largely available in the digestate which makes suitable for soil amendment. Digestate in the form of liquor can be used again in digestion process or it can be used in irrigation. The nutrient content available in different animal excreta before and after digestion is given in Table 12. As per MNRE reports, the digestate obtained after digestion process contains 80% carbon, 1.8% nitrogen, 1% phosphorus and 0.9 % potash.

Table 12. Nutrient content of different animal excreta in percentage [16, 81]

Animal excreta	Before digestion					After digestion				
	N	P_2O_5	K_2O	Mg	Ca	N	P_2O_5	K_2O	Mg	Ca
Cow dung	1.74	1.7	0.60	0.53	0.37	1.92	1.9	0.66	0.58	0.40
Camel dung	1.51	0.35	1.8	0.69	0.70	1.60	0.38	2.0	0.75	0.76
Pig dung	2.27	3.1	1.8	0.54	0.21	2.20	3.4	2.0	0.59	0.23
Horse dung	1.07	2.1	3.6	0.49	0.26	1.34	2.2	3.9	0.52	0.28
Poultry litter	2.17	2.0	4.2	1.39	2.28	2.32	2.3	4.8	1.58	2.59

9. Enrichment of Biogas

Presence of unwanted gases present in the raw product is a major challenge to fuel efficiency, gas storage and transportation. Carbon dioxide is commonly produced in considerable quantities due to interfering bacterial species, affects the combustion efficiency of the biogas. Hydrogen sulfide present in the biogas affects the life of the storage vessel linings and during the combustion it releases sulphur dioxide which makes the exhaust gas very corrosive and pollutes the environment. Purification of biogas is not that much essential when it is used for cooking or lighting purpose. However removal of such gases is mandatory, if it is used as fuel in automobiles and fuel cells.

9.1. Water Scrubbing

Water scrubbing is the most common technique used to remove carbon dioxide and hydrogen sulfide from biogas because these gases are more soluble in water than methane. The absorption process is purely physical. Usually the biogas is pressurized and fed to the bottom of a packed column and water is fed from the top and so the absorption process is operated counter currently [82]. The water which exits the column with absorbed carbon dioxide and/or hydrogen sulfide can be regenerated and recirculate back to the absorption column. Regeneration is accomplished by de-pressuring or by stripping with air in a similar column.

9.2. Polyethylene Glycol Scrubbing

Polyethylene glycol scrubbing works on the same mechanism as water scrubbing, with a physical absorption process that works because both carbon dioxide and hydrogen sulfide are more soluble than methane in the solvent. Selexol is the trade name for one of the common solvents used for this process. The big difference between water and solvents is that carbon dioxide and hydrogen sulfide are more soluble in Selexol which results in a lower solvent demand and reduced pumping. In addition, water and halogenated hydrocarbons (contaminants in biogas) are removed when scrubbing biogas with Selexol. Due to formation of elementary sulfur stripping, the polyethylene glycol scrubbing is normally done with inert gas or steam rather than with air.

9.3. Pressure Swing Adsorption

Pressure Swing Adsorption (PSA) technique used to separate some gas species from a mixture of gases under pressure according to the species molecular characteristics and affinity for an adsorbent material. This technique separates different substances on the basis of physical forces and molecular size, which decides their penetration ability in a material. The upgrading is performed at increased pressure and the adsorption material is regenerated by reducing the pressure. Using zeolites or activated carbon at different pressure levels, carbon dioxide, oxygen and nitrogen are removed from the raw gas by adsorption. The fixed adsorption material also adsorbs hydrogen sulfide, but unlike adsorption of carbon dioxide this is done irreversibly. Therefore, PSA has high security against hydrogen sulfide in the upgraded gas. Usually, hydrogen sulfide is removed in a separate cleaning step. More than 97% methane enrichment is possible by this technique.

9.4. Cryogenic Separation

In a cryogenic separation technique, the biogas is compressed and chilled. As the methane, carbon dioxide and contaminants present in the biogas liquefy at different temperature-pressure domains, it is possible to remove carbon dioxide and other gases by cooling and compressing the biogas. Since no chemical has been used to make this separation, this technique is termed as an environmental friendly technique, though the process uses considerable energy [82]. The main advantage of this technique is that large quantity of gas can be efficiently processed with high purity of the upgraded biogas (99% CH_4). This technique is very costly due to the use of process equipment like compressors, heat exchangers, turbines, etc.

9.5. Membrane Separation

A membrane is a layer of material which acts as a selective barrier between two phases and remains impermeable to specific particles, molecules, or substances when exposed to the action of a driving force [83]. The most common membrane separation process uses pressure and a selective membrane, which allows preferential passage of one of the gases. Several stages of membrane can be used for perfect separation of gases. There are 2 membrane separation techniques: high pressure gas separation and gas liquid adsorption. The high pressure separation process selectively separates hydrogen sulfide and carbon dioxide from methane. Gas liquid adsorption uses micro porous hydrophobic membranes to separate the gaseous from the liquid phase. The carbon dioxide and hydrogen sulfide are dissolved into liquid, were methane (remains as gas) is collected separately.

Conclusion

The commercial energy demand is currently increasing exponentially in the present global and local scenario. A look beyond the fossil fuel resources is crucial for long term energy security and economic growth, as there are several uncertainties about the fresh fossil fuel supplies, coupled with the greater environmental risks. This necessitates the development of alternative techniques for fresh energy resource. Among various alternate techniques available, biogas production by anaerobic digestion seems significant, in terms of energy supply, environmental protection and long term sustenance. Anaerobic digestion process causes number of beneficial utilities including production of biogas (a renewable source of energy) and the spent slurry (a rich organic manure) there by eliminating the emission of greenhouse gas.

References

[1] Chynoweth D.P., Owens J.M and Legrand R., Renewable methane from anaerobic digestion of biomass, *Renew Energ.*, 22, 1-8 (2000).

[2] Gurung A., Ginkel V.S.W., Chang K.W., Qambrani N.A and Sang-Eun O.H., Evaluation of marine biomass as a source of methane in batch tests: A lab-scale Study, *Energy*, 43, 396-401 (2012).

[3] Bapat DW, Kulkarni SV, Bhandarkar VP. Design and operating experience onfluidized bed boiler burning biomass fuels with high alkali ash. In: Preto FDS, editor. *Proceedings of the 14th international conference on fluidized bed combustion.* New York: Vancouver ASME; 165–174 (1997).

[4] Zeng XY, Ma YT, Ma LR. Utilization of straw in biomass energy in China. *Renew Sust Energ Rev* 11, 976–87 (2007).

[5] McKendry P. Energy production from biomass (part 2): conversion technologies. *Bioresource Technology,* 83, 47–54 (2002).

[6] Liew L.N., Jian S and Yebo L., Methane production from solid-state anaerobic digestion of lignocellulosic biomass, *Biomass Bioenergy*, 46, 125-132 (2012).

[7] Nijaguna BT, Biogas Technology. New Age International (P) Limited Publishers, New Delhi; 2012.

[8] Appels L, Lauwers J, Degreve J, Helsen L, Lievens B, Willems K, Impe JV, Dewil R. Anaerobic digestion in global bio-energy production: Potential and research challenges. *Renewable and Sustainable Energy Reviews,* 15, 4295–4301, (2011).

[9] Jensen, J. and A. Jensen, Biogas and Natural Gas Fuel Mixture for the Future, 1st World Conference and Exhibition on Biomass for Energy and Industry, Sevilla (2000).

[10] Khanal S.K., *Anaerobic biotechnology for bioenergy production: principles and applications*, Wiley-Blackwell, Singapore (2008).

[11] Gerardi M.H., *The microbiology of anaerobic digesters*, John Wiley & Sons Inc., USA (2003).

[12] Nayono S.E., *Anaerobic digestion of organic solid waste for energy production*, KIT Scientific Publishing, Karlsruhe (2009).

[13] Abbasi T., Tauseef S.M and Abbasi S.A., Anaerobic digestion for global warming control and energy generation-An overview, *Renew Sust Energ Rev.*, 16, 3228-3242 (2012).

[14] Yebo Li, S.Y. Park, Jiying Zhu, Solid-state anaerobic digestion for methane production from organic waste, *Renewable and Sustainable Energy Reviews* 15 (2011) 821–826

[15] Chanakya H.N and Malayil S., Anaerobic digestion for bioenergy from agro- residues and other solid wastes—An overview of science, technology and sustainability, *J Indian Inst Sci.*, 92, 111-143 (2012).

[16] Khoiyangbam R.S., Gipta N and Kumar S., *Biogas Technology: Towards Sustainable Development*, TERI Press, New Delhi (2011).

[17] Deepanraj B., Sivasubramanian V and Jayaraj S., Solid Concentration Influence on Biogas Yield from Food Waste in an Anaerobic Batch Digester, *International Conference and Utility Exhibition 2014 on Green Energy for Sustainable Development (ICUE 2014,* Thailand, 19-21 March, 2014.

[18] Deublein D and Steinhauser A., *Biogas from wastes and renewable resources: An introduction*, John Wiley & Sons, Germany (2008).

[19] Sivakumar P., Bhagiyalakshmi M and Anbarasu K., Anaerobic treatment of spoiled milk from milk processing industry for energy recovery – A laboratory to pilot scale study, *Fuel*, 96, 482–486 (2012).

[20] Mata A.J., Biomethanization of the organic fraction of municipal solid wastes, IWA Publishing, London (2002).

[21] Ji-shi Z., Ke-Wei S., Man-Chang W., Lei Z., Influence of temperature on performance of anaerobic digestion of municipal solid waste., *Journal of Environmental Sciences,* Vol.18, 2006, pp.810-815.

[22] Rai G.D., Non-conventional energy sources, Khanna Publishers, New Delhi (2010).

[23] San Thy, T R Preston and J Ly, Effect of retention time on gas production and fertilizer value of biodigester effluent, *MSc Thesis, MEKARN-SLU* (2003).

[24] Fricke K., Santen H and Wallmann R., Comparison of selected aerobic and anaerobic procedures for MSW treatment, *Waste Manage.*, 25, 799-810 (2005).

[25] Chen Y., Cheng J.J and Creamer K.S., Inhibition of anaerobic digestion process: A review, *Bioresour. Technol.,* 99, 4044-4064 (2008).

[26] Yadvika., Santosh., Sreekrishnan T.R., Kohli S and Rana V., Enhancement of biogas production from solid substrates using different techniques—a review, *Bioresour. Technol.*, 95, 1-10 (2004).

[27] Babaee A and Shayegan J., Effect of organic loading rates (OLR) on production of methane from anaerobic digestion of vegetables waste, World Renewable Energy Congress 2011, Sweden, 8-13 May 2011.

[28] Dhevagi, P., Ramasamy, K. and Oblisami, G., in Biological Nitrogen Fixation and Biogas Technology, Tamil Nadu Agricultural University, Coimbatore, 149–153 (1992).

[29] Karim K., Klasson T., Hoffmann R., Drescher S.R., Depaoli D.W and Dahhan M.H.A., Anaerobic digestion of animal waste: Effect of mixing, *Bioresour. Technol.*, 96, 1607-1612 (2005).

[30] Keanoi N., Hussaro K and Teekasap S., Effect of with/without agitation of agricultural waste on biogas production from anaerobic co-digestion-a small scale, *American Journal of Environmental Science* 10 (1): 74-85 (2014).

[31] Budiyono I. N. et al., The Influence of Total Solid Contents on Biogas Yield from Cattle Manure Using Rumen Fluid Inoculum, *Energy Research Journal*, 1, 6-11 (2010)

[32] Raheman H and Mondal S., Biogas production potential of jatropha seed cake, *Biomass and Bioenergy*, 37, 25-30 (2012).

[33] Sambo A.S., Garba B and Danshehu B.J., Effect of some operating parameters on biogas production rate, *Renewable Energy*, 6, 343-344 (1995).

[34] Jayaraj S., Deepanraj B and Sivasubramanian V., 'Study on the effect of pH on biogas production from food waste by anaerobic digestion' 9th International Green Energy Conference, Tianjin, China (2014).

[35] Kim JK, Oh BR, Chun YN and Kim SW., Effects of temperature and hydraulic retention time on anaerobic digestion of food waste, *Journal of Bioscience and Bioengineering,* 102, 328-332 92006).

[36] Achmad K.T.B., Hidayati Y.A., Fitriani D and Imanudin O., The effect of C/N ratios of a mixture of beef cattle feces and water hyacinth (Eichornia crassipes) on the quality of biogas and sludge, *Lucrari Stiintifice*, 55, 117-120 (2011).

[37] Moset V., Cerisuelo A., Sutaryo S and Moller H.B., Process performance of anaerobic co-digestion of raw and acidified slurry, *Water Res.*, 46, 5019-5027 (2012).

[38] Fang S., Ping L., Yang Z and Mao J., A review of different pretreatment techniques for enhancing biogas production, *Proceedings of International Conference on Materials for Renewable Energy & Environment (ICMREE)*, Shanghai (2011).

[39] Deepanraj B., Sivasubramanian V and Jayaraj B., Enhancement of biogas production by pretreatments: *A review, Proceedings of IV the International Conference on Advances in Energy Research*, December 10-12, 2013, pp.309-319, India.

[40] A. Mudhoo, Biogas production: Pretreatment methods in anaerobic digestion, *John Wiley & Sons* (2012).

[41] P. Kumar, D.M. Barrett, M.J. Delwiche, P. Stroeve, Methods for pretreatments of lignocellulosic biomass for efficient hydrolysis and biofuel production, *Ind. Eng. Chem. Res.,* 48, 3713-3729, (2009).

[42] Sharma S.K., Mishra I.M., Sharma M.P and Saini J., Effect of particle size on biogas generation from biomass residues, *Biomass*, 17, 251-263 (1988).

[43] R. Dewil, J. Baeyens, R. Goutvrind, Ultrasonic treatment of waste activated sludge, *Environ. Prog.* 25, 121-128, (2006).

[44] Q. Wang, M. Kuninobu, K. Kakimoto, H.I. Ogawa, Y. Kato, Upgrading of anaerobic digestion of waste activated sludge by ultrasonic pretreatment, *Bioresour. Technol.,* 68, 309-313 (1999).

[45] Mudhoo A., Biogas Production: Pretreatment Methods in Anaerobic Digestion, John Wiley & Sons (2012).

[46] H. Carrere, C. Bougrier, D. Castets, J.P. Delgenes, Impact of initial biodegradability on sludge anaerobic digestion enhancement by thermal pretreatment, *J. Environ. Sci. Health, PartA-Toxic/Hazard. Subst. Environ. Eng.* 43, 1551-1555 (2008).

[47] A.T.W.M. Hendriks, G. Zeeman, Pretreatments to enhance the digestibility of lignocellulosic biomass, *Bioresour. Technol.,* 100, 10–18 (2009).

[48] A.U. Ofoefule, E.O. Uzodinm, O.D. Onukwuli, Comparative study of the effect of different pretreatment methods on biogas yield from water Hyacinth (*Eichhornia crassipes*), *Int. J. Phys. Sci.,* 4, 535-539 (2009).

[49] W.C. Neely, Factors affecting the pretreatment of biomass with gaseous ozone. *Biotechnol. Bioeng.,* 26, 59-65 (1984).

[50] I.T. Yeom, K.R. Lee, K.H. Ahn, S.H. Lee, Effects of ozone treatment on the biodegradability of sludge from municipal wastewater treatment plants, *Water Sci.Technol.* 46, 421-425 (2002).

[51] Carrere H., Dumas C., Battimelli A., Batstone D.J., Delgenes J.P., Steyer J.P and Ferrer I., Pretreatment methods to improve sludge anaerobic degradability: a review, *J. Hazard. Mater.*, 183, 1-15 (2010).

[52] P. Kumar, D.M. Barrett, M.J. Delwiche, and Pieter Stroeve, Methods for pretreatment of lignocellulosic biomass for efficient hydrolysis and biofuel production, *Ind. Eng. Chem. Res.* 43, 3713-3729 (2009).

[53] B.L.M. Dermott, A.D. Chalmers, J.A.S. Goodwin, Ultrasonication as a pre-treatment method for the enhancement of the psychrophilic anaerobic digestion of aquaculture effluents, *Environ. Technol.* 22, 823-830 (2001).

[54] Saifuddin N and Fazlili S.A., Effect of microwave and ultrasonic pretreatments on biogas production from anaerobic digestion of palm oil mill effluent, *Am. J. Eng. Applied Sci.*, 2, 139-146 (2008).

[55] I. Toreci, K.J. Kennedy, R.L. Droste, Evaluation of continuous mesophilic anaerobic sludge digestion after high temperature microwave pretreatment, *Water Res.* 43, 1273-1284 (2009).

[56] Y. Lin, D. Wang, S. Wu, C. Wang, Alkali pretreatment enhances biogas production in the anaerobic digestion of pulp and paper sludge, *J Hazard Mater.* 170, 366-373(2009).

[57] F. Monlau, E. Latrille, A.C.D. Costa, J.P Steyer, H. Carrere, Enhancement of methane production from sunflower oil cakes by dilute acid pretreatment, *Applied Energy*, 102, 1105-1113 (2013).

[58] J. Perkowski, L. Kos, S. Ledakowicz, Application of ozone in textile wastewater treatment, *Ozone Sci. Eng.* 18, 73-85 (1996).

[59] Dhouib A., Ellouz M., Aloui F and Sayadi S., Effect of bioaugmentation of activated sludge with white-rot fungi on olive mill wastewater detoxification, *Lett. Appl. Microbiol.*, 42, 405-411 (2006).

[60] Ward AJ, Hobbs PJ, Holliman PJ, Jones DL. Optimisation of the anaerobic digestion of agricultural resources. *Bioresource Technology*, 99, 7928–7940 (2008).

[61] Corral M., Samani Z., Hanson A., Smith G., Funk P., Yu H and Longworth J., Anaerobic digestion of municipal solid waste and agricultural waste and the effect of co-digestion with dairy cow manure, *Bioresour. Technol.*, 99, 8288-8293 (2008).

[62] Amon T., Amon B., Kryvoruchko V., Machmuller A., Sixt K.H., Bodiroza V., Hrbek R., Friedel J., Potsch E., Wagentristl H., Schreiner M and Zollitsch W., Methane production through anaerobic digestion of various energy crops grown in sustainable crop rotations, *Bioresour. Technol.*, 98, 3204-3212 (2007).

[63] Lehtomaki A., Viinikainen T.A and Rintala J.A., Screening boreal energy crops and crop residues for methane biofuel production, *Biomass Bioenergy*, 32, 541-550 (2008).

[64] Raposo F., Borja R., Rincon B and Jimenez A.M., Assessment of process control parameters in the biochemical methane potential of sunflower oil cake, *Biomass Bioenergy*, 32, 1235-1244 (2008).

[65] Lei Z., Chen J., Zhang Z and Sugiura N., Methane production from rice straw with acclimated anaerobic sludge: Effect of phosphate supplementation, *Bioresour. Technol.*, 101, 4343–4348 (2010).

[66] Kusch S., Schumacher B., Oechsner H and Schafer W., Methane yield of oat husks, *Biomass Bioenergy*, 35, 2627-2633 (2011).

[67] Gunaseelan V.N., Biochemical methane potential of fruits and vegetable solid waste feedstocks, *Biomass Bioenergy*, 26, 389–399, (2004).

[68] Nasir I.M., Ghazi T.I.M and Omar R., Anaerobic digestion technology in livestock manure treatment for biogas production: A review, *Eng. Life Sci*, 12, 258–269, (2012).

[69] Kalle G.P., Menon K.K.G., Inhibition of methanogenesis and its reversal during biogas formation from cattle manure, *J. Biosci*, 6, 315–324, (1984).

[70] Klasson K.T and Nghiem N.P., Energy Production from Zoo Animal Wastes, OAK Ridge National Laboratory, Febraury 2003.

[71] Schenk P.M., Thomas Hall S.R., Stephens E., Marx U.C., Mussgnug J.H., Posten C., Kruse O and Hankamer B, Second generation biofuels: high-efficiency microalgae for biodiesel production, *Bioenerg. Res*, 1, 20-43, (2008).

[72] Vaidyanathan S., Kavadia K.M., Shroff K.C., Mahajant S.P., Biogas production in batch and semicontinuous digesters using water hyacinth, *Biotechnol. Bioeng*, 27, 905-908, (1985).

[73] Mussgnug J.H., Klassen V., Schluter A and Kruse O., Microalgae as substrates for fermentative biogas production in a combined biorefinery concept. *J Biotechnol*, 150, 51–56, (2010).

[74] Gunaseelan N.V, Anaerobic digestion of biomass for methane production: a review, *Biomass Bioenergy*, 13, 83-114, (1997).

[75] Nizami A.S and Murphy J.D., What type of digester configurations should be employed to produce biomethane from grass silage, *Renew Sust Energ Rev*, 14, 1558–1568, (2010).

[76] https://energypedia.info/index.php/Types_of_Biogas_Digesters_and_Plants

[77] http://www.fao.org/docrep/010/ah810e/AH810E13.htm

[78] http://www.cd3wd.com/cd3wd_40/biogshtm/en/appldev/design/digestypes.html

[79] http://floatingcovers.wordpress.com/

[80] Appels L., Lauwers J., Degreve J., Helsen L., Lievens B., Willems K., Impe J.V and Dewil R., Anaerobic digestion in global bio-energy production: Potential and research challenges, *Renew Sust Energ Rev.*, 15, 4295-4301 (2011).

[81] Kumar S and Biswas TD., Biogas production from different animal excreta, *Indian Journal of Agricultural Science,* 52, 513-520(1982).

[82] Q. Zhao, E. Leonhardt, C. MacConnell, C. Frear and S. Chen, Purification Technologies for Biogas Generated by Anaerobic Digestion, *CSANR Research Report* 2010–001.

[83] J.I. Huertas, N. Giraldo, and S. Izquierdo, Removal of H_2S and CO_2 from Biogas by Amine Absorption, in *Mass Transfer in Chemical Engineering Processes* (2011).

In: Renewable Energy and Sustainable Development
Editors: R. T. Durai Prabhakaran, S. A. Kale et al.
ISBN: 978-1-63463-429-8

Chapter 5

RSM Based Optimization of Biodiesel Production from Tobacco Seed Oil

***M. A. Waheed*[1], *O. D. Samuel*[1*], *B. O. Bolaji*[1] *and O. U. Dairo*[2]**

[1]Department of Mechanical Engineering,
[2]Department of Agricultural Engineering,
Federal University of Agriculture,
Abeokuta, Nigeria

Abstract

Biodiesel is a monoalkyl ester of fatty acid of long chain derived from vegetable oil. Response surface methodology based on Central Composite Design is used to analyze the interaction effect and to optimize the transesterification reaction variables such as reaction duration, catalyst amount and oil to methanol ratio on biodiesel yield. A quadratic model developed based on the CCD, correlating the yield of biodiesel with reaction variables. The predicted yield by optimization is found very close to the experimental values. The optimum conditions for biodiesel production is found to be at the reaction duration of 78 min, oil to methanol ratio of 1:6.09 and catalyst amount of 1.13%. The fuel properties of tobacco methyl ester at the optimized parameters are determined as per ASTM standard and the results of the TSO biodiesel are within biodiesel specifications.

* Corresponding author: E-mail: olusegundavidsamuel@gmail.com, samolud@yahoo.com.

1. INTRODUCTION

Transesterification is the mostly commonly used method of reducing the viscosity of vegetable oil and it is the reaction of a lipid with alcohol to form esters and a by-product, glycerol. Transesterification seems to be the best choice as the physical characteristics of fatty acid (m) ethyl esters (biodiesel) are very close to those of diesel fuel and the process is relatively simple. Different methods to produce biodiesel abound but transesterification is the most commonly used method of reducing the viscosity of vegetable oil. Transesterification is the reaction of a lipid with alcohol to form esters and a byproduct, glycerol. By transesterifying vegetable oils, alkyl monoesters of the fatty acids present in the vegetable oil are obtained. These esters are commonly referred to as biodiesel. A catalyst is usually required to speed up the reaction; this may be basic, acidic or enzymatic in nature. The reaction is affected by several parameters such as the concentration of catalyst, oil to methanol ratio, reaction temperature, moisture, presence of free fatty acids (FFA) and agitation intensity [1]. This process involves many parameters that effect the reaction and optimizing so many reaction factors require large number of experiments, which is laborious, time consuming, and economically non-viable. Response surface methodology (RSM) is a useful statistical technique for the optimization of complex processes, as it reduces the number of experiments required to achieve ample data for a statistically pertinent result [2]. Optimization of process parameters is the key step in response surface methods to achieve high quality without cost inflation [3].

Biodiesel a renewable fuel with lesser exhaust emissions is one of the alternative sources of energy that is presently of interest worldwide. Hence, researchers have focused on the development and the optimization of the processes of biodiesel production to meet the standards and specifications needed for the fuel to be used commercially without compromising on the durability of engine parts. However, the global inevitability of fossil fuels, ever increasing demands for diesel and uncertainty in their availability, depletion rate of energy source, emission of very dangerous pollutants, rapid industrialization and increasing price that make petroleum no longer economically sustainable have necessitated the search for cheap raw material. A sustainable and economical supply of raw material is the key factor for biodiesel to be competitive commercially [4-8]. Tobacco seed oil is one of the promising alternative as the seeds are not commercial products, they are collected from fields and they are a by-product of tobacco leaf product. Veljkovic et al. [9] remarked that tobacco seed, as agricultural wastes and with quite high oil content, might be a cheap and valuable renewable raw material for biodiesel production and the authors produced tobacco biodiesel with maximum yield of 91%. They reported that tobacco biodiesel obtained had fuel properties within the limits prescribed by international standard. Tobacco (Nicotiana tabacum L.) seed oil (TSO), a by-product of tobacco leaf production, has been shown to be an appropriate substitute for a diesel fuel in a raw [10] or chemically modified [11] form.

Mukhtar and Mukhtar [12] highlighted that tobacco seed oil is comparable to the other commonly used vegetable oil. Usta [13] reported that properties and quality verification of biodiesel produced from tobacco seed oil. Murthy and Babu [14] concluded that B5 tobacco biodiesel is the most suitable blend in view of the engine performance and emission. Parlak et al. [15] investigated parameters that affect the yield of tobacco biodiesel and their interactions on emissions of a diesel engine. They concluded that choosing right catalyst and the tobacco methyl ester blend rate are important two factors in view of minimization of pollutant emissions. Mohanrao et al. [16] reported that tobacco seed oil can be used as an alternate fuel in cases of emergency and short term use with proper precautions and tuning of the engine for best performance level.

The Response Surface Methodology (RSM) is a combination of statistical and optimization methods that can be used to model and optimize designs [17-19]. It has many applications in design improvement of products and process operation.

So far, few or no work has been documented on literature on optimization of experimental conditions for biodiesel production from methanolysis of esterified tobacco seed oil; therefore this chapter was aimed at optimizing the condition of transesterification of esterified tobacco seed oil into biodiesel through reaction variable such as alcohol to oil molar ratio, catalyst amount and reaction time. This paper is prepared with following objectives:

1. To study the effect of process factors such as oil-methanol ratio, catalyst amount and reaction time at various levels on yield of tobacco seed oil methyl ester using central composite experimental design
2. To determine the optimum process factors conditions for methanolysis of tobacco methyl ester
3. To characterize the tobacco methyl ester produced at the optimum process factor levels according to ASTM standard.

The world is faced with challenges of minimizing emission of green house gases and other pollutants as well as overcoming the shortage of power supply. Hence, there is a need to develop alternative cost-effective biodiesel feedstock in form of tobacco seed oil that are capable of powering everyday life without bringing harmful environmental changes. Production of biodiesel from tobacco seed oil would not compete with food uses of edible oils and thus there would be value addition as well as curtailing smoking related challenges. This chapter is restricted to optimization of methanolysis/alkaline transesterification parameters of pre-treated tobacco seed oil. The tobacco seed oil methyl ester produced at the optimum process factor levels is characterized and compared with ASTM standard. Also, effort will be made to develop model for biodiesel produced from Nigerian tobacco seed oil.

2. Materials and Methods

Tobacco seed was obtained from Agricultural farm, Olabisi Onabanjo University, Ibogun campus, Ogun State, Nigeria. Methanol, sulphuric acid, potassium hydroxide, phenolphthalein, ethanol, acetic acid and other chemicals/reagents were obtained from Uche Scientific Company Limited, Lagos, Nigeria. All chemical/reagents were of analytical grade with purity > 99%.

2.1. Seed Preparation and Tobacco Seed Oil Extraction

Tobacco seeds were separated from foreign matters, weighed and grinded using blender in order to expose large surface area of the cells containing oil for the extraction of the oil. Due to unavailability of tobacco seed oil in markets, the oil was initially extracted from the collected seeds in laboratory conditions. The extraction procedures entailed drying tobacco seed until a constant weight was obtained and the seeds were grounded to a fine powder. A soxhlet extraction apparatus was used to extract the oil from seed and hexane was used as solvent in extraction process and the oil was separated from the hexane using a Buchi R114 rotary vacuum evaporator (Figure 1). The percentage of oil yield is given by expression in Equation 1.

$$Oil\ yield\ (\%) = \frac{Mass\ of\ oil\ obtained\ (kg)}{mass\ of\ seed\ used\ (kg)} \times 100\% \quad (1)$$

Figure 1. Soxhlet extraction set-up.

2.2. Acid-base Catalyzed Transesterification of High FFA TSO

Raw TSO was filtered to remove all the insoluble impurities followed by heating at 100°C for 15 min to remove all the moisture. TSO had high FFA (15.98%), which is far above the 1% limit suitable for alkaline catalyzed transesterification reaction. FFAs were, therefore, first converted to esters in a pretreatment process for the production of esterified tobacco seed oil The reaction was carried out at a temperature of 60°C for 60 min. using conc. H_2SO_4 (2.5% w/w) as acid catalyst with methanol/oil ratio of 7.5:1 in author's laboratory. The mixing solution of methanol with catalyst (sulphuric acid) is also heated prior to be added into the reactor. After the reaction completing, the products are allowed for 2 h settling down in a separation funnel and then, the methanol-water fraction at the bottom layer is removed. The high FFAs were reduced from 15.98% to 0.98% after second pre-treatment and the oily phase of reaction mixture (FAME) was subjected to alkaline catalyzed transesterification process for biodiesel production using KOH as base catalyst (Figure 2).

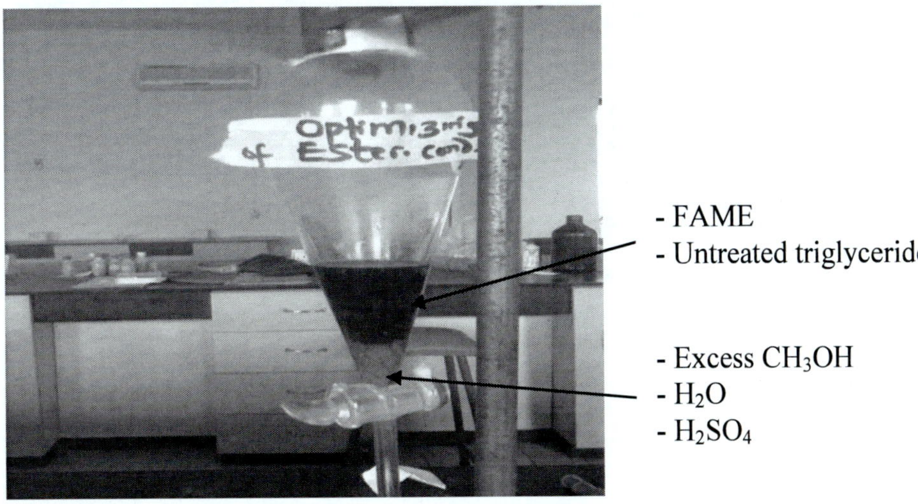

Figure 2. Product from acid esterification during the separation process.

2.3. Alkaline Transesterification Procedures

Esterified tobacco seed oil was subjected to base catalyzed transesterification for biodiesel production at specified reaction conditions according to the Central Composite Response Surface Design (CCRD) using conical flask equipped with a condenser, thermometer, chiller, heating plate and stirrer. The pre-heated oil was trans-esterified with methanol and potassium hydroxide to yield TSO biodiesel and glycerol. The resulting product was poured into a separating funnel mounted on a clamp stand and was allowed to settle down overnight. It was observed that the resulting mixture from the reaction had settled into yellow biodiesel on the top with the black glycerol at the bottom of the separating funnel (Figure 3). The ester was washed with water three times. Small amount of acetic acid (2.5ml/l of the oil) was used in the first washing. At the end of the process, the ester was heated to 90°C at vacuum to remove any water left from the oil. The final ester became clear straw yellow. The procedure was replicated three times and the transesterification of TSO has been optimized using RSM for the maximization of TSB yield which was calculated using the following Equation (2):

$$Yield\ of\ TSOB\ (\%) = \frac{Total\ weight\ of\ methyl\ ester}{Total\ weight\ of\ oil\ in\ the\ sample} \times 100 \tag{2}$$

Figure 3. Two-layers separation after alkali catalyzed transesterification.

2.4. Fuel Properties Tests

The fuel properties of TSOME produced from the optimum reaction variables of TSO transesterification were determined in accordance to biodiesel standard test methods. The standards employed were by the American ASTM D 6751 and European Union EN 14214 biodiesel standards [20]. Fuel properties such as kinematic viscosity (ASTM D 445), density (ASTM D 4052), flash point (ASTM D 93) and acid value (EN 14104) were determined. Values obtained were compared with the ASTM D 6751 and EN 14214 standards.

2.5. Design of Experiment

The alkaline transesterification of tobacco seed oil and optimization of the methyl ester was studied using Design Expert 6.0.6 (Stat-Ease Inc., USA). Response Surface Methodology (RSM) and Central Composite Design (CCD) were used to find the interactions between the three variables and to predict the optimum condition for biodiesel yield. Selection of these parameters was based on the literature studies reported on the alkaline solutions. These parameters were catalyst (KOH) amount, methanol to oil molar ratio and time. Methyl ester yield was used as the response. The reported data in this work were average of triplicate measurements to ensure accuracy. Table 1 shows the levels of the variables used and the ranges. Selection of the levels was based on the preliminary study and literature research. The reaction time, methanol to oil molar ratio catalyst (KOH) amount, were varied from 40 to 80 minutes, 4 to 8 mol/ mol and 0.5 to 1.5 wt. %, respectively. The reactant mixture temperature was held constant below the boiling point of methanol (65°C) in order to prevent mass transfer limitation and bubble formation of methanol. A three-level-three-factors CCD (20 experiments) was adopted, consisting of 8 factorial points, 6 axial points and 6 replicates at the center points design. The yield of methyl ester was analyzed and optimized using Equation (3) as follows:

$$y=\beta_o+\sum_{i=1}^{3}\beta_i x_i+\sum_{i=1}^{3}\beta_{ii} Xi^2+\sum_{j=1}^{3}\sum_{j=1}^{3}\beta_{ij}X_iX_j \quad (3)$$

where β_o, β_i and β_{ii} are regression coefficients for intercept, linear, quadratic and interaction coefficients, respectively are x_i and x_{ij} are coded independent variables.

3. Results and Discussion

3.1. Percentage Oil Content of Tobacco Seed Oil

The oil content of the collected seeds was determined as approximately 35.6% (in weight basis) using hexane as extraction solvent.

3.2. Development of Regression Model

In this chapter, the correlation between response (yield) and three parameters, reaction time, methanol to oil molar ratio and amount of catalyst, KOH, were analyzed using alkaline transesterification. The experimental data of the three parameters and the response yield were shown in Table 1. The Response Surface Methodology was conducted to find the relationship between the time, methanol to oil molar ratio and catalyst dosage. This investigation was employed to obtain high yield of biodiesel without adding co-solvent. A model for the experimental design was fitted using Design Expert 6.0.6 (Stat-Ease Inc., USA). Final equation in terms of (coded factors) experimental data was:

$$Yield=+86.40+3.70A+0.73B+7.45C-0.81A^2-$$
$$5.71B^2-18.17C^2-1.83AB+2.01AC-1.39BC \quad (4)$$

where, A, B and C denote reaction time (min), methanol to oil molar ratio (M), and catalyst (KOH) amount (wt.%), respectively, and Zabeti et al. [21] reported that positive sign in front of each term denotes synergetic effect and negative sign shows the antagonic effect. Figure 4 shows the predicted value and experimental value using the model equations in Equation (1) and the graph shows that the predicted values are quite close to the experimental value and confirmed by the R^2 value, 0.9828. Analysis of variance (ANOVA) was employed to investigate the significance and fitness of the model. ANOVA also represents the interaction of variables on the respond and the effect of individual parameters. Table 3 indicates that the model was very significance because the p value < 0.0001. There was no lack of fit since the value was not significant. Lack of fit implies that the model was fitted to the data Adequate precision measures the signal to noise ratio. This model can be utilized to navigate the design region as ratio greater than 4. Therefore it is desirable and in this case, the ratio was 23.801.

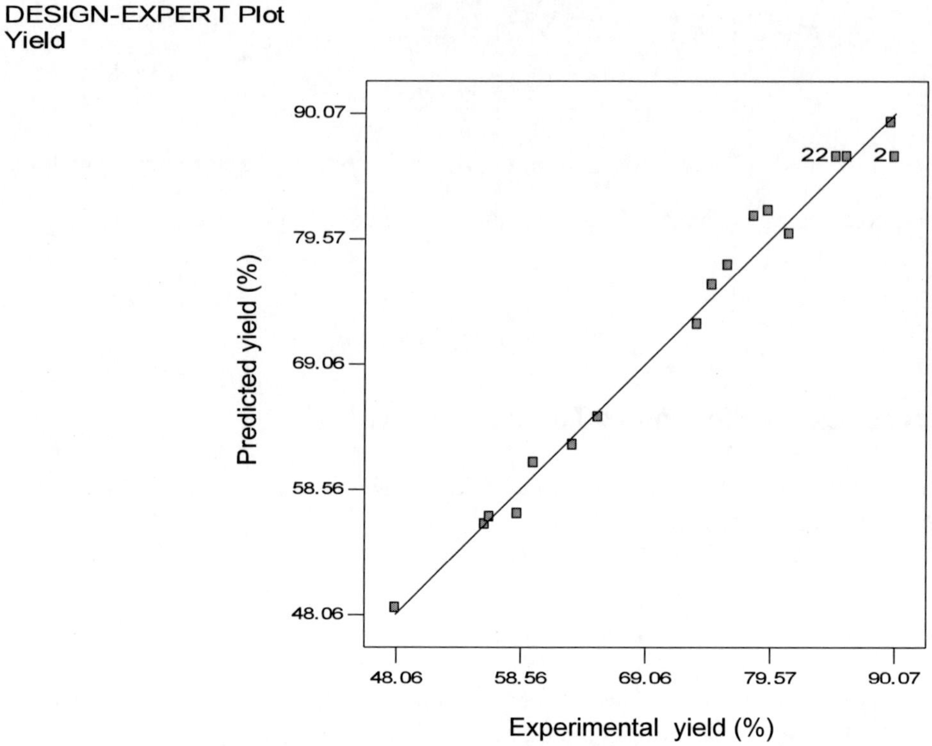

Figure 4. A parity plot between actual experimental values and predicted values of TSO biodiesel.

Table 1. Levels for Biodiesel Production of Tobacco seed Oil

Variables	Coding	Unit	Lower level	High Level
Time	A	Min	40 (-1)	80 (+1)
Methanol to oil Molar ratio	B	M	4(-1)	8 (+1)
Catalyst amount	C	wt. %	0.5 (-1)	1.5 (+1)

Table 2. Central composite design for optimizing NaOH-catalyzed transesterification

	Coded factors			Actual factors			Biodiesel Yield (%)		
Run	A	B	C	Reaction time (min.)	Methanol to Oil molar ratio	Catalyst (w/w. %)	Experimental	Predicted	Residual
1	-1	-1	-1	40	4	0.5	48.06	48.62093	-0.56093
2	1	-1	-1	80	4	0.5	55.63	55.64193	-0.01193
3	-1	1	-1	40	8	0.5	58.36	56.51893	1.841068
4	1	1	-1	80	8	0.5	56.04	56.23493	-0.19493
5	-1	-1	1	40	4	1.5	62.97	62.26493	0.705068
6	1	-1	1	80	4	1.5	76.01	77.34093	-1.33093
7	-1	1	1	40	8	1.5	65.13	64.60793	0.522068
8	1	1	1	80	8	1.5	73.45	72.37893	1.071068

	Coded factors			Actual factors			Biodiesel Yield (%)		
Run	A	B	C	Reaction time (min.)	Methanol to Oil molar ratio	Catalyst (w/w. %)	Experimental	Predicted	Residual
9	-1	0	0	40	6	1	79.38	81.88727	-2.50727
10	1	0	0	80	6	1	89.75	89.28327	0.466727
11	0	-1	0	60	4	1	81.15	79.95127	1.198727
12	0	1	0	60	8	1	78.18	81.41927	-3.23927
13	0	0	-1	60	6	0.5	59.71	60.78327	-1.07327
14	0	0	1	60	6	1.5	74.71	75.67727	-0.96727
15	0	0	0	60	6	1	90.07	86.39982	3.670182
16	0	0	0	60	6	1	86.04	86.39982	-0.35982
17	0	0	0	60	6	1	86.04	86.39982	-0.35982
18	0	0	0	60	6	1	90.07	86.39982	3.670182
19	0	0	0	60	6	1	85.13	86.39982	-1.26982
20	0	0	0	60	6	1	85.13	86.39982	-1.26982

Table 3 illustrates the time (A), catalyst (C), quadratic term methanol to oil molar ratio (B^2), quadratic term catalyst amount (C^2) and time- catalyst amount (AC) were significant as the p- value < 0.05. There exist four insignificant values which were higher than p- value > 0.005, linear methanol to oil ratio (B), quadratic term time (A^2), time- methanol to oil molar ratio (AB) and molar ratio – catalyst amount (BC). This indicates that the model could be utilized to predict biodiesel yield. Figure 4 again shows that the regression model equation provided a very accurate description of the experimental data, indicating that it was successful in capturing the correlation between the process parameters to the yield of TSO biodiesel. This is further supported by the correlation coefficient, R^2 which was found to be very close to unity, 0.961.

Table 3. ANOVA table for response surface quadratic model

Source	Prob.> F	Remark
Model	< 0.0001	Significant
A	0.0007	Significant
B	0.3594	Not significant
C	< 0.0001	Significant
A^2	< 0.5884	Not significant
B^2	0.0029	Significant
C^2	< 0.0001	Significant
AB	0.0582	Not significant
AC	0.0401	Significant
BC	0.1351	Not significant
Lack of fit	0.4553	Not significant
Pure error		Not significant
R^2	5.53	
Adj R^2	0.9672	
Pred R^2	0.9138	

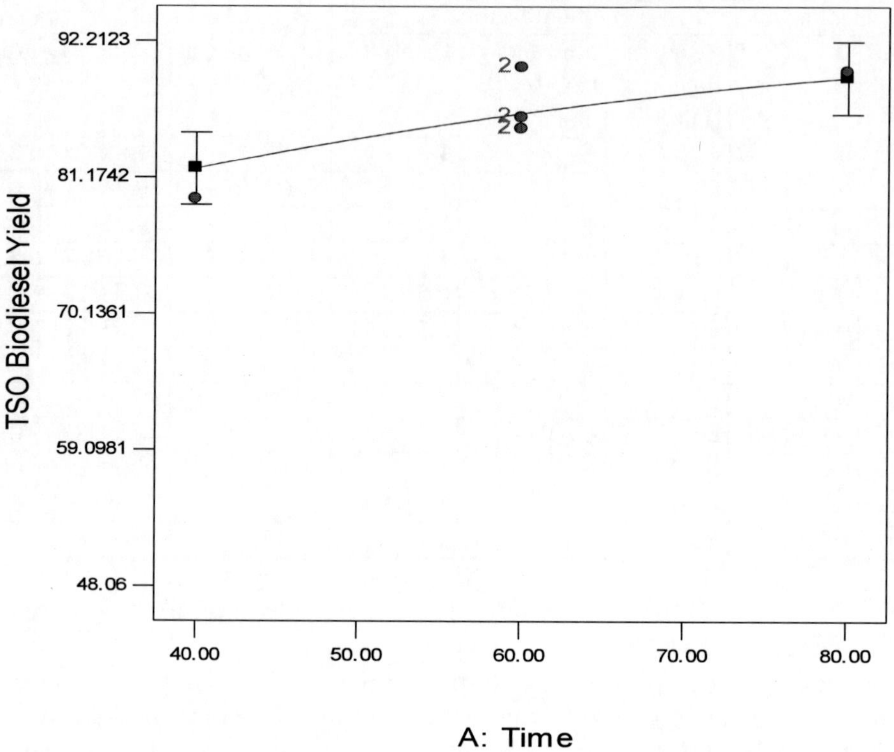

Figure 5. Effect of reaction time on TSO biodiesel yield.

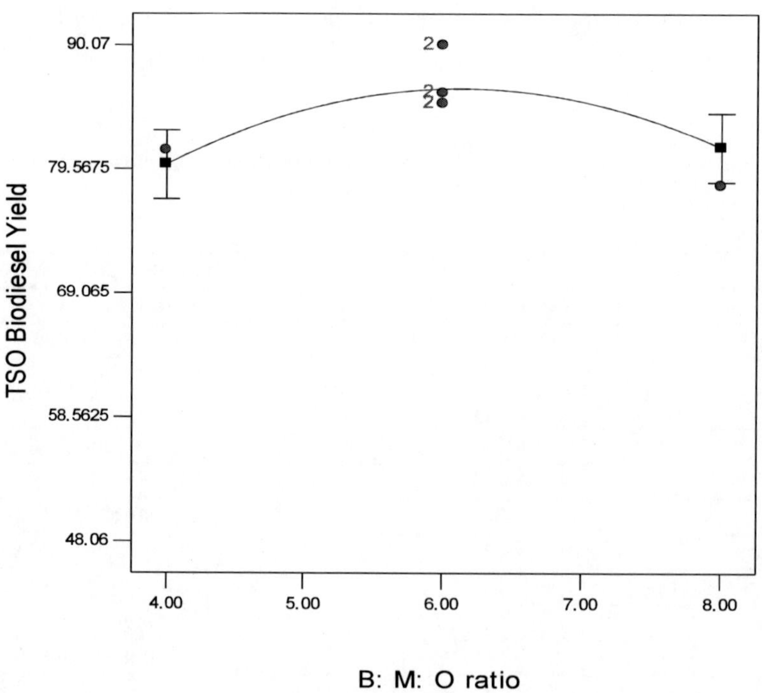

Figure 6. Effect of methanol-to oil molar ratio on TSO biodiesel yield.

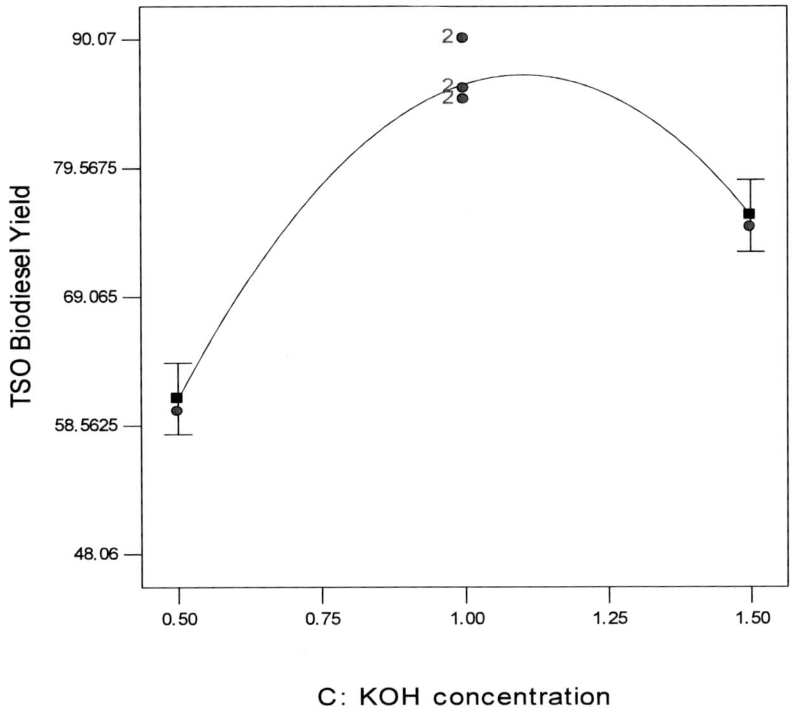

Figure 7. Effect of amount of catalyst on TSO biodiesel yield.

3.3. Effect of Transesterification Process Variables

Based on the analysis of variance, the transesterification reaction was remarkable influenced by various interactions between the process variables. However, significant individual process variables that affect the transesterification reaction is reaction time (A) methanol- oil molar ratio (B) and catalyst amount (C). The effect of reaction time on biodiesel yield is shown in Figure 5. It is seen from one factor plot that the yield of biodiesel increased with increase in reaction time from 40 to 80 minutes. Above 40 minutes, the biodiesel yield greatly increased and maximum yield was obtained at a reaction time of 80 minutes.

The ratio of methanol to oil is one of the important factors that affect the conversion of triglyceride to fatty acid methyl ester (FAME). The results obtained in this chapter as revealed in Figure 6 shows that the yield of biodiesel increased with increase in methanol- oil molar ratio from 4 to 6 mol/mol. Above 6mol/mol, the biodiesel yield slightly decreased. Therefore, maximum yield is obtained at a molar ratio of 6, showed that biodiesel yield increased as methanol-to-oil molar ratio increases up till an optimum value after which it decreased. Aworanti et al. [22] reported that higher ratio of methanol used could also minimize the contact of excess triglyceride molecules on the catalyst's active sites which could decrease the catalyst activity. They further stressed that an increase in the ratio of methanol could also lead to the increase in the purity of the biodiesel layer which would also be responsible for the observed increased in biodiesel yield.

Figure 7 illustrates influence of the dosage of KOH catalyst on biodiesel yield. When the mass of catalyst was increased from 0.5% to 1%, the active sites of KOH was increased; accelerating transesterification reaction, thereby enhancing biodiesel yield. However, optimum yield was obtained at a mass of 1%.

3.4. Effect of Interaction between Process Variables

Three-dimensional surface plots of the predicted yield of the biodiesel production are given in Figures 8-10. Each surface was generated by keeping one factor constant at the central point of CCD and varying the other two within their experimental ranges. The figures show the yield of biodiesel increases with the increase of time, oil to methanol molar ratio and catalyst loading.

Figure 8 shows the effect of interaction of methanol-to- oil molar ratio and reaction time on methanolysis of TSO to biodiesel at a constant amount of KOH dosage (1%). There was a high significant effect of reaction time on alkali-catalyzed transesterification. Under such conditions, methanol-to-oil molar ratio of about 6: 1 gave the suitable conversion to biodiesel. Also, there was negative significant interaction between methanol-to oil molar ratio and reaction time. This shows that TSO biodiesel yield decrease beyond an optimum value, whereas it increases up till an optimum value (6:1).

Figure 9 shows the interaction effect of catalyst amount and reaction time on TSO biodiesel production at a constant methanol to-oil molar ratio of 6:1 dosage of KOH and reaction temperature significantly influences TSO biodiesel production. There was a positive significant interaction between catalyst and reaction time. This indicated that TSO biodiesel increases with increase in temperature an increase of catalyst. However, at longer reaction time (80min), an increase of catalyst amount from 0.5wt. % to 01.5 wt. % only slightly increases in the yield of TSO biodiesel.

Effects of interaction of methanol-to oil molar ratio and catalyst amount on transesterification of TSO to biodiesel at a constant reaction duration is showed in Figure 10. At lower methanol-to-oil molar ratio, the conversion to biodiesel increased with increase in KOH amount. It appears increase in methanol-to-oil molar ratio had remarkable influence on the conversion at different catalyst dosage.

3.5. Model Evaluation of Alkaline Transesterification

The optimum condition of these three parameters was obtained by optimization of the Equation (3) using Design Expert 6.0.6. Table 4 shows the summary of optimization conditions with high and low limit to optimize based on the previous model. In this case, reaction time and catalyst amount were in the range. The optimal condition for this study was 1.13 wt. % of KOH catalyst, 6.09 methanol/ oil molar ratio and time of 78 min (Table 5).

To evaluate the accuracy of the developed model, base catalyzed transesterification was conducted according to the optimum condition as in Table 5. Table 5 shows the results obtained and the yield was 90.11% if compared to the predicted yield, 90.17%. Both of the values are close with smaller error of 0.06% and this shows that this model was in agreement with the actual value.

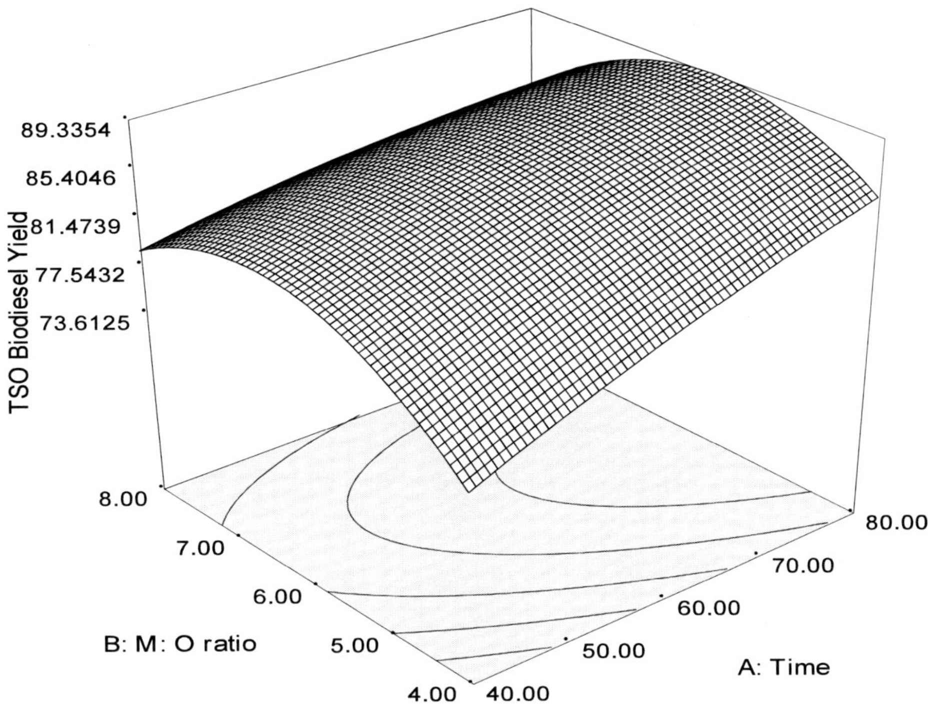

Figure 8. Response surface plot of the interaction of time and methanol-to oil ratio on TSO biodiesel yield when the amount of catalyst is 1 percent.

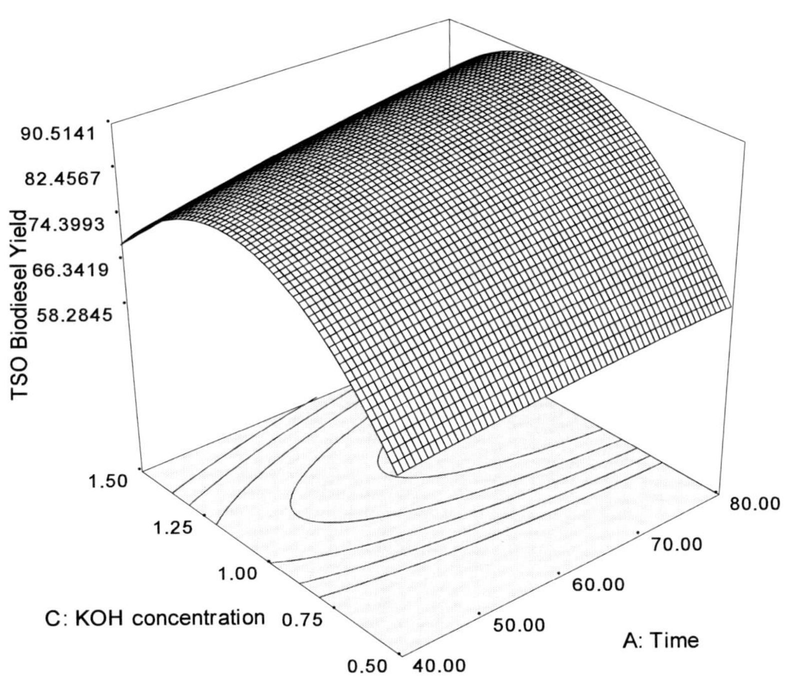

Figure 9. Response surface plot of the interaction effect of catalyst amount and reaction time on TSO biodiesel yield when the molar ratio of methanol to oil is 6:1.

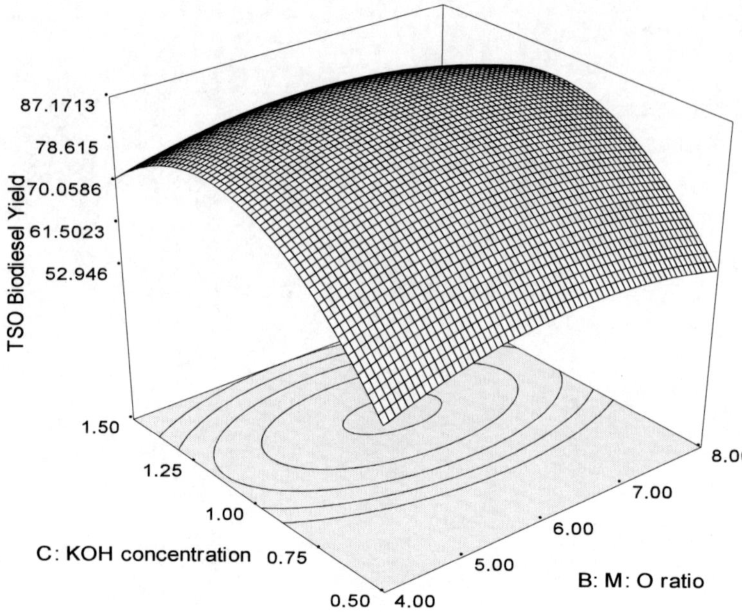

Figure 10. Response surface plot of the interaction effect of methanol-to oil molar ratio and catalyst amount on TSO biodiesel yield when the reaction time is 60 minutes.

Table 4. Summary of optimization conditions

Parameter	Objective of optimization	Lower limit	Upper limit
Reaction time	Is in range	40	80
M: O ratio	Is in range	4	8
KOH amount	Is in range	0.5	1.5
Yield	To maximize	48.06	90.07

Table 5. Results of modeling/optimization of biodiesel yield

Experiment	Time, A (min)	Methanol/ oil ratio, B (mol)	Catalyst amount, C (wt. %)	Yield (%) (predicted)	Yield (%) (Experimental)
1	78	6.09	1.13	90.17	90.11

Table 6. Properties of TSO biodiesel

Fuel Characteristics	Unit	TSO methyl ester	ASTM Standard D6751	EU Standard EN 14214 Corrected value	IS 15607
Specific gravity@ 50°F/50°F	g/cm^3	0.891	0.880	0.86-0.90	0.860-0.900
Density	Kg/m^3	890		860-900	
Kinematic viscosity @40°C	mm^2/s	3.87	1.9-6.0	3.5-5.0	2.5-6.0
Flash point	°c	126	>120	> 120	-
Acid value	Mg KOH/g	0.420	<0.8	<0.50	-

Validation of the Final Model

Through the *t*-test, no significant differences are observed between the experimental and predicted values ($P \geq 0.05$). Moreover, the closeness of actual and predicted values in Figure 11 shows that the regression equation is adequate (Mirhosseini et al., 2008).

3.6. Biodiesel Properties

Table 6 shows the summary of the properties of the biodiesel of tobacco seed oil The properties of TSO biodiesel were within the mentioned biodiesel standards, ASTM D6751, EN 14214 and IS 15607 The result of few standard fuel test such as specific gravity, viscosity, flash point and acid value of the TSO biodiesel are within biodiesel specification.

CONCLUSION

Central Composite Design was used for optimization of methanol/oil ratio, catalyst amount, and reaction time on the transesterification of Nigerian tobacco seed oil. Thus, research work gave an optimal value of 90.17% biodiesel, in 78 minutes of reaction time with 1.13 wt.% of catalyst loading and methanol/oil ratio of 1:6.09. Quadratic polynomial models were obtained to predict yield of biodiesel. On the basis of ANOVA; reaction time, catalyst amount, quadratic term methanol to oil molar ratio, quadratic term catalyst amount and time-catalyst amount were significant reaction time, catalyst amount, quadratic term of methanol/ oil ratio, quadratic term catalyst amount, reaction time - catalyst amount had a significant effect on TSO biodiesel yield. The result showed that the predicted value was in agreement with the experimental value, which was established with additional experiments to confirm the optimized parameters. The prepared TSO biodiesel gave promising results as alternative diesel fuel with fuel properties in good agreement within limit set by international biodiesel standard. Therefore production of biodiesel from Nigerian tobacco seed oil is a feasible process.

ACKNOWLEDGEMENT

The authors are grateful to Pastor Akinola Omotayo, Biotechnology Central Laboratory, Federal University of Agriculture, Abeokuta, Nigeria for his assistance during the research work.

REFERENCES

[1] Encinar, J. M., Gonzalez, J. F. and Rodryguez-Reinares, A. 2005. Biodiesel from used frying oil. Variables affecting the yields and characteristics of the biodiesel. *Ind Eng Chem Res*, 44, 5491–5499.

[2] Jeong, G. T. & Park, D. H. 2009. Optimization of Biodiesel Production from Castor Oil Using Response Surface Methodology. *Appl Biochem Biotechnology*, (156): 431–441.

[3] Abhang, L. B. and Hameeddullah, M. 2012. Response surface modeling and grey relational analysis to optimize turning parameters with multiple performance characteristics. *International Journal of Manufacturing, Materials, and Mechanical Engineering (IJMMME), 2*(2), 12-45.

[4] Fazal, M. A., Haseeb, A. S. M. A, and Masjuki, H. H. 2011. Biodiesel feasibility study: an evaluation of material compatibility; performance; emission and engine durability. *Renew Sustain Energy Rev,* (15), 1314-24.

[5] Zeng, J., Wang, X., Zhao, B., Sun, J., and Wang Y. 2008. Rapid in situ transesterification of sunflower oil. *Ind Eng Chem Res,* 48, 850-6.

[6] Su, E., You, P., & Wei, D. 2009. In situ lipase-catalyzed reactive extraction of oilseeds with short-chained dialkyl carbonates for biodiesel production. *Bioresour Technol.,* (100), 5813-7.

[7] Lei, H., Ding, X., Zhang, H., Chen, X., Li, Y., & Zhang, H. 2010. In situ production of fatty acid methyl ester from low quality rice bran: an economical route for biodiesel production. *Fuel,*(89): 1475-9.

[8] Samuel, O. D., Waheed, M. A., Bolaji, B. O., and Dairo, O. U. (2013). Production of biodiesel from Nigerian Restaurant Waste Cooking oil using Blender. *International Journal of Renewable Energy Research*, 3, 976-979.

[9] Veljkovic, V.B., Lakic´evic S. H., Stamenkovic, O. S., Todorovic, Z. B, Lazic, M. L. 2006. Biodiesel production from tobacco (Nicotiana tabacum L.) seed oil with a high content of free fatty acids. *Fuel,* 85: 2671-2675.

[10] Giannelos, P. N., Zannikos, F., Stournas, S., Lois, E., and Anastopoulos, G.2002. Tobacco seed oil as an alternative diesel fuel: physical and chemical properties. *Ind Crop Prod*., 16:1-9.

[11] Usta N. 2005. Use of tobacco oil methyl ester in turbocharged indirect injection diesel engine. *Biomass Bioenergy,* 28:77-86.

[12] Mukhtar, A., and Mukhtar, H. 2006. Extraction and characterization of tobacco seed oil. *Asian Journal of Chemistry*, 18(1):20-24.

[13] Usta, N., Aydogan, B., Con, A. H., Uguzdogan, E. and Ozka, S. G.2011. Properties and quality verification of biodiesel produced from tobacco seed oil. *Energy Conversion and Management,* 52:2031-2039.

[14] Murtyhy, S. Y. V. V., and Babu, A. V. 2011. Performance and emission evaluation of diesel engine fueled with methyl esters of tobacco seed oil. *Int. Journal of Mechanical Engineering,* 3(1): 37-41.

[15] Parlak, A., Karabaş, H., Zsert, I., Ayhan, V. and Cesur, I. 2012. Application of Taguchi's methods to investigate factors affecting emissions of a diesel engine running with tobacco oil seed methyl ester, *Int. J. of Vehicle Design*, 59(2/3): 196-211.

[16] MohanRao, C. V., Sita, A. V. Rama Raju, R., and Govindarajulu, K. 2013. Experimental investigations on a four Stroke compression ignition engine using neat tobacco seed oil as an alternate. *Fuel Global Journal of Researches in Engineering Mechanical and Mechanics Engineering,* 13(4):1-11.

[17] Montgomery, C. C. 2005. *Design and Analysis of Experiments,* 6th Edition, John Wiley & sons, USA.

[18] Myers, R. H., and Montogomery, D. C. 2002. *Response Surface methodology: Process and Product optimization using designed Experiments* 2nd Ed, John Wiley & Sons, USA.

[19] Mohammed, J. K., Aziz, H. M., Yosoff, M. S., and Adlan, N. 2010. Application of response surface Methodology (RSM) for optimization of ammonical nitrogen removal from semi-aerobic landfill leachate using ion exchange resin, *Desalination,* 254: 154-161.

[20] Knothe, G., Gerpen, J.V, Krahl, J., and Ramos, L. P. 2006. *Manual do Biodiesel*, São Paulo: Blücher p. 352.

[21] Zabeti, M., Daud W. M. A. W, Aroua M. K. 2010. Biodiesel production using alumina supported calcium oxide: an optimization study. *Fuel Process Technol.;* 91:243-8.

[22] Aworanti, O. A., Agarry, S. E., and Ajani, A. O. 2013. Statistical optimization of Process variables for biodiesel production from waste cooking oil using heterogeneous base catalyst. *British Biotechnology Journal,* 3(2): 116-132.

In: Renewable Energy and Sustainable Development
Editors: R. T. Durai Prabhakaran, S. A. Kale et al.
ISBN: 978-1-63463-429-8

Chapter 6

BIODIESEL FROM RUBBER SEED OIL: A SUSTAINABLE SOLUTION FOR BIOFUEL PRODUCTION

Uma Krishnakumar[1] and V. Sivasubramanian[2,*]
Department of Chemical Engineering, National Institute of Technology Calicut, Kozhikode, India

ABSTRACT

The initial surge of biofuels was driven by the concerns regarding energy security, environmental pollution, global warming and rising fossil fuel prices. However, in terms of sustainability perspective, biofuels offers the advantages like energy security, reduction of greenhouse gases and reduced air pollution and have some risks like the intensive use of resources, monocultures, reduced biodiversity. The established feedstocks like corn, rapeseed, soybeans are found to be efficient, but are seems to have certain sustainability issues in terms of food and water security as most of these are originally food crops and they need lots of water for irrigation. So in order to address the sustainability issues second generation of biofuels and related feed stocks are in force based on waste residues and other non-crop biomass. Large scale research activities are involved in non-edible vegetable oils such as jatropha seed oil as a possible alternative for sustainable biodiesel feedstock. However, this also has some issues since we require a lot of land for its cultivation. In this chapter, we examine the possibilities of using the rubber seeds since at present it is being an underutilized by-product from the rubber plantations. The oil extractions from the rubber seeds are found to be rich in fatty acid contents and hence it can be employed as a potential feedstock for biodiesel production. Also, this feed stock will have the minimum sustainability issues since rubber seed is the essential by-product of rubber plantation and the major product the natural rubber has a high economic value. However, the potential of converting rubber seed oil into biodiesel must be well examined because physical and chemical properties of biodiesel produced from any feedstock must comply with the limits of ASTM and DIN EN specifications for biodiesel fuels.

* Department. of Chemical Engineering, National Institute of Technology Calicut, Kozhikode, India-673601. Email: siva@nitc.ac.in, p110069ch@nitc.ac.in.

In this chapter, we examine the various physical- chemical properties of the biodiesel produced from rubber seed oil as feedstock. Here we present an overview of biofuels, biodiesel production techniques, biodiesel standard and characteristics. We are also presented in detail the engine performance and emission analysis. As conclusion, it has been found that rubber seed oil has the potential to boost the future biodiesel production.

1. INTRODUCTION

"Biofuels" is a collective term for liquid fuels derived from renewable sources, including ethanol, biodiesel, and other renewable liquid fuels. It is a type of fuel whose energy is derived from biological carbon. These include fuels derived from biomass conversion, as well as solid biomass, liquid fuels and various biogases. Biodiesel is one among the promising renewable fuels, holding various advantages compared with fossil fuel.

Most of the countries in the world are keen on reducing their dependence on coal and petroleum to meet their increasing energy demand and are encouraging the use of biofuels in their energy policies. First-generation biofuels are created largely from feedstocks that have traditionally been used as food. For example, ethanol from corn, biodiesel from vegetable oil & animal fats, etc. Despite the considerable potential of biodiesel, the production of biodiesel is found to be expensive from edible oils owing to the growing demand for edible oils and the high cost of the feedstock. Hence the future of the biodiesel is going to be limited unless biodiesel can be extracted from other non edible raw materials. To avoid these, large scale plantations of non edible oil plants were also thinking of. Jatropha oil produced from the seeds of the Jatropha curcas, a plant that can grow in wastelands are considered to be an excellent source of bio-diesel for the next generation. However, this requires a large scale utilization of land for its cultivation and raise some issues of sustainability.

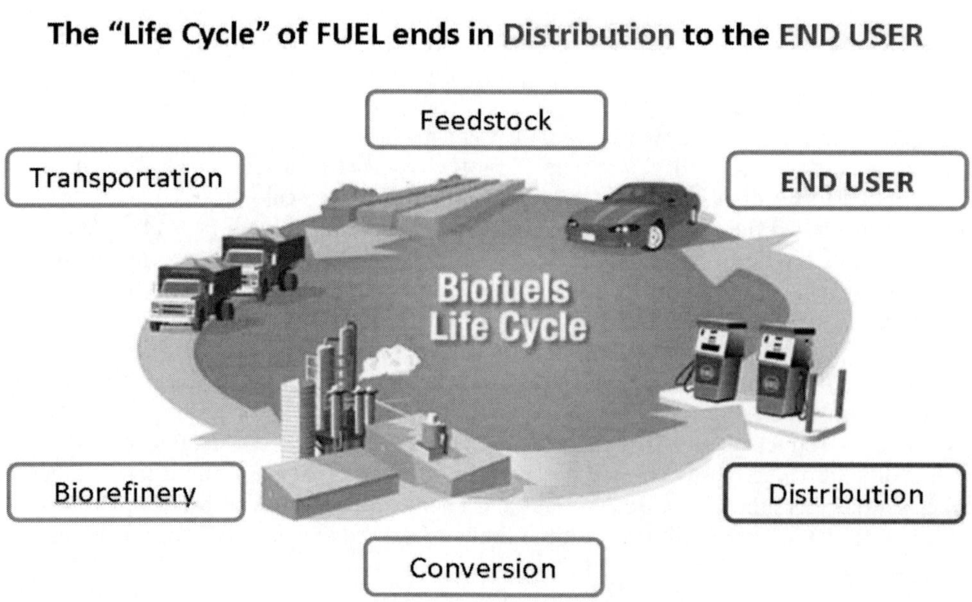

Figure 1. Biofuels life cycle.

Figure 2. Rubber seeds and Jatropha seeds.

Another possible feedstock identified was the oil produced from the rubber seeds. Rubber seeds are the byproduct of the rubber plantations. The major rubber producers in the world are Thailand (35%), Indonesia (23%), Malaysia (12%), India (9%), and China (7%). Rubber seeds are the essential by the product in this plantation which are not being utilized properly at present. Normal seed production yields from the rubber plantation vary from 70 to 500 kg/hectare/year and these are normally regarded as waste. Some efforts have been initialized in some part for extracting oils from these seeds. In present rubber seed oil does not find any major application and hence even the natural production of seeds it remains underutilized. In this situation, analyzing the possibilities of biodiesel production utilizing the rubber seed oils has great relevance. In this chapter, we are describing a single step transesterification process developed using a solid metal oxide catalyst for the efficient conversion of rubber seed oil to biodiesel. The performance of this newly developed fuel is evaluated by fueling a four stroke diesel engine and examines its performance and emission characteristics. It could be observed from this study that the rubber seed oil can be used as an effective feedstock for the production of biodiesel. The engine performance study reveals that the performance does not differ greatly from that of diesel fuel, which confirms that the rubber seed oil based biodiesel can be used as an effective substitute to Petro Diesel. This study also highlights that rubber seeds have a potential to give more income to the rubber farmers and can generate more revenue to the state by its effective use for biodiesel production. [1-3]

2. BIODIESEL

Biodiesel consists of the mono alkyl esters formed by a catalyzed reaction of the triglycerides in the oil or fat with a simple monohydric alcohol. It is a biodegradable and non-toxic diesel fuel substitute that can be used in diesel engines without any modification of the engine architecture. It has been proved that the diesel engines operated on biodiesel have lower emissions of carbon monoxide, unburned hydrocarbons, particulate matter and air toxics than when operated on petroleum-based diesel fuel. It can be used as a fuel for vehicles in its pure form, but it is usually used as a diesel additive to reduce levels of particulates, carbon monoxide, and hydrocarbons from diesel-powered vehicles. [10]

The use of biodiesel is found to be good for diesel engines as it lubricates better than petroleum- based diesel fuel and has excellent solvent properties. Conventional diesel fuel can leave deposits inside fuel lines, storage tanks and fuel delivery systems over time. Biodiesel dissolves this sediment while adding no deposits of its own, resulting in cleaner, more trouble-free fuel handling systems. Some of the major advantages and properties of biodiesel can be summarized as below:

a) Requires no engine modifications (except replacing some fuel lines on older engines)
b) Can be blended in any proportion with petroleum diesel fuel
c) High cetane number and excellent lubricity

Table 1. Technical properties of biodiesel

Chemical name	Fatty acid (m)ethyl ester
Chemical formula	$C_{14} - C_{24}$ Methyl ester or $C_{15\_25}H_{28\_48}O_2$
Kinematic viscosity (mm^2/s at 313 K)	3.3 -5.2
Density (kg/m^3 at 288 K)	860 – 894
Reactivity	Stable, but avoid strong oxidizing agents

3. METHODS AVAILABLE FOR BIODIESEL PREPARATION

Biodiesel is produced by a chemical reaction called transesterification. Three basic routes have been identified for biodiesel production from oils and fats:

a) Base catalyzed transesterification of the oil
b) Direct acid catalyzed transesterification of the oil
c) Conversion of the oil to its fatty acids and then to biodiesel

Transesterification is the reaction of a triglyceride (fat/oil) with an alcohol to form esters and glycerol. A triglyceride has a glycerin molecule as its base with three long chain fatty acids attached. In most of the production methods, methanol or ethanol is the alcohol used (methanol produces methyl esters, ethanol produces ethyl esters) and is catalyzed by a base, either potassium or sodium hydroxide.

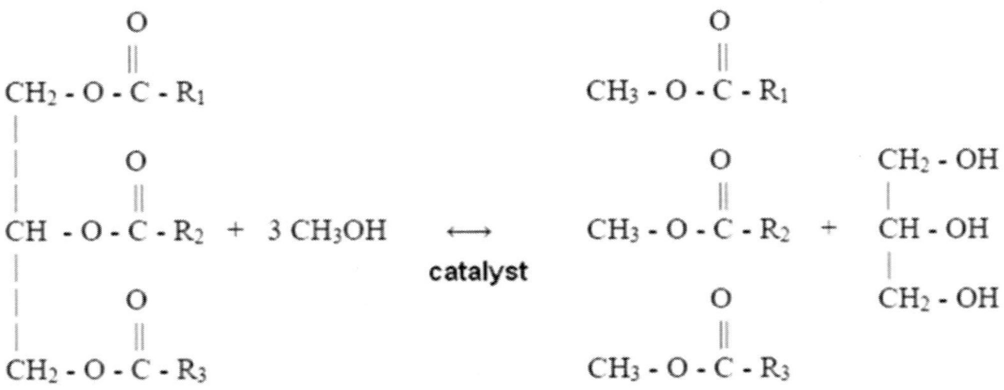

The characteristics of the fat are determined by the nature of the fatty acids attached to the glycerin. After the transesterification reaction, the solution is kept for a specified period of time to settle the biodiesel and sediment layers clearly. The biodiesel was separated from sedimentation by separating funnel carefully. The by- product glycerol can be purified and sold for commercial purposes.

The alcohol/catalyst mix is charged into a closed reaction vessel and the oil or fat is added. The system is totally closed to the atmosphere to prevent the loss of alcohol. Once the reaction is complete, two major products exist: glycerin and biodiesel. The glycerin phase is much denser than biodiesel phase and the two can be gravity separated with glycerin simply drawn off from the bottom of the settling vessel. The most important aspects of biodiesel production to ensure trouble free operation in diesel engines are: complete reaction, removal of glycerin, removal of catalyst, removal of alcohol and the absence of free fatty acids. [4]

4. Major Characteristics and Standards of Biodiesel

Biodiesel is different from Petro diesel in its physical and chemical properties. The quality of biodiesel highly depends on many factors like the quality of feedstock, fatty acid composition of the feedstock, type of production and refining process employed and post production parameters. Therefore, it is required to have some standard characteristic of the biodiesel fuel. The internationally accepted standards for biodiesel properties are ASTM 6571; EN 14214 etc. these standards describe the physical and chemical characteristics of variety biodiesel produced from non-edible and edible oil resources. Some of these properties include; calorific value (MJ/kg), cetane number, density (kg/m^3), viscosity (mm^2/s), cloud and pour points (°C), flash point (°C), acid value (mg KOH per g-oil), ash content (%), copper corrosion, carbon residue, water content and sediment, distillation range, sulfur content, glycerin (% m/m), phosphorus (mg/kg), oxidation stability.

4.1. Density

Density can be found from specific gravity. Initially empty specific gravity bottle is weighed and the weight is represented as W_1. The specific gravity bottle is filled with water and weighed again, W_2. Oil is taken in the bottle and weighed which is noted as W_3. Biodiesel usually has density higher than conventional diesel. Density indicates the long term storage, handling and combustion of automotive fuels. This property helps in splash blending of biodiesel with conventional diesel.

$$Specific\ gravity = \frac{[W_3 - W_1]}{[W_2 - W_1]} \qquad (1)$$

Density of the oil = Specific gravity of the oil × density of water.

4.2. Viscosity

This is the most important property of biodiesel since it affects the operation of fuel injection equipment, particularly at low temperatures when an increase in viscosity affects the fluidity of the fuel. The lower the viscosity of the biodiesel, the easier it is to pump, atomize and achieve finer droplets. The viscosity of the oil samples were generally being determined using Ostwald viscometer which comprises of a capillary through which the samples are allowed to flow and the time measurements are made. The law controlling the flow of a liquid through a capillary is expressed mathematically by Poiseuille's equation, which forms the basis for deriving the expression to calculate viscosity of unknown samples with reference to water. The viscosity of the oil samples was calculated by:

Density of water (ρ_1)	:	kg/m^3
Time of flow for water (t_1)	:	seconds
Viscosity of water (μ_1)	:	8.517 mP
Density of oil (ρ_2)	:	kg/m^3
Time of flow for oil (t_2)	:	seconds

The kinematic viscosity of the oil samples (μ_2) was determined using the Equation (2).

$$\mu_2 = \frac{[\mu_1 \times \rho_2 \times t_2]}{[\rho_1 \times t_1]} \qquad (2)$$

4.3. Flash and Fire Point

The flash point is a measure of the lowest temperature at which application of the flame causes the vapor above the sample to ignite, i.e., it is a measure of the tendency of a sample to form a flammable mixture with air. Fire point of a fuel is the temperature at which it will continue to burn for at least 5 seconds after ignition by an open flame. Flash point is used in safety regulations to define "flammable" and "combustible" materials. Higher values indicate materials that are less likely to ignite accidentally. The flash point of the oil samples was determined using a Penskey Closed / Open Cup apparatus. With the heating device set to the proper level, the thermometer inserted in the testing apparatus and the stirring device attached, the test was started.

4.4. Cloud and Pour Point

Cloud point of a fluid is the temperature at which dissolved solids are no longer completely soluble, precipitating as a second phase giving the fluid a cloudy appearance. The wax crystals typically form on the lower circumferential wall with the appearance of a whitish or milky cloud. The cloud point is the temperature at which these crystals first appear. The pour point is the lowest temperature at which the fuel continues to flow. It is an indication of the temperature at which fuel can still be pumped. Alternately, pour point can also be described as the lowest temperature at which a fuel performs satisfactorily and beyond this temperature, the fuel stops flowing and starts to freeze.

4.5. Acid Value

Acid value or neutralization number is an indication of the corrosive nature of the fuel. Acid value is calculated by equation

$$Acid\ value = \frac{[No\ of\ ml\ of\ N/10KOH\ used \times 5.6]}{Weight\ of\ sample(g)} \tag{3}$$

where 5.6 represents the amount of KOH in mg present per each ml of N/10 KOH solution.

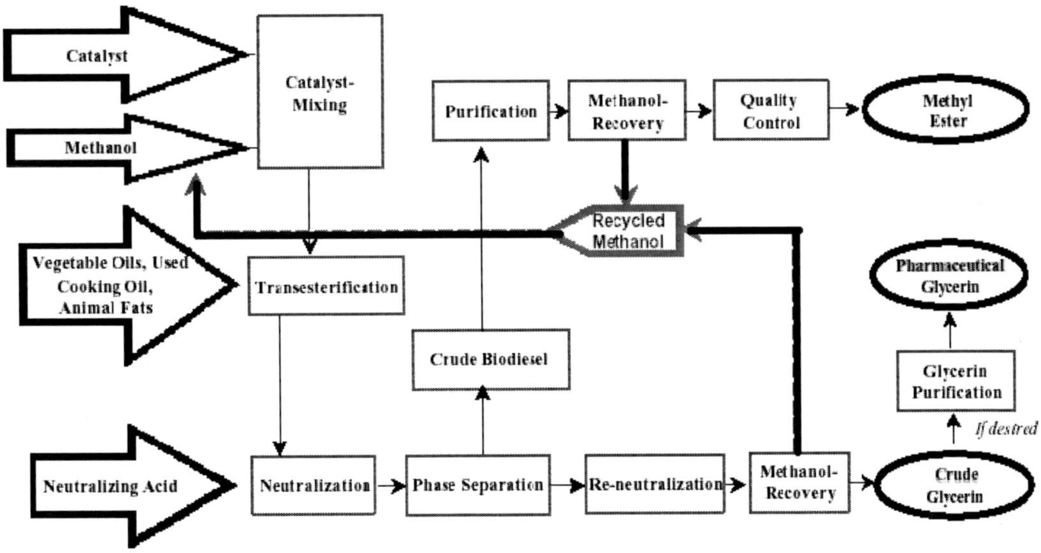

Figure 3. Biodiesel Production Process.

4.6. Cetane Number

Cetane number (CN) is a measure of a fuel's ignition delay, the time period between the start of injection and the first identifiable pressure increase during combustion of the fuel. Cetane number is a good indication of performance of a fuel since fuels with high cetane number could provide short ignition delay and good combustion. Low cetane number could lead to poor combustion, smoking, and excessive engine deposits.

4.7. Iodine Number

Iodine number is an indication of the amount of unsaturated fatty acids in the mixture. A higher number of iodine will lead to fuel thickening leading to plugging of fuel filters. When iodine is introduced in the oil, the iodine attaches itself over a single bond to form a double bond. This iodine number refers to the amount of iodine required to convert unsaturated oil into saturated oil. Hanus method determines the iodine value of oil which is the number of g expressed as iodine which will combine with 100g of oil or fat under the condition of the determination at 20^{0}C. Iodine value is calculated by the equation:

$$Iodine\ value = \frac{[1.269(X_1 - X_2)]}{w} \tag{4}$$

where W is the weight of the oil, X_1 is the ml of 0.1 N, phosphate required for blank determination, X_2 is the ml of 0.1 N thiosulphate required for titration.

4.8. Saponification Value

Saponification value represents the number of milligrams of KOH or NaOH required saponifying 1g of fat under the conditions specified. The saponification value of oil is determined by refluxing a known quantity of the sample with a known excess of standard KOH solution and determining the alkali consumed by titrating the unreacted alkali.

The weight of the sample taken	=	Xg
The volume of alcoholic KOH added to both the flasks	=	5 ml.
Volume of HCl used in sample determination	=	V_1 ml
Volume of HCl used in blank determination	=	V_2 ml

Saponification value of the sample is calculated by the equation:

$$Saponific\ ation\ value = \frac{[(V_2 - V_1) \times 28]}{X} \tag{5}$$

Table 2. Fatty acid composition of the crude rubber seed oil

Sr. No	Component	Structure	Percentage
1	Palmitic acid	C 16:0	0.23
2	Linoleic acid	C18:2	52.84
3	Oleic acid	C18:1	12.70
4	Stearic acid	C18:0	35.85

Table 3. Physico-chemical properties measured

Oil/Fuel	Kinematic Viscosity [mm2/s]	Flash Point [0C]	Acid Value [mg KOH/g]	Cloud Point [0C]	Density [kg/m3]	Calorific Value [MJ/kg]
Crude RSO	6 ± 0.2	208 ± 0.5	1.68 ± 0.3	-5 ± 0.2	882.4 ± 0.5	36.1±1.08
Biodiesel from RSO	3.7 ± 0.2	110 ± 0.5	0.22 ± 0.3	-6 ± 0.2	857.1 ± 0.5	39.78 ± 1.08

Table 4. Comparison of properties of the biodiesel produced with commercial petro diesel

Fuel property	Biodiesel from RSO	Petro diesel
Kinematic Viscosity [mm^2/s]	3.7	2.0-5.5
Flash Point [^{0}C]	110	54
Acid Value [mg KOH/g]	0.22	0.835
Cloud Point [^{0}C]	-6	2max
Calorific Value [MJ/kg]	39.78	43.73
Cetane number	50	45

Table 5. Optimized conditions

Sr. No	Parameter	Condition
1	Alcohol to oil molar ratio	4.5:1
2	Catalyst to oil weight ratio	0.6:1
3	Reaction time	4 hrs
4	Reaction temperature	65 ^{0}C

5. Preparation of Biodiesel and Property Analysis

Here we are describing the development of a single step transesterification process using a solid metal oxide catalyst for the efficient conversion of rubber seed oil to biodiesel. The performance of this fuel is evaluated by fuelling a four stroke diesel engine and testing emission characteristics. The physico-chemical properties of this fuel are also analyzed and compared with the petro diesel.

Table 6. Engine test data for diesel

Load (kg)	Time taken for 10cc fuel consumption (s)	Manometer reading (mm)
No load	59.65	13.5
5	56.97	13.0
8	49.59	13.5
12.5	43.29	12.5
17	39.41	13.5
21	30.36	13.4

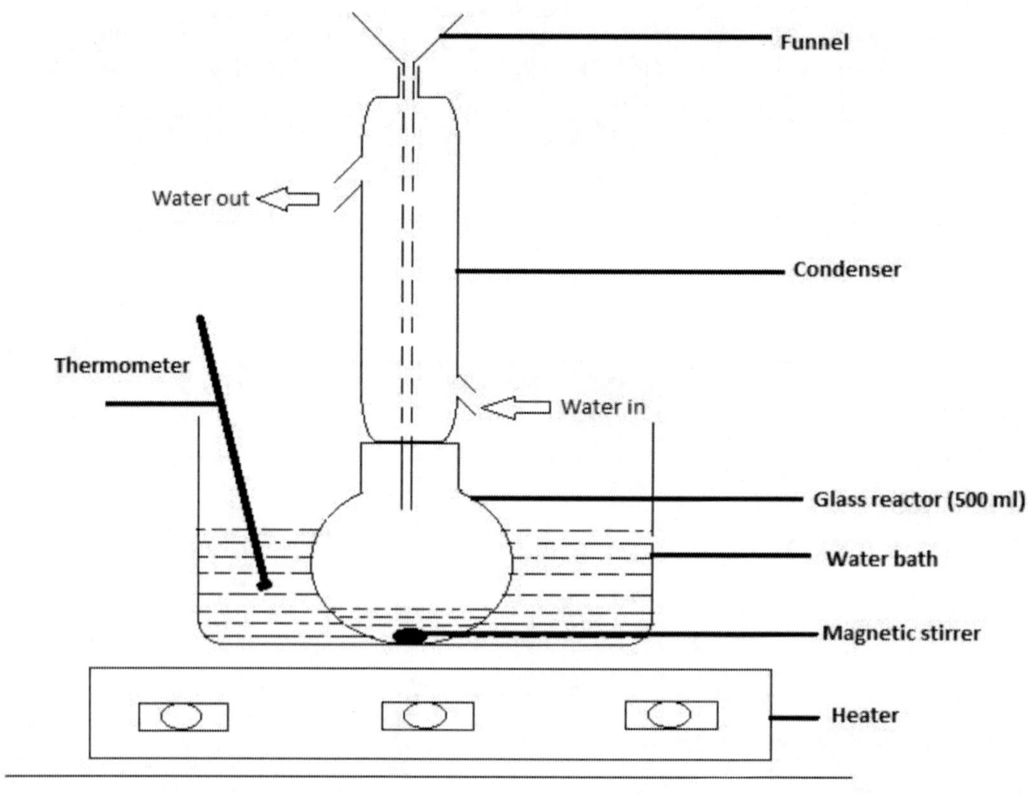

Figure 4. Experimental set up for biodiesel preparation.

5.1. Feedstock for Biodiesel Preparation

Crude rubber seed oil had been selected as a potential feedstock for biodiesel preparation. Oil used is non-edible and is commercially obtained after extraction from rubber seeds. Methanol with a purity of 99.5% and calcium oxide, CaO are also commercially obtained. Molecular weight of crude rubber seed oil was measured by using gas chromatography analysis from which the approximate molecular weight of oil can be calculated. The results of the gas chromatographic analysis of the fatty acid composition of the crude rubber seed oil are shown in Table 2. From this table it is clear that the rubber seed oil used in this present study mainly consist of linoleic, oleic and stearic acids with a minor component of palmitic acid. The ability of biodiesel to meet ASTM D 6751 standard criteria is dependent on the fatty acid composition [2]. Petroleum diesel is largely made of hydrocarbon with carbon chain length of 8 to 10 carbon atoms compared to rubber seed oils that contained fatty acids comprising mainly of 18 carbon atoms. Cetane number which is a prime indicator of fuel quality for diesel engines is increased with the increased carbon number [3]. Branched chains and double bonds improve low temperature, flow properties [4-5]. So from this analysis it is clear that the biodiesel produced from this rubber seed oil will have a higher cetane number as it contain fatty acids with 18 carbon atoms [6-7].

Figure 5. Biodiesel and Glycerin.

5.2. Catalyst Preparation

Calcium oxide was dipped in an ammonium carbonate solution (12% by weight) then the mixture was stirred for 30 minutes at room temperature. After filtration and drying at 110^0C till constant weight, the dried solid was milled and sieved then calcined at 900^0C for 1.5 hrs. After cooling in a dessicator to room temperature the base CaO is ready to use. High temperature calcinations of CaO improved the performance of the catalyst.

5.3. Experimental Set-up

As shown in the figure, the bench scale set-up consisted mainly of a round glass reactor placed in an adjusted temperature bath. This flask was provided with a reflux condenser, magnetic stirrer and thermometer for temperature follow up and funnel for methanol addition.

The measured amount of methanol + calcium oxide (CaO) stirred by a magnetic stirrer for 10 minutes, after which it was preheated to about 60 – 65^0C below the boiling point of methanol. Thereafter, 150 ml of RSO was taken and heated to 60 – 65^0C on a temperature water bath before it was poured into a blender. Methanol + CaO were then poured gently into the RSO in the blender. The entire content was allowed to blend for the required reaction time. After the whole reaction, the reaction product was decanted from the blender & centrifuged at specified rpm for a particular time, where the CaO catalyst was separated from the reaction product. Then the supernatant product mixture was exposed to open air for 30 minutes to evaporate excess methanol. Thereafter, it was poured into a separating funnel and allowed to settle overnight so as to separate the glycerin from the biodiesel. The lower glycerin layer was drawn off and the upper biodiesel layer was then removed, dried and weighed.

Percentage yield is given by:

$$\%\ Yield = \frac{[Weight\ of\ fuel\ produced\,/\,Weight\ of\ oil]}{100} \tag{6}$$

5.4. Property Analysis

Physical and chemical properties of crude rubber seed oil had been analyzed. Physical properties include density, viscosity, flash & fire point, cloud & pour point and calorific value. Chemical properties include acid value, iodine value and saponification value. After the preparation of biodiesel, physical and chemical properties of the product had been analyzed. Physical properties include density, viscosity, flash & fire point, cloud & pour point, calorific value, cetane number and total glycerol. Chemical properties include the acid value of the sample prepared. The quality of biodiesel is important for end-users, engine manufacturers and petroleum companies to ensure that it can be used directly or blended with petroleum diesel. So a detailed analysis about the qualities of the biodiesel as well as the feedstock is important. Table 2 presents various physico-chemical properties measured for the crude rubber seed oil as well as that of the biodiesel produced using the present method.

From the data presented in Table 5, it can be seen that the measured acid value of the rubber seed oil is not so high in comparison with that of the ASTM standard 0.8 mgKOH/g for non-edible oils. This is an interesting result since the acid value measures the presence of corrosive free fatty acids and oxidation products which is an important variable in considering the quality of oil because the lower the free fatty acid content, the better the quality of oil. The acceptable limit for edible oils is $\leq$10. Non edible vegetable oils generally contain high free fatty acids and have significant effects on the transesterification with methanol using alkaline catalyst. It also interferes with the separation of fatty acid ester and glycerol. This means that these types of oils would be better converted to biodiesel using the two-stage process of esterification and trans-esterification. But the measured acid value of the rubber seed oil shows that it is possible to convert rubber seed oil to bio diesel with a single stage process. Similarly the trans-esterification process brings down the flash point and kinematic viscosity to an acceptable level. This shows that its volatile characteristics had improved and it is also safe to handle.

As we know the fuel's viscosity is an important factor since it has the effect on pumps and influence on injector system's spray pattern. In comparison, of the viscosity of the biodiesel produced from the rubber seed oil it is clear that the kinematic viscosity of the biodiesel produced is in comparison with that of petro diesel and it is having a moderate value. This is quite interesting because the high viscosity will cause poor atomization and high spray jet penetration. The jet tends to be a solid stream. The fuel spray may impinge upon the cylinder walls, washing away the lube oil film and causing dilution of the crankcase oil. Low viscosity fuel does not penetrate far enough in the combustion chamber. Low viscosity leads to excessive leakage past the injection pump plunger. In either case the results are poor combustion, loss of power, inaccurate metering and increased wear in both the fuel system and engine.

Similarly the cloud point is relatively low compared to that of petro diesel. The cloud point relates to the tendency of filter plugging at cold temperatures. The cloud point is a guide as to what temperature the fuel may plug filter systems and restrict flow. The higher cloud point indicates a higher precipitation temperature of the fuel. Which indicates the fuel is less suitable for low temperature operation. The flash points of the obtained bio diesel extract are found to be higher than that of petro diesel. The flash point temperature is the minimum temperature at which the fuel will ignite (flash) on application of an ignition source under specified conditions. Flash point varies inversely with the fuel's volatility. Flash point minimum temperatures are required for proper safety and handling of fuels. Note that the biodiesel component must meet flash point criteria, prior to blending. In general fuels with flash point above 66^0C are considered as safe fuels which can be blended with petro diesel. The obtained results show that the bio diesel extract from rubber seed oil can be safely blended with petro diesel.

Another important parameter which quantifies the quality of a fuel oil is its cetane number. Cetane number is a measure of ignition quality of the fuel. A high cetane number indicates shorter ignition delay and improved combustion. For most engines, high cetane number fuels generally cause a shorter ignition delay and therefore pressure will rise only at lower rates. This tends to lessen combustion noise, improve control of combustion, and results in increased engine efficiency and power input. The present bio diesel extract has got a cetane number of 50 which shows it will improve the cetane number of petro diesel on blending. [5-10]

6. Process Optimization

The important parameters that influence the transesterification reaction mechanism are:

a) Alcohol to oil molar ratio
b) Catalyst to oil weight ratio
c) Reaction time
d) Reaction temperature

Optimization of each parameter was done by fixing all the other variables. By trial and error method, most favourable conditions were selected for each of the parameters. Temperature is one of the factors that influence the reaction and yield of the biodiesel product. A higher reaction temperature can decrease the viscosities of oils and results in an increased reaction rate and a shortened reaction time. However, high reaction temperature can decrease biodiesel yield because a higher reaction temperature accelerates the saponification reaction of triglycerides. Reaction temperature must be lower than the boiling point of alcohol in order to ensure that the alcohol will not leak out through vaporization. The results show that the reaction temperature influences the formation of methyl ester (ME). The formation of ME increases with the increase of reaction temperature but drops drastically at 70°C. The initial increase in the yield may be due to the vaporization of the fatty acid contents in the mixture which may encourage the forward reaction to form ME. At high temperatures the reaction may slow down due to the alcohol evaporation. [11-14]

The effect of reaction time on the transesterification reaction of RSO using CaO catalyst is shown in Figure 6. As reaction time increases from 1 to 5 hours, the percentage yield found to be increasing initially and decreasing at higher reaction times. A reaction time of 4 hours gives a maximum yield of 96.45%.

The effect of four process parameters on biodiesel yield in the transesterification reactions of rubber seed oil using the solid metal oxide catalyst (CaO) has been analysed. Based on this study a set of optimized conditions were identified which has been summarized in Table 5.

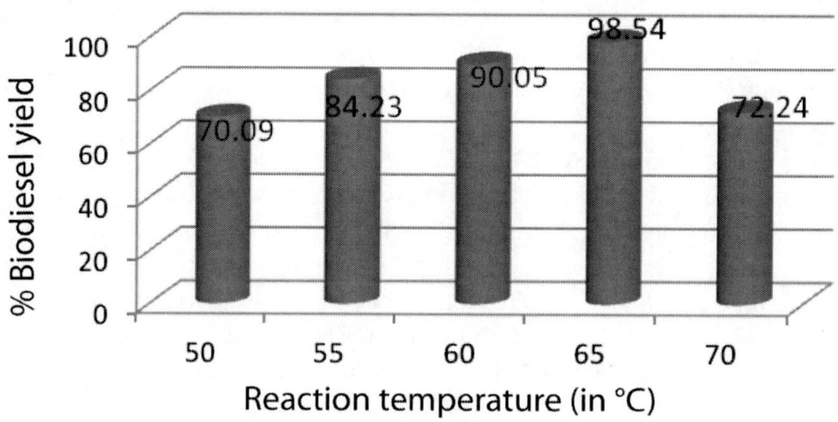

Figure 6. Influence of reaction temperature on biodiesel yield.

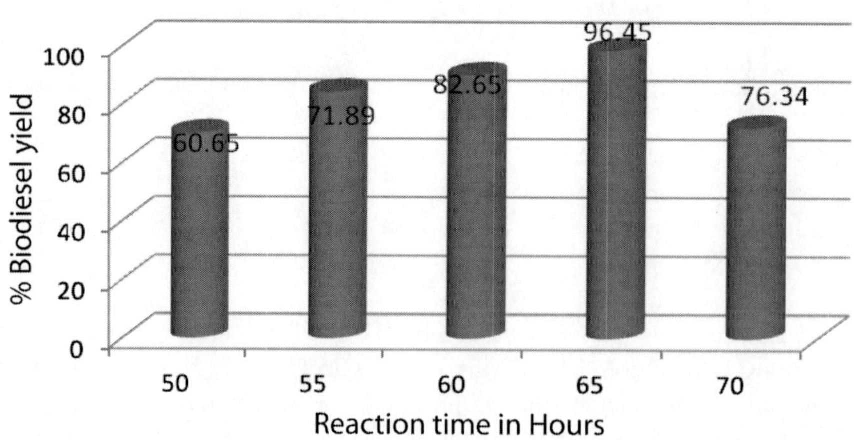

Figure 7. Influence of reaction time on biodiesel yield.

7. Engine Test and Emission Analysis

Vehicles that run on alternative fuels such as biodiesel are the most efficient in terms of mileage and exhaust emissions. Biodiesel runs in any conventional diesel engine without any major modifications. A single stage cylinder, 4-stroke water cooled diesel engine developing 8 HP at 850 rpm was used for the engine test. The fuel tank is connected to a graduated burette to measure quantity of fuel consumed in unit time. The trials were run on conventional diesel, pure biodiesel and blends of biodiesel blended with conventional diesel in different proportions. The thermal energy generated is known as the calorific value of the fuel. A part of this energy alone gets converted into mechanical power of the piston movement, known as the indicated power (I.P). A little amount of indicated power is used to overcome the frictional losses of the engine known as frictional power (F.P) and the remaining is delivered as the useful power output of the engine known as Brake power (B.P). Engine performance is analysed by many parameters especially specific fuel consumption (SFC), Brake thermal efficiency (BTE) and Indicated thermal efficiency (ITE). [15-20]

Table 7. Engine test data for B20

Load (kg)	Time taken for 10cc fuel consumption (s)	Manometer reading (mm)
No load	60.05	13.4
4.5	59.31	13
7.5	51.58	14
12	39.55	13.2
16	32.17	13.1
22	31.14	13.2

Figure 8. Four Stroke Diesel Engine.

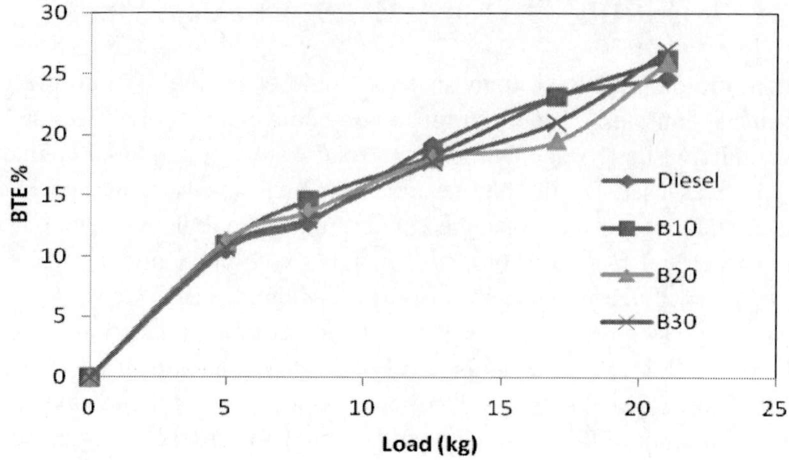

Figure 9. BTE Vs Load.

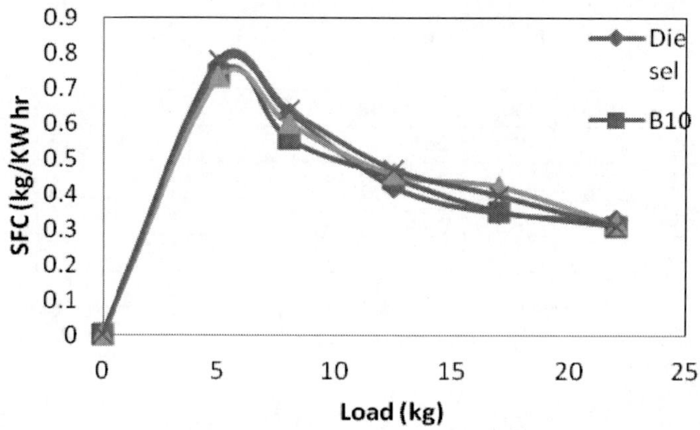

Figure 10. SFC Vs Load.

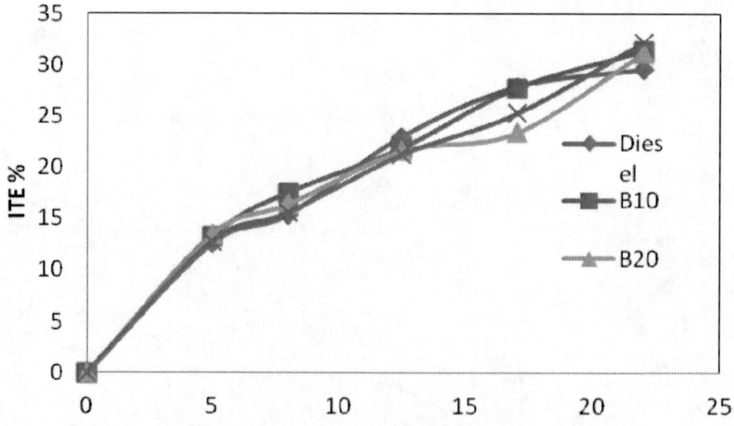

Figure 11. ITE Vs Load.

A single stage cylinder, 4-stroke water cooled diesel engine developing 8 HP at 850 rpm was used for the engine test. Biodiesel can be blended in many proportions with conventional Diesel. B100 is pure biodiesel. B10 blend is 10 volume percentage pure biodiesel blended with 90 volume percentage of conventional Diesel. Similar blends of B20 and B30 were prepared and engine tests were conducted. From the specifications of the engine, the effective maximum load that can be applied to the brake drum is calculated. When there is no load applied to the engine, the engine is started and run at the rated speed. Cooling water is allowed to the brake drum and engine is loaded. The time taken for 10 cc of fuel consumption for various loads is noted. At the end of the experiment, the engine is stopped by operating the fuel cut off lever. Emission analysis was done on a four- stroke, single cylinder air-cooled Diesel engine fitted with Gas analyser. The engine was started and the emissions of Carbon monoxide, Carbon dioxide, Oxides of nitrogen and hydrocarbons were monitored for various loads. [13-15]

For diesel, B10, B20 & B30 blends of biodiesel, engine test is conducted and the data obtained are given in the following tables.

CONCLUSION

It can be concluded that the biodiesel derived from the rubber seed oil is most suitable to use in diesel engines given that its kinematic viscosity, flash point, cloud point, and calorific value conform to the recommended international standards. This means that CaO can be effectively used as a solid metal oxide catalyst in the transesterification reaction of the rubber seed oil for biodiesel production. By the usage of CaO catalyst savings could be made from the expensive refining processes admissible in the case of conventional acid/alkali based tranesterification reactions. More over rubber seed oil can be converted into biodiesel directly by a single stage transesterification process while most of the non edible vegetable oils required two stage transesterification reactions for the bio diesel conversion. This is also an added advantage in this process. 98.54% of methyl ester was formed from transesterification of RSO with methanol. The main factors affecting the methyl ester yield during transesterification reaction were the catalyst content, molar ratio of oil to alcohol and reaction time. High methyl ester yield and fast reaction rate could be obtained even if reaction temperature was relatively low, which is quite favourable to the industrial production of biodiesel from the rubber seed oil. Results of optimization test showed complete formation of methyl ester under the following circumstances; reaction duration: 4 hours, temperature: 65 ^{0}C, catalyst to oil weight ratio of 0.6: 1 and with an alcohol to oil molar ratio of 4.5:1. Engine test has been conducted and performance analysis is also being done in a 4-stroke diesel engine. Results show that the performance of the engine can be improved by blending the pure diesel with different blends of biodiesel produced from crude rubber seed oil.

REFERENCES

[1] Agarwal A. K., Biofuels (alcohols and biodiesel) applications as fuels for internal combustion engines, *Prog. Energy Combust. Sci.* 33 (2007) 233–271.

[2] Highina K. et al., Biodiesel production from Jatropha caucus oil in a batch reactor using ZnO as catalyst, *Journal of Petroleum Technology and Alternative Fuels* Vol. 2(9), pp. 146-149, September 2011.

[3] Ezeanayanso S. et al., Evaluation of homogeneous catalysis and supercritical methodology for biodiesel production from Jatropha curcas, Azadirachta indica and Hevea brasiliensis oil seeds, *Journal of Petroleum Technology and Alternative Fuels* Vol. 2 (10), pp. 190-199, October 2011.

[4] Demirbas Ayhan, Progress and recent trends in biodiesel fuels, *Energy conversion and management* 50 (2009) 14-34.

[5] Benavides Pahola T. & Diwekar Urmila, Optimal control of biodiesel production in a batch reactor: stochastic control, *Fuel* April 2012 Vol 94 pp 218-226.

[6] Refaat A., Biodiesel production using solid metal oxide catalysts, *Int. J. Environ. Sci. Tech.,* 8 (1), 203-221, Winter 2011.

[7] Gopinath et al., Effect of unsaturated fatty acid esters of biodiesel fuels on combustion, performance and emission characteristics of diesel engine, *International Journal of Energy & Environment* Vol. 1 Issue 3, 2010 pp. 411-430.

[8] Gimbun Jolius et al., Biodiesel production from rubber seed oil using a limestone based catalyst, Advances in Materials Physics and Chemistry Supplement: *2012 world Congress on Engineering and Technology.*

[9] Gui M. M., Lee K. T., Bhatia S., Feasibility of edible oil vs. non-edible oil vs. waste edible oil as biodiesel feedstock, *Energy* 33 (2008) 1646-1653.

[10] Mathur Y. B, Economics, formulation techniques and properties of biodiesel, *Universal Journal of Environmental Research & Technology,* Vol, 1 Issue 2: 124-134.

[11] Ikwuagwu O. E., Ononogbu I. C., Njoku O. U., Production of biodiesel using rubber [Hevea brasiliensis (Kunth. Muell.)] seed oil, *Industrial Crops and Products* 12 (2000) 57-62.

[12] Aworanti O. A., Statistical optimization of process variables for biodiesel production from waste cooking oil using heterogeneous base catalyst, *British Biotechnology Journal* 3(2): 116-132, 2013.

[13] Chopade Shruthi G. et al., Solid heterogeneous catalysts for production of biodiesel from transesterification of triglycerides with methanol, *Acta Chim. Pharm. Indica:* 2(1), 2012, 8-14.

[14] Hawash S. et al., Optimization of biodiesel production from jatropha oil by heterogeneous base catalyzed transesterification, *International Journal of Engineering Science & Technology* Vol. 3 No. 6 June 2011.

[15] Chincholkar S. P., Srivastava Saurabh, Rehman A., Dixit Savita and Lanjewar Atul, Biodiesel as an alternative fuel for pollution control in diesel engine, *Asian J. Exp. Sci.,* Vol. 19, No. 2, 2005, 13-22.

[16] Buyukkaya Ekrem, Effect of biodiesel on a DI diesel engine performance, emission and combustion characteristics, *Fuel* 89 (2010) 3099-3105.

[17] Atadashi I. M., Aroua M. K., Aziz A. Abdul, High quality biodiesel and it diesel engine application, *Renewable and Sustainable Energy* 14 (2010) 1999-2008.

[18] Demirbas Ayhan, Importance of biodiesel as transportation fuel, *Energy Policy* 2007, Vol 35 Issue 9 4661-4670.

[19] Karuppasamy K. et al., The effect of biodiesel blends on single cylinder DI Diesel engine and optimization using response surface methodology, *European Journal of Scientific Research* Vol.84 No.3 (2012), pp.365-376.

[20] Ramadhas A. S., Jayaraj S., Muraleedharan C., Performance and emission evaluation of a diesel engine fueled with methyl esters of rubber seed oil, *Renewable Energy* 30 (2005) 1789–1800.

In: Renewable Energy and Sustainable Development
Editors: R. T. Durai Prabhakaran, S. A. Kale et al.
ISBN: 978-1-63463-429-8

Chapter 7

EVALUATION OF BIOGAS GENERATION AND GHG EMISSIONS FROM POULTRY WASTE

Paria Sefeedpari, Shahin Rafiee*, Asadollah Akram and Seyyed Hassan Pishgar Komleh
Department of Agricultural Machinery Engineering,
University of Tehran, Karaj, Iran

ABSTRACT

This chapter estimates biogas production potential from poultry manure in 40 poultry farms of Karaj city, Iran. Biogas technology is quite immature however sustainability assessment of biogas production is a relatively documented topic in Iran. Resulting from this chapter, there is the potential of 7 m^3 day^{-1} (1000 $birds^{-1}$) biogas production which is equal to 4.7 m^3 day^{-1}(1000 birds) methane (CH_4) gas and the electricity potential of about 47 kWh. Based on these results, educating farmers through extension programs, making new policies for extending the idea of extracting biogas from poultry manure to provide the energy requirements and facilitating the establishment of biogas plants in Iran are suggested.

NOMENCLATURE

AD	Anaerobic Digestion
BDV	Biogas digester volume
CH_4	Methane gas
CO_2	Carbon dioxide
GHG	Greenhouse gas
GWP	Global warming potential
PEM	Potential Energy of Methane

* Corresponding author: Email: shahinrafiee@ut.ac.ir, Tel: +98 26 32801011.

PVB	Potential volume of biogas
PVM	Potential volume of methane
PWM	Potential weight of methane
TS	Total solid

1. Introduction

One important characteristic of a sustainable waste management program includes the recovery of materials and energy from generated wastes. There are various alternative sources of energy, which may compensate for the future scarcity of conventional energy resources. Biogas is one of those renewable high-quality energy resources which should be taken under consideration, particularly in developing countries where energy is costly and a lot more activities are developing [1]. In the not-too-distant future, many countries will be facing another problem due to a shortage of fossil fuels. One of the alternative sources of energy, which may in the future compensate for the shortage of conventional fuels, is biogas which is naturally produced from biogradable wastes [2].

Biogas is the by-product of anaerobic decomposition of organic matter by micro-organisms, mainly bacteria [3]. The process is referred to as anaerobic digestion (AD). Biogas contains about 60% methane (CH_4) and 40% carbon-dioxide (CO_2). Thus, the anaerobic digester has been widely used in the production of biogas on a worldwide scale. Containing this amount of greenhouse gas percentage, there is a great potential to reduce the GHG emissions from livestock farms by using anaerobic digesting approach of animal manure. Moreover, the resulting methane can be contributed to the electrical energy generation, lighting, heating and cooking gas as the world would face the energy shortage crisis.

Heat and power energy are the top two of those energy sources in agricultural production technologies required for heating of homes, drying of harvest, heating of greenhouses, powering of electric equipment, etc. Undeniably, generating heat from burning fossil fuels and biomass causes the greenhouse gas (GHG) emissions. On the other hand, unused biomass residues can produce gases that contain up to 50 % methane (CH_4) during its natural degradation or 5-10 % CH_4 during its open burning [4]. Consequently, the most environment friendly technology is biogas (mixture of CH_4 and CO_2) production from biomass, as it is usable for a relatively “clean” heat and power energy production.

The above ‘greenhouse’ gases have differing abilities to absorb the heat radiated from the Earth and are assigned factors (their ‘global warming potential’) based on their absorbing ability relative to that of carbon dioxide.

Electricity is inevitable in poultry industry either for the production of eggs or the growth of the broilers as well. Lamps to provide proper lighting in the poultry sheds, fans to maintain the required temperature, brooder to brood up the chicks and the water pumps to supply water are the main electricity consumers in this industry [5].

The potential for producing biogas from animal manure has been addressed by several researchers [6-10]. In literature, Koelsch et al. reported that the total solids content range is about 8–13% and 80% of the solids are volatile solids. One-half of the volatile solids (the biodegradable ones) are converted to methane and carbon dioxide [11]. Fantozzi and Buratti designed a single-stage, batch, mixed, laboratory anaerobic digester to evaluate biogas production from different substrates. One was poultry wastes. It was concluded that pig and chicken manure were the highest methane production resources [12]. Gelegenis et al. concluded that the biogas can not only be utilized for the generation of heat and/or electricity, but co-digestion results in liquid and solid effluents are also valuable, as they retain all their nutrient constituents (nitrogen, phosphorus, trace elements, etc.) and so they can be used as fertilizers and soil improvers [13]. A greenhouse gas emission for a North American beef feedlot operation, which includes biogas production by AD with subsequent electricity generation, was calculated by Ghafoori et al. [14]. In their study, AD, biogas production and electricity production are the major sources of differences in emissions. This author also calculated the cost of power and gas and the carbon emission reduction from AD plants receiving manure from multiple sources in another study [15]. Since no investigation has been carried out in the potential area of biogas production using poultry waste in Karaj city of Iran, the need to invest energy in studying this fertile area is justified. In the study area, there are two types of poultry farms, layer farms and broiler farms. In this chapter, all the poultry farms with the potential of biogas production were selected from layer farms. One of the reasons that broiler farms do not produce biogas is the nature of rearing the birds which produces the droppings in batch not continuous. Therefore, this chapter was conducted among different layer poultry farms. Lack of governmental attentions and research in expanding and establishing biogas production sites has caused little improvement in this industry. Hence, the principal focus of this chapter is to calculate the biogas generation potential from poultry wastes, the corresponding amount of diminished GHG emissions, the global warming potential (GWP) of the estimated GHG and finally the energy equivalent potential of the obtained biogas.

2. Material and Methods

2.1. Data Collection

This chapter was carried out as a case study in Karaj city of Iran due to high number of poultry layer farms. Data were collected from 40 poultry farms of egg production, and the farmers were interviewed in order to gather the required information on manure generation and electricity use of the target farms with structured questionnaires. It is believed that these poultry farms have a huge potential to produce biogas which can later be used to generate electricity. All the quantitative data were analyzed using Microsoft Excel program.

2.2. Methane Calculation

The modeling comprised three sets of calculations: manure generation, biogas generation and baseline methane emissions prior to biogas. Manure generation was calculated for poultry using the following equations [12]:

Total raw manure (kg/day)=No. of animals (head) × *Raw manure per animal per day(kg)* (1)

Manure total solids (kg/day) = Total raw manure (kg/day) × Percent total solids (%) (2)

Collectible manure total solids (kg/day) = Manure total solids (kg/day) × % collectible (3)

The input data including some assumptions for calculations were derived from previous studies and are given in Table 1.

Table 1. Input data for biogas generation [16]

Average raw manure (kg/day/head)	Total solids (% of fresh manure)	Collectible manure (%)	Typical volume of biogas (L/kg TS)	Methane in biogas (%)	Density of CH_4 (kg/L)	Energy content of Methane (MJ/m^3)
12	25	50	300-450 (375)	57-70 (65)	0.00067	36.7

It is worth mentioning that in the studied region, manure collecting operation is done by human labor. As layers are housed in the cages, 100% of the manure quantity is collectible. While chicken manure is being used in farming applications, as well; it was assumed that at least 50% of the collectible manure can be utilized in biogas production. Moreover, in order to facilitate the calculations, all equations were evaluated per 1000 birds. Following equations were applied in this chapter [17]:

Potential Volume of Biogas (PVB) (Litres/day) = Collectible manure total solids (kg/day) × typical volume of biogas produced per kg total solids (L/kg TS) (4)

Potential Volume of Methane (PVM) $(LCH_4\ /day)$ *= PVB × Methane in biogas* (5)

Potential Weight of Methane (PWM) $(Kg\ CH_4\ /day)$ *= PVM × Density of Methane (Kg/L)* (6)

Weight of CH_4 *(kg/year) = PWM × 365 days* (7)

Weight of CH_4 *(tons/ year) = Kg* CH_4 */year /1000* (8)

In addition to the analysis of potential biogas production as an alternate resource of electricity supply, the CO_2 equivalent quantity of produced methane was also reported.

Weight of CO_2 *equivalent (tons/year) = Weight of* CH_4 *(tons/ year) × GWP* CH_4 (9)

Where GWP CH_4 is the Global Warming Potential for CH_4 (t CO_2/CH_4) and equals to 21 [18]. Assuming that retention time of 20 days is needed, the volume of biogas plant was estimated as follows:

Biogas Digester Volume (BDV) (m^3/day) *= 20 × Total Collectible Manure volume (kg/d) /1000* (10)

Potential Energy of Methane (PEM) (MJ/day)=PVM (L/day) $\times$ *36.7* (11)

The result of the last equation (Eq. 11) can be converted to kWh/day (1000 birds) to calculate the potential of electricity generation.

3. RESULTS

Input data to estimate the potential of biogas generation in some poultry farms of Iran were daily layers' manure amount and the number of birds in each farm. Biogas production potential was estimated due to some assumptions and previous works and the results are given in Table 2.

Table 2. Calculated results of biogas production potential in poultry layer farms of Iran per 1000 birds

Calculated Results	PVB* (L)	PVM* (LCH_4)	PWM* ($KgCH_4$)	Weight of CH_4 (tons/year)	Weight of CO_2 equivalent (tons/year)	BDV* (m^3)	PEM* (MJ)	PEM* (kWh)
Average	7,179.5	4,666.7	3.1	1.1	24.0	20.1	171.3	47.6
Maximum	20,270.3	13,175.7	8.8	3.2	67.7	150.0	483.5	134.3
Minimum	3,472.2	2,256.9	1.5	0.5	11.6	4.0	82.8	23.0

* per day.

The results of the chapter indicated that capture and management of methane from collected poultry wastes is significantly an effective tool for providing energy in both current and future production systems which can ensure generations against the term "Sustainability".

The results showed that out of 38 kg of raw material (chicken wastes) per day per 1000 birds (measured in the target poultry farms), the possible extracted methane gas was estimated to be 4.6 m^3 which is approximately equivalent to1 tonne of CH_4 per year per 1000 birds. This amount of biogas, containing 65% methane, is expected to generate about 24 tonnes $year^{-1}$ (1000 $birds^{-1}$) of carbon dioxide. Regarding this much of GHG, the need for having control on it becomes apparent. Furthermore, biogas energy potential was calculated to be 171.3 MJ day^{-1} (1000 $birds^{-1}$) which has the potential of generating electricity as about 47.6 kWh. The electricity would be sufficient either for poultry demand (an average of 10 kWh day^{-1} (1000 $birds)^{-1}$ in the sample farms) or running biogas systems and any excess electricity can be sold back into the local grid at wholesale prices.

CONCLUSION

The chapter looked at the present potential of biogas production, GHG emissions and responding to electricity demand from poultry wastes in some poultry for egg production farms of Iran. The chapter results have revealed that there is a potential to produce electricity from poultry waste and fortunately there is a high interest among farmers to take advantages of the produced electricity power as a resource of renewable energy. This interest has come due to the fact that all the poultry farms experience load shedding throughout the day mostly in the evening which hampers the production of the farm. According to the findings electricity can be produced from poultry waste for the total daily consumption of most of the poultry farms and in addition electricity can also be produced for the peak hour only to save farms from being cut off. Moreover, the excess can be sold back into the local grid at wholesale prices.

At present, high average yields of biogas 7 m^3 day^{-1} (1000 $birds^{-1}$) and methane 4 m^3 day^{-1} (1000 $birds^{-1}$) could be obtained from poultry wastes through anaerobic fermentation process.

The drawn results will be of importance regarding the elimination of GHG emission to the atmosphere (environmental impacts). As it was calculated and under the assumptions used in this chapter, high amount of CO_2 and CH_4 was produced which would lead to great global warming concerns for our planet, the Earth.

On the basis of findings and analysis of the chapter the authors recommend the followings:

1) To make the technology available to equip farms with modern wastes collection systems in case of caged houses for layers, regarding the point that now wastes are collected by human labours in this area.
2) Awareness development program is required to build up the knowledge about the benefit of using animal wastes, especially poultry as a rich source of biogas production, among the poultry farmers and policy makers.
3) Awareness development program is required to make the poultry farmers aware about the use of energy efficient appliances which can cause reduction in electricity consumption of the farm.
4) To facilitate research conditions for executing proposals and also installing biogas plants in sample farms concerning its efficiency and economical justifications.
5) Further works can be done to find out the potential of producing biogas by the best suitable biogas technology for other applications rather than electricity demand, in a sample farm.
6) Regarding the previous recommendation, it is suggested to carry out economic and environmental analyses after each potential metering of biogas generation project in order to show whether it has economic and environmental justifications or not.

REFERENCES

[1] Morini, M., Pinelli M. and M. Venturini. (2009). Analysis of biogas compression system dynamics. *Appl. Energy* 86: 24-66.

[2] Hvid-Hanson, K., Angelidaki, I. and B.K. Ahring. (1998). Anaerobic Design for Swine Manure in Habitation by Ammonia. *Water Resour.* 32(1): 5-12.

[3] Demirbas, M.F. and Balat, M. (2009). Progress and Recent Trends in Biogas Processing. *Int. J. Green. Energy* 6(2): 117-42.

[4] Dubrovskis, V., Plūme, I., Koteļeņecs, V. and I. Straume. (2009). Investigation of biogas production from relatively dry biomass. *Engineering for rural development* 2: 239-42.

[5] Zaman, S.A.U. (2007). The Potential of Electricity Generation from Poultry Waste in Bangladesh: A Case Study of Gazipur District. M.Sc. Thesis, International Institute of Management University of Flensburg, Germany.

[6] Sommer, S.G., Møller, H.B. and S.O. Petersen. (2002). Reduction in methane and nitrous oxide emission from animal slurry through AD. *Proceedings of the 3rd International Symposium, Maastricht, the Netherlands*, 21–23 January 2002.

[7] Hashimoto, A.G., Chert, Y.R., Varel V.H. and R.L. Prior. (1979). Anaerobic fermentation of animal manure. Paper No. 79.4066, Proceedings of Summer Meeting of ASAE and CSAE, University of Manitoba, Winnipeg, Canada, 24-27 June 1979.

[8] Hashimoto, A.G., Chen, Y.R. and V.H. Varel. (1980). Theoretical aspects of methane production: state-of-the-art. In: livestock waste: A renewable resource. *The Proceedings of the 4th International Symposium of Livestock Wastes.* ASAE, St. Joseph, Michigan 49085; 1980.

[9] Batzias, F.A., Sidiras, D.K. and E.K. Spyrou. (2005). Evaluating livestock manures for biogas production: a GIS-based method. *Renew. Energ.* 30: 1161-76.

[10] Börjesson, P. (1996). Energy analysis of biomass production and transportation. *Biomass Bioenerg.* 11: 305-18.

[11] Koelsch, R.K., Fabian, E.E., Guest, R.W. and J.K. Campbell. (2001). Anaerobic digesters for dairy farms. *Agricultural and biological engineering extension bulletin* 458. Ithaca, NY: Cornell University, 2001.

[12] Fantozzi, F. and C. Buratti. (2009). Biogas production from different substrates in an experimental Continuously Stirred Tank Reactor anaerobic digester. *Bioresource.* Technol. 100, 5783-89.

[13] Gelegenis, J., Georgakakis, D., Angelidaki, I., Christopoulou N. and M. Goumenaki. (2007). Optimization of biogas production from olive-oil mill wastewater, by codigesting with diluted poultry-manure. *Applied Energ.* 84: 646-63.

[14] Ghafoori, E., Flynn, P.C. and M.D. Checkel. (2006). Global Warming Impact of Electricity Generation from Beef Cattle Manure: A Life Cycle Assessment Study. *Int. J. Green. Energy* 3(3): 257-70.

[15] Ghafoori, E., Flynn, P.C. and M.D. Checkel. (2007). Carbon Credits Required to Make Manure Biogas Plants Economic. *Int. J. Green. Energy* 4(3): 339-49.

[16] Anonymous. (1985). Energy: Biogas Plant Construction: Design and Operation (Ministry of Agriculture and Fisheries, *New Zealand MAF AgLink* FPP663 (Attached), 1985).

[17] Anonymous. (2008). *Methane from animal waste management systems* (Final report, Ministry of Agriculture & Forestry of New Zealand, 2008).

[18] Anonymous. (2006). *Guidelines for Greenhouse Gas Inventories* (IPCC (Intergovernmental Panel for Climate Change), Chapter 10, Emissions from Livestock and Manure Management, p.87, 2006).

In: Renewable Energy and Sustainable Development
Editors: R. T. Durai Prabhakaran, S. A. Kale et al.
ISBN: 978-1-63463-429-8

Chapter 8

WOODY BIOMASS FOR SUSTAINABLE BIOENERGY HARVESTING

Carmen Rocío Rodríguez Pleguezuelo[1], Víctor Hugo Durán Zuazo[2*] and José Ramón Francia Martínez[3]

[1]Université Catholique de Louvain, Louvain-la-Neuve, Belgium
[2]IFAPA Centro Las Torres-Tomejil, Spain
[3]IFAPA Centro Camino de Purchil, Spain

ABSTRACT

Biomass and bioenergy potential estimation is crucial for gaining a comprehensive understanding of bioenergy development, especially under the imminent energy-crisis scenario. Biomass production has the capacity for providing a major portion of overall energy, revitalizing local economies, promoting energy self-sufficiency, and reducing environmental problems. The main challenge in this research area will be to point out a new direction in sustainable energy farming by applying new genetic and innovative agricultural approaches for improving energy-farm functioning by boosting yields, lowering costs, and maximizing sustainability. Therefore, sustainable biomass harvesting for producing renewable energy provides significant opportunities for reducing greenhouse-gas emissions, a major contributor to climate change. Also, agricultural lands, especially those that are abandoned, offer an alternative for new energy farming. In this sense, woody biomass has proved to offer broad choices, particularly those produced under short-rotation systems, based mainly on re-growth after each harvest, allowing several crops without having to replant. Currently energy crops utilized as raw materials for bioenergy production, and their potential, have been analyzed both in Spain and elsewhere in Europe. In this chapter, a compilation of data from the literature confirms the considerable potential for implementing woody-energy crops under selective environmental conditions, and thus, the feasibility for sustainable bioenergy harvesting.

[*] Corresponding author: E-mail: victorh.duran@juntadeandalucia.es.

1. Introduction

The current system based on fossil fuels as the main suppliers for energy production is unsustainable. Therefore, it is of priority to search for alternative and sustainable energy sources for palliating global warming while supplying present and future energy demands. In this sense, according to the International Energy Agency, the global demand for energy is expected to increase by 36% between 2008 and 2035 [1], since energy is a basic part of the world's present and future development. Worldwide annual energy demand has considerably grown, from 10,000 to 13,000 Mtep in the last 15 years [2]. However, 40% of this consumed energy comes from non-renewable liquid fuels such as gasoline and diesel, transport being almost fully dependent on these types of fuels. Nevertheless, fossil fuels will still be the main energy source in the coming decades, used primarily by the emerging countries. However, according to IEA [1] the use of renewable energy will triple for 2035 and its share in the electricity supply will rise from 19% in 2008 to 32% in 2035.

On the other hand, the net reduction in the total amount of greenhouse gas (GHG) emissions is one of the most important objectives in energy and environmental policies being implemented in the European Union (EU). In this sense, the awareness of the global impact of GHG emissions on climate change calls for imperative strategies to mitigate the effects. Consequently, energy-conservation policies to reduce electricity consumption, including more efficient appliances, high-efficiency illumination, and more responsible energy practices will be fomented as well. In this context, the EU has established the following objectives to be reached up to 2020: 20% of the EU's total energy supply will come from renewable sources, and the EU's GHG emissions will be reduced 20% with respect to 1990. To achieve these objectives, energy diversification seems to be the surest method to guarantee this increasing demand as well as the promotion of alternative energy sources, which definitely minimize the environmental impact and avoid the dependence on fossil fuels.

Concerning Europe and particularly to Spain, it is necessary to point out their considerable dependence on fossil-fuel importation. Recently, Spanish policies have been focused on a new energy model with legal measures and budgetary allocation to foster the transformation process which is necessary in order to fulfil the European energy requirements by the National Plan of Action for Renewable Energy (Plan de Acción Nacional de Energías Renovables; PANER) of 2010-2020.

In this regard, it is predicted that bioenergy plantations will eventually become the main source of biomass for energy production worldwide. Particularly, forest wood is considered a key source of biomass for energy in several studies (up to about 115 EJ yr^{-1} in 2050) [3]. The increasing interest of biomass has been focused on heat and electricity production; however, the extensive research and the subsequent creation of new technologies, together with the rising prices of common fuels have currently intensified interest in liquid biofuel generation.

The production of bioenergy crops will most likely continue to place significant demands on land resources worldwide. To achieve global demand for biofuels, the IEA has reported that 65 million hectares of land will be necessary by 2030 and 105 million hectares by 2050 [4].

Short-rotation plantations (SRP) have several advantages among other cropping systems, since they utilize shorter return periods than do traditional wood plantations. Short-rotation coppices are wood plantations of high-growth-rate tree species under intensive agricultural practices (weed control, fertilizers application, and irrigation) for high biomass yields. Basically, trees are planted as single trunks or as coppice systems with rotation periods of less than 15 years, i.e., being cut every period of rotation and left to re-sprout.

Comparing biomass crops with conventional energy and some other renewable energy sources, it is necessary to point out that they have many advantages, mainly because of reduced dependence on short-term weather changes, promotion of regional and local economy, and favouring employment in rural areas.

In this context, the most commonly used woody species in SRPs to produce biomass with electrical and thermal energy purposes include poplar (*Populus* spp.), willow (*Salix* spp.), eucalyptus (*Eucalyptus* spp.), locust (*Robinia* sp.), acacia (*Acacia dealbata, Acacia melanoxylon*), and paulownia (*Paulownia fortunei x Paulownia tomentosa*). Concretely in Spain, renewable energy from different sources has enormous growth potential [5], biomass generating positive economic results, given its capacity to promote employment, develops rural areas, and respects to the environment. Therefore, the challenge for the forthcoming decades will be to search for new sustainable energy sources.

In particular, this chapter focuses on the potential of eucalyptus, poplar, and paulownia cultivation for woody biomass production under short-rotation systems in order to meet the growing demand of biofuels in a sustainable context.

2. Capability Expansion of Solid Biomass in Spain and EU Countries

Spain has still an excessive dependence on fossil fuels, with consumption in 2012 being 42% of the total primary energy of the country and much higher than the average European consumption (24%) (www.idae.es). Nevertheless, there has been a major increase in the share of renewable energy sources from 2004 to 2011 (from 0.8 to 5.9%, respectively). Specifically, Spain occupies the sixth position for this quota, after Sweden, Austria, Poland, and Germany (8.8, 7.6, 6.5, and 6.1% for 2011, respectively). In fact, in 2009, for the first time, the energy share from renewable sources was higher than energy generated from coal in Spain. In 2012, this tendency was the same, coal and renewable energy sources representing 11.6 and 12.2%, respectively, of the total of energy consumption (www.idae.es) (Table 1). With respect to renewable energy sources, this share of 12.2% was divided among biomass (30.6%), wind (26.8%), solar energy (15.2%), hydro energy (11.2%), biofuels (13.5%), biogas (1.6%), solid urban waste (1.0%), and geothermal energy (0.1%). In addition, the renewable sector created approximately 130,000 jobs in 2010, most of them in the wind, solar photovoltaic, and biomass sectors (Figure 1).

In this context, Gómez et al. [6] studied the Spanish energy framework, based on nine different scenarios according to different economic and strategic criteria. Their results emphasized that the 20% energy supply objective implies that approximately 45% of the electricity should be renewable by 2020 (assuming that biofuels contribute 10% to the final energy consumption in the transport sector). In line with this, and related to solid biomass, Spain is the sixth producer of primary energy from solid biomass (after Germany, France, Sweden, Finland, and Poland) (www.eurobserv-er.org) (Figure 2A). Nevertheless, if these amounts of energy production are calculated as primary energy production of solid biomass per tonne of oil equivalent (toe) per inhabitant in the EU, Spain occupies the 18th position with a rate of 0.104 in 2011, which is lower than the average rate for EU (0.157 in 2011) and far lower than Finland (1.391), Sweden (0.867) or Latvia (0.784), these latter being the main producers by toe per inhabitant [7]. The gross electricity production from solid biomass in UE-27 is represented in Figure 2B, showing that Spain is the 9th producer (2.937 TWh), quite far from the leading producers: Germany, Finland, and Sweden, with 11,539, 9,968, and 9.641 TWh, respectively. It is important to mention that France is not among the most important positions of gross electricity production from solid biomass since it has promoted nuclear-energy politics.

Table 1. Primary energy consumption in Spain during 2012

Source	Consumption (ktoe)	Respect to total (%)
Coal	14,986	11.6
Fuel	54,108	41.9
Natural gas	28,241	21.9
Nuclear	15,993	12.4
Renewable energies	15,778	12.2
	With respect to the total of renewable energy sources (%)	
Hydro	1,763	11.2
Wind	4,227	26.8
Biomass	4,831	30.6
Biogas	260	1.6
Solid Urban Waste	159	1.0
Biofuels	2,124	13.5
Geothermal	18	0.1
Solar	2,397	15.2

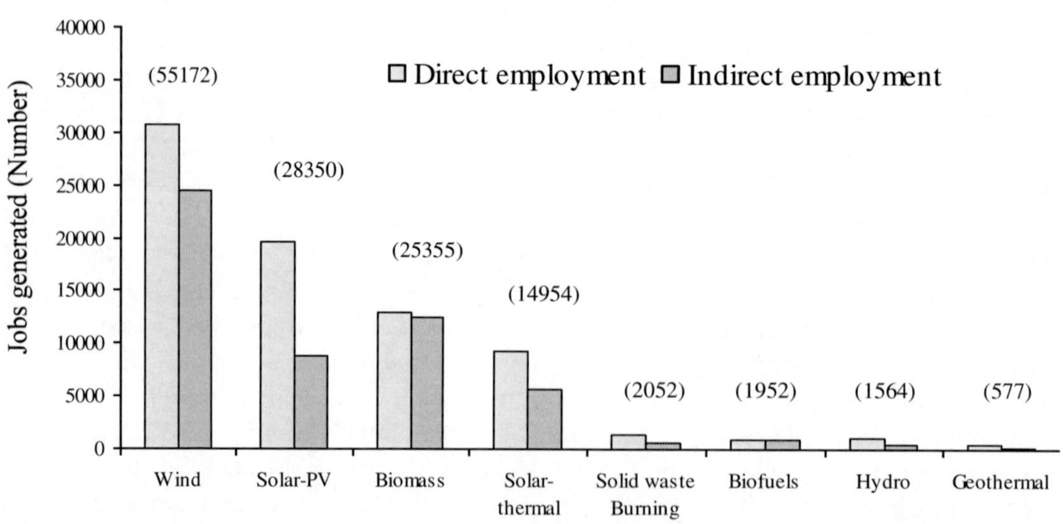

Figure 1. Direct and indirect jobs generated from renewable energy source in Spain during 2010. Values in brackets are total jobs generated.

In the framework of the so-called "Renewable Energy Plan (Plan de Energías Renovables) 2011-2020", several environmental analyses are presented, including the results from a study on biomass potential of non-industrial origin in Spain: energy crops, herbaceous or woody, forest, and agricultural wastes (Table 2). In the research for this plan, available areas for potential biomass energy crops were determined under certain criteria, seeking non-interference with the food market, promoting the sustainability of production markets, and encouraging the optimisation of the use of irrigation water. According to that report, potential biomass in Spain for energy purposes is approximately 11-fold the current energy used. This potential is shared among agricultural biomass (36.9%), biomass-crop establishment in agricultural areas (24.4%), forest biomass (21.4%) and biomass installation in forest areas (17.3%). Despite this considerable potential, woody energy crops under short-rotation management are not widespread in Spain. Particularly in Andalusia (S Spain) such crops are limited almost exclusively to experimental plots [8, 9].

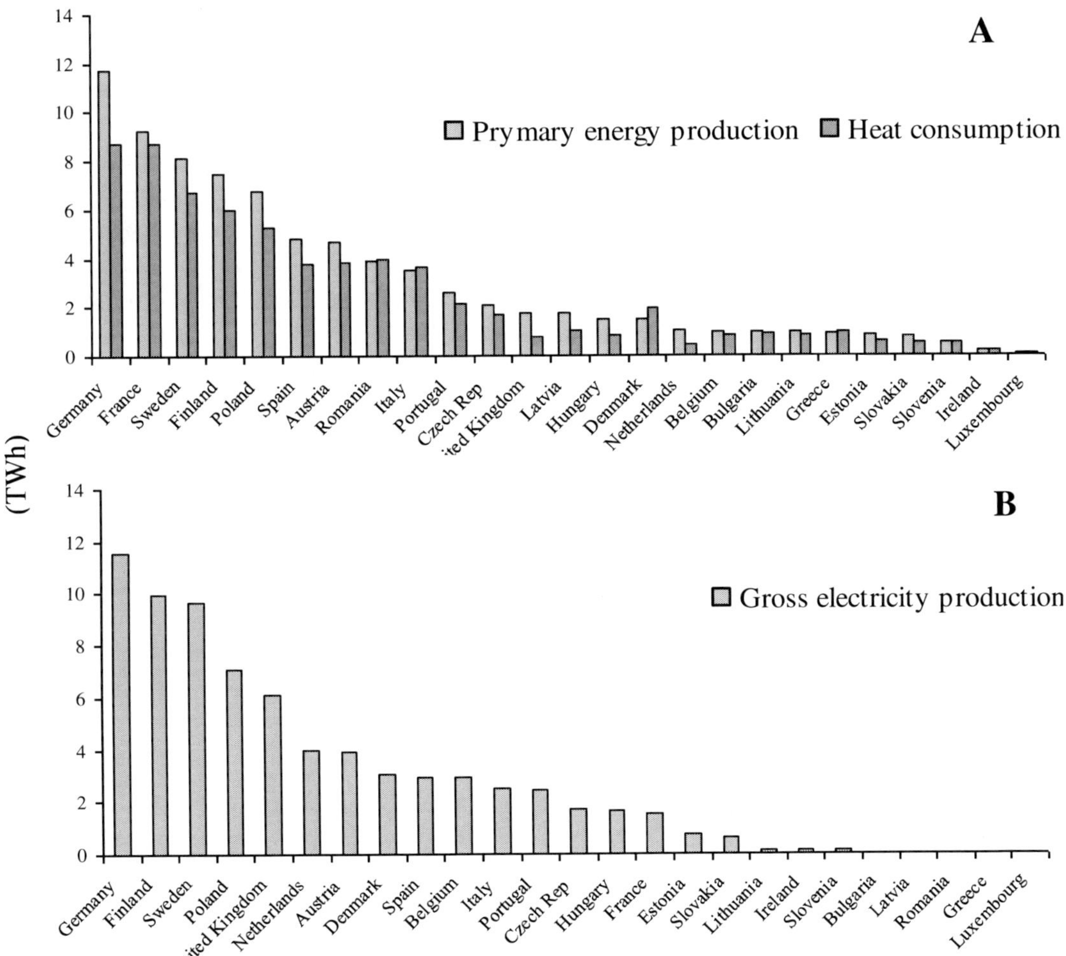

Source: Solid Biomass Barometer 2012 (www.eurobserv-er.org).

Figure 2. Primary energy production and heat consumption (A), and gross electricity production (B) from solid biomass in 2011.

Table 2. Available potential of expanding biomass in Spain (Plan de Energías Renovables 2011-2020)

Source		Potential available biomass (t yr^{-1})	Average price (€ t^{-1})	Biomass consumption (t yr^{-1})
Existing forest areas	Forest wastes	2,984,243	26.59	5,545,287
	Whole tree	15,731,116	43.16	
Agricultural wastes	Herbaceous	13,586,759	19.98	478,011
	Woody	18,605,756	19.98	1,912,046
Grasslands available for agricultural uses		15,874,572	45,62	0
Woody areas available for agricultural uses		5,457,812	34.73	0
Woody areas available for forest purposes		15,072,320	42.14	0
Total biomass potential in Spain		87,312,398	--------	7,912,046

In the Spanish renewable-energy context, the most suitable fast-growing tree crops for irrigated SRP are poplar (*Populus* spp.), paulownia (*Paulownia fortunei* x *Paulownia tomentosa*), and eucalyptus (*Eucalyptus* spp.). This latter species can also be established under rainfed conditions. Concretely, the most suitable areas for poplar crops are the fertile plains next to rivers, being periodically flooded. Many species adapted to Spain as energy crops for solid biofuels are suited to SRP that are generally recommended to be grown in low-lying areas without cold winters. Poplar and eucalyptus have traditionally been used for wood production, pulp, and veneer, but for some years it has been used for energy. The available poplar clones are those traditionally used for veneer: NNDV, I0MC, Triplo, I0214, Viriato, Beaupré, and others that could be used for energy farming under SRP system.

In Spain, more than 135,710 ha are devoted to poplar-wood production [10]. In relation to eucalyptus, according to Veiras and Soto [11], during the last few decades its area devoted to plantations has declined in Spain amounting to 760,000, most for pulp production. Moreover, RWE Innogy Iberia, the subsidiary of RWE Innogy on the Iberian Peninsula, is operating its first energy-crop plantation in Villamartín, close to Cadiz in Spain [12]. On the 235-ha field more than 390,000 paulownia trees were cultivated in 2009. This plantation is probably the largest devoted to paulownia short-rotation coppices cultivated in Spain. However, there is a strong potential to establish also willow and locust as woody energy crops, these trees being naturally widespread on Spanish riverbanks and river valleys [13].

3. Woody Biomass as Feedstock for Bioenergy Production

It is predicted that short-rotation woody crops for energy purposes will presumably become the most important source for biomass production and will reduce GHG emissions by some 80-90% compared to the current fossil-fuel baseline utilization [14]. In this context, several studies have undertaken their financial viability for bioenergy: Mitchel et al. [15] concluded that a stable market for wood chips and government aid are necessary in the UK for SRP to be feasible at a commercial scale. Ericson et al. [16] found that willow is a profitable crop in Poland; basically because the cost production is lower (diesel prices, labour, and fertilizers).

As stated above, biomass as a raw material can be utilized for producing heat and electricity. However, without a structural change in the energy system, adverse environmental, economic, or social impact could result from the production of biomass-energy crops and removal of biomass waste from forest and agricultural systems. Furthermore, unsustainable biomass production would erode the climate-related environmental advantage of bioenergy. Consequently, it is of high priority to demonstrate that the advantages of biofuels made from biomass exceed the potential cost of environmental damage originated by their production, making sustainability the key to produce biomass for use as biofuels and thereby allowing the expansion of energy farming.

High standards are needed to incorporate for investments in making biomass energy sustainable: i) reducing life-cycle GHG emissions, ii) protecting the environment, including all natural resources; and iii) promoting the general public acceptance of sustainably produced and renewable fuels. The emerging programmes must explain their environmental implications, and their corresponding measures for the reduction of greenhouse-gas emissions, better management practices for water use, and collection methods of feed-stocks used in the production process. As a consequence, new measuring methodologies are required to make evaluations of indirect land-use changes resulting from the production of woody energy crops. Also, the development of new varieties and clones for the respective local sites will be necessary for optimising biomass production, since there is a strong dependency between variety and growth behaviour in many SRPs. Particularly, improved clones and varieties with low water requirements and suitable for establishment on marginal lands could be developed through plant breeding. In the following sub-sections, the main species and their potential for establishment as biomass crops are described.

3.1. Eucalyptus (*Eucalyptus* sp.)

Currently there are about 14,000 km^2 of eucalyptus in southern Europe, *E. globulus* being the most common species but with an increasing proportion of *E. nitens*, which is grown successfully as a frost-tolerant species. The main objective of these plantations is the production of wood pulp or fibreboard. However, logging waste and the bark derived from the harvesting operations are increasingly used as biofuel for thermal and electricity purposes.

According to Johnson et al. [17], eucalyptus species are well adapted for energy production, since they produce higher yields and have lower water and nutrient demand than in the case of poplars and willows. These plantations in southern Europe were established at initial densities of 1,000 to 2,400 trunks per ha to supply the bioenergy industry as the main plantation objective or through the use of logging debris for energy purposes. Figure 3 shows the field performance for eucalyptus SRP in Granada (SE Spain). In this context, eucalyptus plantations in Spain are handled by intensive crop management, including mechanical soil preparation, fertilization at establishment and planting of 1,000-1,500 containerized seedlings per ha. Brush weeding is applied frequently before canopy closure, and clear cutting is carried out at 10-12 years. Logging debris and small-diameter logs are also used for energy purposes, through the use of bundles and further chipping once the bundles are air dried.

Recently, Jiménez et al. [18] evaluated biomass yield potential under SRP management for the Mediterranean Spanish Iberian Peninsula for *E. camaldulensis* and *E. Dunnii*. They reported that the production ranged from 19.0 to 20.5 t_{DM} ha^{-1} after three-year and two-year coppicing cycle, respectively (Table 3).

Research related to its environmental viability is still under way. For example, Gabriele et al. [19] have assessed the environmental impact of producing 1 GJ of heat from eucalyptus SRP using life-cycle assessment (LCA). All eucalyptus scenarios offered savings of fossil energy and GHG emissions of 80-90%, and had generally lower impact compared with the equivalent fossil-fuel production chain. González et al. [20] concluded the same results on assessing the environmental effects of eucalyptus-based ethanol production in comparison to conventional gasoline.

Figure 3. Irrigated eucalyptus plantation under short-rotation management in Granada (SE Spain).

Table 3. Biomass yield biomass from different species in EU countries

Country	Specie	Yield (t_{DM} ha^{-1})	Rotation (years)	Production costs (€ GJ^{-1})	Calculation period (years)	Reference
Portugal	Eucalyptus	20.0	-	-	8-10	[52]
Spain	Eucalyptus	19.0-20.5	3-2	-	10-15	[18]
Spain	Eucalyptus	10.0-15.0	-	-	8-10	[53]
Spain	Poplar	13.5	5	0.8-0.85	16	[54]
Italy	Poplar	10.0	2	4.1-4.9	8	[55]
Italy	Poplar	13.0-10.0	3-2	-	10	[56]
Italy	Poplar	18.0	5	3.27	10	[33]
Czech R.	Poplar	10.0	3	3.3	21	[57]
Spain	Paulownia	7.0-14.0	2-10	-	-	[39]
Spain	Paulownia	2.6-6.0	3-5	-	-	[58]
Spain	Paulownia	35.0-45.0	3	-	21	[59]

It is also important to point out that fuel properties are related to the age and height of eucalyptus [21]. According to these authors, the calorific value was substantially higher with the age of tree (2-6 years) with a significant variation, and the fuel properties of younger tress were compared with those of mature trees (20 years). In general, the fuel properties of mature trees were better than those of younger trees.

In the same context of sustainability, Perez et al. [22] have developed a tool for estimating biomass and bioenergy production and carbon sequestration in *E. globulus* and *E. nitens* planted at the observed range of initial stocking densities in southern Europe, in order to propose two crop-management systems for both species. These authors found higher yield in *E. nitens* than in *E. globulus* for all variables because of faster diameter increment at similar densities. The total yield in terms of biomass was from 13.9 to 14.6 t ha^{-1} yr^{-1} for *E. globulus* and from 20.4 to 21.5 t ha^{-1} yr^{-1} for *E. nitens*. The evaluated energy in aboveground biomass was between 233 and 245 GJ ha^{-1} yr^{-1} for *E. globulus* and 345 and 364 GJ ha^{-1} yr^{-1} for *E. nitens*. In this sense, Sochacki et al. [23] developed equations that relate stem diameter over bark at 10 cm (D_{10}) and tree height (ht) to total tree biomass, leaves, stems and roots by combining data with different planting densities (500, 1000, 2000, and 4000 stem ha^{-1}) and landscape positions (upper-slope, mid-slope, and lower-slope). These researchers used *E. globulus, E. occidentalis* and *Pinus radiata* for this purpose. Mean oven-dried yields of the three species studied, in the high-planting-density treatment did not significantly differ, ranging from 12.0 to 14.0 t ha^{-1} (3 years)$^{-1}$. Biomass increased with higher planting density. Productivity also varied with slope position: *E. globulus* and *E. occidentalis* had the highest yield in lower landscape positions with initial planting densities of 4,000 stem ha^{-1}, with 16.6 and 22.2 t ha^{-1} (3 years)$^{-1}$ of total biomass produced, respectively. These authors confirmed that biomass productivity can be optimised by means of high initial planting densities and accepting the interaction of different species with *in situ* hydrology.

3.2. Poplar (*Populus* sp.)

On average, poplar trees in SRC systems under temperate European conditions have dry-mass yields of 10.0-15.0 t ha^{-1} y^{-1} [24]. *Populus* spp. has been widely studied in terms of ecophysiology and productivity and has a considerable capacity to protect the soil against erosion while controlling nitrate leaching [25]. Moreover, poplar has been used for studies of phytoremediation of contaminated soils by SRP cultures. Many studies have been made on cultivation, production, and technological aspects of SRP with poplar, and it is widely known that the success of these systems depends on the establishment and the first-year performance of stands [26]. Poplar is a light-demanding species, and weed control in a SRP is crucial during the establishment period.

In this context, Broeckx et al. [27] have conducted a large-scale experiment in Belgium with SRP system, selecting 12 *Populus* genotypes, and characteristics of production have been assessed during the first two years, reaching a volume index ranging from 1.00 (± 0.68) to 1.93 (± 0.97) dm^3 in the first, and from 2.75 (± 1.70) to 11.91 (± 0.63) dm^3 in the second season. Average growth rates were 247 cm of tree height and 445 cm of trunk diameter and 25.1 mm and 40.7 mm after the first and second growing seasons, respectively. These authors found high potential of SRP with poplar on agricultural land for bio-energy purposes.

Under European Mediterranean conditions, species from *Populus* are economically viable as bioenergy crops purposes. Fernández et al. [28] evaluated the yield in Spain from poplar based on data for potential evapotranspiration, with results ranging from 10.1 to 14.4 t ha^{-1} yr^{-1}. Also, other studies have shown that biomass from poplar SRP has a positive energy balance and a more beneficial effect on the environment than alternative biomass sources [29]. However, a possible drawback of SRPs on a large scale is the water irrigation demand under Mediterranean climatic conditions, as well as the water footprint in other EU areas [30].

More recently, Pérez et al. [31] have studied the potential biomass production for mainland Spain. For this purpose, they used the 'I-214' clone of *Populus* spp. under several management systems and land-availability scenarios. Empirical models were fitted to the data from a network of 144 experimental plots located at 12 different sites in Mediterranean climate areas; specifically, four models were developed with the following parameters: the average maximum temperature of the hottest month, length of drought, intensity of drought, and soil pH. Predictions were made for irrigated agricultural land, and the capacity of energy production was evaluated considering the different alternatives for converting biomass of poplar short-rotation crop: heat, bio-ethanol, and electricity. These authors calculated that the mean poplar productivity at between 15.3 and 10.9 t ha^{-1} yr^{-1} for standard management scenario, and the poorly irrigated and weeded management scenario, respectively. Therefore, poplar SRPs could take on an important role in the Spanish energy sector by providing stability to the biomass for energy market, as well as helping to achieve greater energy self-sufficiency in the Mediterranean climate areas of Spain. It would also prove useful to other entities involved in the biomass industry, facilitating the identification of the most appropriate sites to establish biomass-consumption industries.

On the other side, Deckmyn et al. [32] evaluated the benefits for C sequestration of afforestation with a multifunctional oak-beech forest *vs.* a poplar SRP by running the models SECRETS (Stand to Ecosystem Carbon and Evapotranspiration Simulator) and GORCAM (Graz Oak Ridge Carbon Accounting Model). They estimated the net primary production of oak-beech forest of 2.5 t C ha^{-1} yr^{-1} after 150 years, compared to 6.2 t C ha^{-1} yr^{-1} for a poplar SRP. When considering the energy substitution, coppice cultures reduce emissions with 24.2-29.3 t CO_2 ha^{-1} yr^{-1} while the oak forests reduce only 6.2-7.1 t CO_2 ha^{-1} yr^{-1}. In the same context, Fiala and Bacenetti [33] have studied the economic, energy, and environmental impact in SRP harvesting operations in two different poplar-plantation systems. The economic cost, energy input, and GHG emissions depended on yield, rate global use of the machines, and the scheduling of operations. These authors have concluded that the optimum systems are achieved when harvest and transport are undertaken in an area larger than 400 ha, with an effective plantation design, a proper-sized transport system, and without mechanical failures. For these conditions, the productivity of the harvest transport system could be up to 65 t_{WB} h^{-1} (wet basis), and the economic cost, the energy input, and GHG emissions could reach, 15 MJ per t_{DM}, 212 MJ per t_{DM}, and 16 kg CO_2 eq per t_{DM}, respectively.

On the other hand, especially for EU Mediterranean countries, it is essential to gather fuller knowledge and develop better background for water-use efficiency (WUE) for SRP poplar, since it is expected to be expanded intensively, making it necessary to perform spatial predictions and up-scaling of yield for minimizing environmental impacts and optimising economic benefits. Figure 4 shows the field performance for assessing the poplar biomass yield in Granada (SE Spain). Along this line, Tallis et al. [34] have recently promoted a process-based model for SRP poplar and willow (ForestGrowth-SRC) and exposed its suitability for predicting yield and WUE. This model has predicted annual biomass yield for poplar of 1.46 odt (oven-dried tonnes) ha^{-1} yr^{-1} and higher WUE for poplar than for willow (9.5 and 5.5 g kg^{-1}, respectively). This model has been demonstrated to be suitable to forecast present and future SRP yields at a regional scale. Table 3 shows the fixed wood-yield biomass under SRP systems in different EU producer countries.

3.3. Paulownia (*Paulownia* sp.)

Paulownia is attracting increasing attention as an energy crop, and considerable interest as an industrial raw material [35]. The genus is composed of nine species, which have very fast growth, being plausible to be harvested only 15 years after establishing the plantation to achieve high added-value products [36].

In China, paulownia is being used for some types of agroforestry systems [37], providing agricultural benefits, which is important in maintaining the ecological system and its sustainability. In addition, paulownia intercropping systems have been analysed, these studies being related to crop yield, rotation period, tree densities, and light and heat distribution. In this context, the economic and energy features in developing paulownia intercropping systems have been reported by Jianbo [38]. This author reported an energy output: input ratio of paulownia intercropping systems of 1.39 (compared with non-paulownia intercropping system or conventional systems of 1.27). In economic terms, the output: input ratio of paulownia was 2.45 (compared with 2.25 from the non-paulownia intercropping system). Consequently, the intercropping system has higher energy efficiency and also provides better financial returns.

Figure 4. Field performance for different poplar clones in Granada (SE Spain).

Figure 5. Paulownia plantation under two-year coppicing cycle in Granada (S Spain).

In the Spanish context, Durán et al. [39] have assessed the potential biomass yield of *Paulownia elongate* x *fortunei* with two of their clones (Cotevisa 2 and Suntzu 11) at different sites in Andalusia, S Spain (Figure 5). They found that Cotevisa 2 was the more productive (1.8 fold higher) in terms of biomass than Suntzu 11. Also, significantly higher woody-biomass yield was found for both clones, ranging between 7.2 and 14.0 tDM ha-1 in Villanueva del Río y Minas (Sevilla province, Spain). By contrast, significantly lower paulownia biomass production was found at Palma del Río (Córdoba province, Spain), registering between 1.7 and 2.3 t_{DM} ha^{-1} (Table 3). Consequently, it is feasible to introduce *Paulownia* into a Mediterranean climate, although data are still insufficient and further research is needed in relation to management, water efficiency, and carbon sequestering.

In addition, López et al. [40] have analysed paulownia as a raw material for solid biofuel production, finding a lower ash content (8.9 g kg^{-1}) and higher cellulose content (440 g kg^{-1}) than for other woody energy crops. Also, a low content in S and N was found in comparison with poplar or willow, with a gross heating value of 20.3 MJ kg^{-1}, this being higher than for hardwood, slightly higher than for *Pinus pinaster* and softwood, and much higher than those for residues of food plants and agricultural crops. In the same context, according to Jiménez et al. [41] in southern Spain, a higher calorific value has been attained using paulownia (18.6 MJ kg^{-1}) and poplar (18.5 MJ kg^{-1}), biomass with lower chlorine content in relation to herbaceous biomass, and paulownia biomass also produced lower ash content with respect to poplar (1.45 vs. 2.50%, respectively).

4. Environmental Impact and Challenges for Sustainable Woody-Energy Farming

SRP systems for producing biomass for heat and electricity is one of the promising means of helping to reach European targets of increasing the amount of renewable energy, and has been identified as the most energy-efficient C-conversion technology to reduce GHG emissions. However, intensive cultivation can exert a potentially negative impact that could complicate their development or that, at least, should be considered. Therefore, biomass generation will depend on the sustainability of the production-resource base: water, soil, air, and biodiversity of the ecosystem as a whole.

It is well known that trees grown with intensive weed control promote higher erosion rates than non-tillage agricultural systems; however, this problem may be ameliorated by planting cover crops between the tree rows, as demonstrated in cultivating rainfed-tree crops [42]. Moreover, Malik et al. [43] studied the effects of varying strip widths of four species of cover crops on the growth of sweetgum (*Liquidambar styraciflua* L.) seedlings planted as a short-rotation bioenergy woody crop. This study reported that cover crops of lespedeza, tall fescue, crimson clover, and ryegrass reduced the biomass of sweetgum by approximately 41, 37, 27, and 15%, respectively, compared to the control. Furthermore, several studies have evaluated the influence of biomass-energy production on biodiversity. Pedroli et al. [44] discussed the compatibility of the objectives of the 2020 horizon for renewable-energy production with the protection of biodiversity. They concluded that the increasing demand for biomass for energy production will covert valuable landscapes into productive land, and will intensify other areas already in production, with negative effects on biodiversity. Contrarily, in certain areas, biomass will promote biodiversity, since perennial crops may help increase it. Consequently, the effects of the cropping of biomass and/or elimination of biomass on biodiversity depend on specific regional characteristics, on the type of land and land-use changes, and on crop management.

In this sense, discussion concerning the environmental impact of energy crops has usually been focused on the analysis of land-use changes and energy balances. Even the Directive of Renewable Energy requires the analysis of the GHG emissions and impact on biodiversity where they are cultivated. However, there were no demands on evaluating energy balances or water-consumption requirements [29]. Some research has been conducted to analyse the water footprint of biofuels [45, 46]. However, those related to the consumption of water resources in areas with significant water shortages, such as the south-eastern Spain, are still limited. In this context, Sevigne et al. [29] studied the relation between water, energy, and CO_2 emissions for a *Populus* field trial in Spain to evaluate the feasibility of its establishment as a large-scale crop, highlighting the positive energy balance and environmental improvement with respect to other energies, such as natural gas. The water consumption required to avoid a kg of CO_2 is 4.6 m^3 and per unit of energy obtained is 45 m^3 GJ^{-1}. They also concluded that the cultivation of *Populus* spp. should be limited to areas with no water shortage. Also, these authors suggest the use of *Populus* as a complementary raw material to fuel power plants, together with the use of forest debris, agricultural wastes, etc. and the reuse of urban or industrial water for irrigation.

Also in Spain, Butnar et al. [47] evaluated poplar and Ethiopian mustard for power generation. They calculated different theoretical situations of electricity production from biomass in power plants of different capacity (10, 25 or 50 MW), different transport scenarios, and different yields for biomass production. According to these authors, Ethiopian mustard is more environmentally harmful than poplar when used for electricity production and compared to electricity from natural gas or the Spanish electricity mix, the electricity derived from biomass is more damaging in three of six impact categories they assessed (acidification, human toxicity, and photochemical oxidation). Also, better environmental profiles were found for 10 or 25 MW power plants. To guarantee a good environmental profile of the electricity from biomass, the transport distance from the field to the power plant has to be the shortest possible. Moreover, since the use of fertilizers is the most harmful part in the cultivation of biomass, a possible strategy to reduce these effects would be to replace mineral fertilizers by natural fertilizers such as livestock manure, etc.

According to Dinica [48], from a logistical perspective, improvements are needed in raw-material collection, transport, storage, and transformation. These affect both the volume and the reliability of the resource market, as all stages influence the energy quality of resources in time, exerting an effect on the final production costs of biomass power. Also, cultural factors affect mainly the size of the available market. This is a reality for farmers, industrial companies, public agencies managing public lands, and some private sectors [49]. There is still strong opposition of farmers to change to a completely new type of cultivation, such as energy crops [48]. Even farmers who already produce biomass wastes as by-products have been unwilling to respond to contacts for biomass supply to power producers.

Regarding to biofuel environmental analysis, González et al. [50], by using the life-cycle assessment technique, analysed the environmental footprint of ethanol production and use from three energy wood crops in a Spanish context. The alternative with the lowest impact and the best characteristics was from black locust, compared to eucalyptus and poplar biofuels. Also, concerning the production stage of ethanol, black locust has the lowest environmental impact due to the low levels of agricultural inputs during its cultivation. Poplar cultivation has higher impact due to fertilizer application and eucalyptus as a consequence of the use of heavy machinery during harvesting.

In addition, we can report that the development of SRP systems in Europe and Spain may help to generate rural income and employment. Goldemberg [51] recognizes the generation of jobs as one of the main benefits of biomass-producing areas. Energy crops lead to changes in agricultural practices and contribute positively to rural diversification. Therefore, the transformation of land for the production of energy crops should be considered an alternative answer to social and environmental problems such as land abandonment, rising unemployment, and may eventually bring the return of people to agricultural activities in rural areas due to the current economic crisis, especially in the European Mediterranean regions. Additionally, rural development will be promoted from bioenergy guidelines which, in accordance with the rural-development policies, will represent a viable alternative use for large marginal areas of agricultural and forestry land.

Consequently, the development and installation of small power plants should be integrated with efficient biomass-management strategies that include short-rotation coppicing of raw material, forest, agricultural, and urban wastes and well as other types of recycled wastes.

The environmental viability of the use of underused agricultural areas for non-food and lignocellulose bioenergy crops remains controversial and uncertain for several reasons, given that the environmental impact from the use of these lands requires further research, as crops often need significant inputs of water and nutrients to maintain the required high rates of productivity.

CONCLUSION

Existing worldwide bioenergy sources are presumably sufficient to fulfil the predicted biofuel and biomass requirements without competing with food production, although the land-use changes will have to be rigorously projected. The use of energy from renewable sources is supported by falling technology costs, by the shifting and rising prices of fossil fuels, and by continued support subsidies. The bioenergy sector is innovative, and the investment in it may bring considerable strategic advantages to investor companies. However, there are still many gaps in certain sectors related with wood crops, and the necessary research efforts on crop selection, development and management might retard the effective establishment of further SRPs. Consequently, future research should focus primarily on key aspects such as the improvement of genetic material and its adaptation to different regions in order to diminish the potential environmental impact, planting design, and rotation length, and management operations such as weed control, fertilization, and irrigation. Research related to other types of social impact will also be decisive, to promote public participation of the different stakeholders in the process to encourage farmers' interest and engagement to this new economic activity. Sustainable bioenergy harvesting is becoming more important, especially those from energy-crop plantations that may become an increasingly utilized resource.

ACKNOWLEDGMENTS

This publication was supported by the research project: "European regions fostering innovation for sustainable production and efficient use of woody biomass (ROKWOOD)", FP7-REGIONS-2012-2013-1, Grant Agreement Nº 319956.

REFERENCES

[1] IEA. (2010). *World Energy outlook 2010.* International Energy Agency.

[2] García, R., Pizarro, C., Lavín, A. G. & Bueno, J. L. (2012). Characterization of Spanish biomass wastes for energy use. *Bioresource Technol*, *103*, 249-258.

[3] Hämäläinen, S., Näyhä, A. & Pesonen, H. L. (2011). Forest biorefineries. A business opportunity for the Finnish forest cluster. *J. Clean Prod.*, *19*, 1884-1891.

[4] IEA. (2011). *Technology Roadmap: Biofuels for Transport.* International Energy Agency, Paris, France.

[5] Montoya, G. F., Aguilera, J. A. & Manzano, A. F. (2014). Renewable energy production in Spain: A review. *Renewable and Sustainable Energy Reviews*, *33*, 509-531.

[6] Gómez, A., Zubizarreta, J., Dopazo, C. & Fueyo, N. (2011). Spanish energy roadmap to 2020. Socioeconomic implications of renewable targets. *Energy*, *36*, 1973-1985.

[7] Solid biomass barometer 2012. EuObserv'ER. http://www.eurobserv-er.org/pdf/baro212biomass.pdf. Accessed 1 November 2013.

[8] AAE. (2011). La biomasa en Andalucía. Agencia Andaluza de la Energía, Consejería de Economía, Innovación y Ciencia, Junta de Andalucía. Sevilla, Spain.

[9] Junta de Andalucía. (2012). Ensayos con cultivos energéticos. Periodo 2005-2010: Síntesis de resultados y principales conclusiones. Agencia de Gestión Agraria y Pesquera de Andalucía, CAPMA, Sevilla, Spain.

[10] MAGRAMA. (2012). Anuario de Estadística Agraria. Ministerio de Agricultura Alimentación y Medio Ambiente, Madrid, Spain. http://www.magrama.gob.ess/agricultura/estadisticas/Estadisticas-chopo.aspx#section. Accessed 25 November 2013.

[11] Veiras, X. & Soto, A. (2011). La conflictividad de las plantaciones de eucalipto en España y Portugal. Análisis y propuestas para solucionar la conflictividad ambiental y social de las plantaciones de eucalipto en la península Ibérica. Greenpeace. Madrid, Spain.

[12] RWE. (2013). http://www.rwe.com/web/cms/en/398786/rwe-innogy/technologies/biomass/spain/. Accessed 14 October 2013.

[13] Ciria CMP. (2011). Desarrollo de los cultivos energéticos leñosos en España. Vida Rural 329.

[14] Djomo, S. N., Kasmioui, O. E. & Ceulemans, R. (2011). Energy and greenhouse gas balance of bioenergy production from poplar and willow: a review. *Glob Change Biol Bioenerg*, *3*, 181-197.

[15] Mitchell, C. P., Stevens, E. A. & Watters, M. P. (1999). Short rotation forestry-operations, productivity and costs based on experience gained in the UK. *Forest Ecol Manag*, *121*, 123-136.

[16] Ericsson, K., Rosenqvist, H., Ganko, E., Pisarek, M. & Nilsson, L. (2006). An agro-economic analysis of willow cultivation in Poland. *Biomass Bioenerg*, *30*, 16-27.

[17] Johnson, J., Coleman, M. U. F. S., Gesch, R., Jaradat, A., Mitchell, R. & Reicosky, D. (2007). Biomass bioenergy crops in the United States: a changing paradigm. *Am J Plant Sci Biotechnol*, *1*, 1-18.

[18] Jiménez, B. J. A., Perea, T. F., Lobo, G. J., Pavón, P. L. & Durán, Z. V. H. (2013a). Evaluación del cultivo del eucalipto para la producción de biomasa en Andalucía. *Vida Rural*, *366*, 62-66.

[19] Gabriele, B., Nguyen, T. N., Paupu, P. & Vial, E. (2013). Life cycle assessment of eucalyptus short rotation coppices for bioenergy production in southern France. *GCB Bioenerg*, *6*, 30-42.

[20] González, G. S., Moreira, M. T. & Jeijoo, G. (2012a). Environmental aspects of eucalyptus based ethanol production and use. *Sci Total Environ*, *438*, 1-8.

[21] Kumar, R. Pandey, K. K., Chandrashekar, N. & Mohan, S. (2011). Study of age and height wise variability on calorific value and other fuel properties of Eucalyptus hybrid, Acacia auriculaeformis and Casuaria equisetifolia. *Biomass Bioenerg*, *35*, 1339-1344.

[22] Pérez, C. C., Merino, A. & Rodríguez, S. R. (2011). A management tool for estimating bioenergy production and carbon sequestration in Eucalyptus globulus and Eucalyptus nitens as short rotation woody crops in north-west Spain. *Biomass Bioenerg*, *35*, 2839-2851.

[23] Sochacki, S. J., Harper, R. J. & Smettem, K. R. J. (2007). Estimation of woody biomass production from a short-rotation bio-energy system in semi-arid Australia. *Biomass Bioenerg*, *31*, 608-616.

[24] Labrecque, M. & Teodorescu, T. I. (2003). High biomass yield achieved by Salix clones in SRIC following two 3-year coppice rotations on abandoned farmland in southern Quebec, Canada. *Biomass Bioenerg*, *25*, 135-146.

[25] Isebrands, J. G. & Karnosky, D. F. (2001). Environmental benefits of poplar culture. In: Dickmann DE, Isebrands JG, Eckenwalder JE, Richardson J (ed) Poplar culture in North America. Ottawa, NRC Research Press, 207-218.

[26] Otto, S., Loddo, D. & Zanin, G. (2010). Weed-poplar competition dynamics and yield loss in Italian short-rotation forestry. *Weed Res*, *50*, 153-162.

[27] Broeckx, L. S., Verlinden, M. S. & Ceulemans, R. (2012). Establishment and two-year growth of a bio-energy plantation with fast-growing Populus tress in Flanders (Belgium): effects of genotype and former land use. *Biomass Bioenerg*, *42*, 151-163.

[28] Fernández, J., Sánchez, J. & Esteban, B. (2009). Potential lignocellulosic biomass production from dedicated energy crops in marginalized agricultural lands of Spain. 17-th European Biomass Conference, Hamburg, Germany, 131-137.

[29] Sevigne, E., Gasol, M. C., Brun, F., Rovira, L., Pagés, J. M., Camps, F., Rieradevall, J. & Gabarrell, X. (2011). Water and energy consumption of Populus spp. bioenergy systems: A case study in Southern Europe. *Renew Sust Energ Rev*, *2*, 1133-1140.

[30] Gerbens, L. W., Hoekstra, A. Y. & Van Der Meer, T. H. (2009). The water footprint of bioenergy. *P Natl Acad Sci*, *106*, 10219-10223.

[31] Pérez, C. C., Sánchez, R. D., Rodríguez, S. R., Hernández, M. J., Sánchez, M. M., Cañellas, I. & Sixto, H. (2013). Biomass production assessment from Populus spp. short-rotation irrigated crops in Spain. *Bioenergy*. doi: 10.1111/gcbb.12061

[32] Deckmyn, G., Muys, B., García, Q. J. & Ceulemans, R. (2004). Carbon sequestration following afforestation of agricultural soils: comparing oak/beech forest to short-rotation poplar coppice combining a process and carbon accounting model. *Global Change Biol*, *10*, 1482-1491.

[33] Fiala, M. & Bacenetti, J. (2012). Economic, energetic and environmental impact of short rotation coppice harvesting operations. *Biomass Bioenerg*, *42*, 107-113.

[34] Tallis, M. J., Casella, E., Henshall, P. A., Aylott, M. J., Randle, T. J., Morison, J. I. L. & Taylor, G. (2012). Development and evaluation of ForestGrowth-SRC process-based model for short rotation coppice yield and spatial supply reveals poplar uses water more efficiently than willow. *GCB Bioenerg*, *5*, 53-66.

[35] Jiménez, L., Rodríguez, A., Ferrer, J. L., Pérez, A. & Angulo, V. (2005). Paulownia, a fast-growing plant, as a raw material for paper manufacturing. *Afinidad*, *62*, 100-105.

[36] Kalaycioglu, H., Deniz, I. & Hiziroglu, S. (2005). Some of the properties of particleboard made from Paulownia. *J Wood Sci*, *51*, 410-414.

[37] Lu, J., Zhao, X. & Ding, L. (2004). Typical patterns of ecological engineering in southern China. *Korean J Economic*, *27*, 1-7.

[38] Jianbo, L. (2006). Energy balance and economic benefits of two agroforestry systems in northern and southern China. *Agr Ecosyst Environ*, *116*, 255-262.

[39] Durán, Z. V. H., Jiménez, B. J. A., Perea, T. F., Rodríguez, P. C. R. & Francia, M. J. R. (2013). Biomass yield potential of paulownia trees in a semi-arid Mediterranean environment (S Spain). *Int J Renew Energy Res.*, *4*, 789-793.

[40] López, F., Perez, A., Zamudio, M. A. M., d, Alba, H. E. & Garcia, J. C. (2012). Paulownia as raw material for solid biofuel and cellulose pulp. *Biomass Bioenerg*, *45*, 77-86.

[41] Jiménez, B. J. A., Lobo, G. J., Pavón, P. L., Durán, Z. V. H. & Perea, T. F. (2013b). Biomasa de cultivos energéticos para la producción sostenible de energía. *Energética*, *XXI*, 1356, 53-55.

[42] Martínez, R. A., Durán, Z. V. H. & Francia, M. J. R. (2006). Soil erosion and runoff response to plant cover strips on semiarid slopes (SE Spain), *Land Degrad. Dev.*, *17*, 1–11.

[43] Malik, R. K., Green, T. H., Brown, G. F., Beyl, C. A., Sistani, K. R. & Mays, D. A. (2001). Biomass production of short-rotation bioenergy hardwood plantations affected by cover crops. *Biomass Bioenerg*, *21*, 21-33.

[44] Pedroli, B., Elbersen, B., Frederiksen, P., Grandin, U., Heikkilä, R., Henning, P., Izakovicová, Z., Johansen, A., Meiresonne, L. & Spijker, J. (2013). Is energy cropping in Europe compatible with biodiversity? Opportunities and threats to biodiversity from land-based production of biomass for bioenergy purposes. *Biomass Bioenerg*, *55*, 73-86.

[45] Stone, K. C., Hunt, P. G., Cantrell, K. B. & Ro, K. S. (2010). The potential impacts of biomass feedstock production on water resource availability. *Bioresource Technol*, *101*, 2014-2025.

[46] Iriarte, A., Rieradevall, J. & Gabarrell, X. (2010). Life cycle assessment of sunflower and rapessed as energy crop under Chilean conditions. *J Clean Prod*, *18*, 336-345.

[47] Butnar, I., Rodrigo, J., Gasol, C. M. & Castells, F. (2010). Life-cycle assessment of electricity from biomass: case studies of two biocrops in Spain. *Biomass Bioenerg*, *34*, 1780-1788.

[48] Dinica, V. (2009). Biomass power: exploring the diffusion challenges in Spain. *Renew Sust Energ Rev*, *13*, 1551-1559.

[49] IDAE. (2005). Plan de las energías renovables en España 2005–2010, Madrid.

[50] González, G. S., Moreira, M. T., Feijoo, G. & Murphy, R. J. (2012b). Comparative life cycle assessment of ethanol production from fast-growing wood crops (black locust, eucalyptus and poplar). *Biomass Bioenerg*, *39*, 378-388.

[51] Goldemberg, J. (2002). Brazilian energy initiative, world summit on sustainable development; Setembro. Joanesburgo, South Africa.

[52] Sebastián, N. F., García, G. D. & Rezeau, A. (2010). Energía de la biomasa (volumen I). Prensas Universitarias de Zaragoza.

[53] Iriarte, C. (2008). Caracterización del olmo (Ulmus pumila L.) como cultivo energético. Dissertation, Universidad Politécnica de Madrid.

[54] Gasol, C. M., Martínez, S., Rigola, M., Rieradevall, J., Anton, A. & Carrasco, J. (2009). Feasibility assessment of poplar bioenergy systems in the Southern Europe. Renew Sust Energ Rev 13, 801-812.

[55] Manzone, M., Airoldi, G. & Balsari, P. (2009). Energetic and economic evaluation of poplar cultivation for the biomass production in Italy. *Biomass Bioenerg*, 33, 1258-1264.

[56] Di, Matteo, G., Sperandio, G. & Verani, S. (2012). Field performance of poplar for bioenergy in southern Europe after two coppicing rotations: effects of clone and planting density. *iForest-Biogeosciences Forestry*, *5*, 224-229.

[57] Havlickova, K., Weger, J. & Zanova, I. (2007). Short rotation coppice for energy purposes e economy conditions and landscape functions in the Czech Republic. In: Goswami DY, Zhao Y (ed) Solar energy and human settlement. *Proceedings of Ises solar world congress*. Springer, Beijing, China, 2482-2487.

[58] Martínez, G. E., Lucas, B. M. E., Andrés, A. M., López, S. F. R., García, M. F. A. & del Cerro, B. A. (2010). Aprovechamiento energético de Paulownia spp.en el ámbito Mediterráneo. *Revista de Montes*, *102*, 5-11.

[59] Gexbioma. (2013). General de Explotaciones para biomasa. Paulownia, la mejor alternativa para el futuro. http://www.gexbioma.com/paulownia.php. Accessed 28 August 2013.

In: Renewable Energy and Sustainable Development
Editors: R. T. Durai Prabhakaran, S. A. Kale et al.
ISBN: 978-1-63463-429-8

Chapter 9

AN ASSESSMENT ON THE POTENTIAL OF TIDAL POWER GENERATION IN CHITTAGONG, BANGLADESH

Tanzina Akter and Samiran Das*
Environmental Sciences Program, Asian University for Women,
Chittagong, Bangladesh

ABSTRACT

Tidal energy is one of the possible renewable energy sources to fulfil the power demand of a country, especially the coastal areas. This chapter investigates tidal prospects in Sadarghat, Chittagong. Sadarghat is located at the Karnaphuli River estuary of the Bay of Bengal. The chapter estimates average power output that can be obtained from the plant using one-way and two-way method based on a suitable mathematical model. Data are collected from Bangladesh Inland Water Transport Authority and WXTide32 software. Suitability of data obtained from WXTide32 is also assessed. Findings suggest that a tidal power station can be constructed around 3 km away from Sadarghat and the power can be generated with a tidal range of 4.2 m. It is also found that tidal data obtained from the software WXTide32 can be successfully used when observed data are unavailable.

1. INTRODUCTION

Bangladesh, a developing country, is facing a severe power crisis problem and her economic growth is hindered by the shortage of power. Bangladesh mostly relies on natural gas, oil, coal and hydro power stations to manufacture a large portion of its power. The utmost power generation capacity is 6350 MW (as of August 2012). Still, Bangladesh lags behind its projected generation capacity. The government plans to generate 11,500 MW by 2015 and 20,000 MW by 2021 [1]. To meet the increasing demand of electricity by 2030,

* Corresponding Author. Tel:+88 (031)2854980, Email:samiran.das@auw.edu.bd.

around $35-40 billion of new capital investment is needed annually [2]. Being the second largest city of Bangladesh, Chittagong suffers from an acute power crisis problem because of lack of gas and decrease of water level in the Kaptai Lake where country's largest hydropower plant is situated. The power consumers under Chittagong Power Development Board have to face on average 120-140 MW load shedding in daytimes and 200-220 MW in peak hours whereas the total power demand is 440-775 MW.

As a developing country, Bangladesh should focus on renewable energy to meet power demand [3]. Renewable energy is a form of energy that can be generated from various kinds of natural resources including sunlight, wind, rain, tides, bio-diesel, bio-fuel and geo thermal heat. In 2011, renewable energy contributed around 0.5% of power production in Bangladesh where solar and biogas contributed around 20.75% and 14.18% of total electricity generated from renewable energy sources, respectively [4, 5]. According to a comparative study done on the use of different renewable energy sources, compared to other energy sources tidal energy is rightly predictable, has significant efficiency, negligible operating costs and also provides both small and large scale power production [3]. Thus, for the coastal areas of Bangladesh like Chittagong, exploitation of tidal energy would be less challenging and more profitable compared to other forms of unpredictable renewable energy sources [6].

Tidal energy is one of the most promising energy sources due to its continuous, conventional and spatially concentrated traits [7]. According to a study, nearly 1800 TWh/year of electricity can be generated from tidal power in all over the world [8]. There are some worldwide established tidal power plants, such as La Rance, in France since 1967 with 240 MW capacities, Annapolis Royal in Canada since 1984 with 20MW capacities, Lake Shiwa in South Korea with 254 MW and others [9, 10, 11]. South Asian countries are also getting inspired by tidal power. The Indian state of Gujarat has taken the initiative to manufacture electricity of 50 MW in the Gulf of Kutch [12]. Similar to India, Bangladesh has the potential to harness tidal power [13, 14]. Bangladesh possesses a long coastal area of 710 km with 2 to 6 m of tidal head rise and fall and the conventional one-way generation (ebb mode generation) and two-way generation methods for power generation can be applicable to coastal areas of Bangladesh.

Data at Sandwip (a coastal area of Bangladesh) have been analyzed with positive tidal power potential. However, it is recommended to examine the environmental factors carefully, since Sandwip has constructed several engineered structures to control flood and also for fish farming and rice cultivation [15]. Similarly, it is also suggested that installation of a tidal power plant project at Sadarghat, Chittagong will offer sustainable energy for Chittagong that might contribute to reduce power crisis.

This chapter assesses the possibility of establishing a tidal power station at Sadarghat and emphasizes the role of tidal energy to solve the power crisis problem in Chittagong, one of the coastal areas of Bangladesh, located on the edge of Bay of Bengal. It also estimates average power output that can be obtained from the plant using both one way and two way systems based on a suitable mathematical model. Another significant aim of the study is to assess suitability of data obtained from the software WXTide32 when observed data are not available.

2. METHODOLOGY

Tide is generated due to the gravitational pull of the moon mainly and the sun on the world's oceans. Depending on the positions of the sun and the moon associated to the earth, the size of lunar tide increases or decreases. The daily rotation of the earth in its own axis is also a factor in the production of high tides. Tidal power is produced by holding the tidal energy. Tides can be predicted in nature, thus have the potential to play an important role in power generation. One of the recommended procedures to produce energy in coastal areas in Bangladesh is by installing tidal barrage with the use of vertical water movement following the rise and fall of the tides.

2.1. Tidal Barrages and Generation Modes

This system is almost like a hydro electric system. Generally, a huge barrage is built across the shore. Main components for tidal barrage are barrage, turbines, and sluice gates. All of the elements are linked to the shore with embankments. When tides come into the shore, it can be trapped in reservoirs behind dams.

The dam includes a sluice gate that is opened to allow the tide to flow into the basin; the sluice is then closed [16]. Later on, when there is a sufficient head difference in the level of water on opposite sides of the dam, the sluice gate is opened. When the tide drops, the water flows from one side of the barrage to the other side through low-head turbines and using the potential energy of the tides it produces electricity [16]. Mainly, two different modes of operation, ebb generation and flood generation are used to produce electricity from a single basin.

The detail description of these modes with illustrations can be obtained from [6]. The Ebb generation is considered as the simplest mode of operation. In this mode at the time of flood tide, incoming water flows through the sluice gates freely in the barrage. During the flood tide, sluice gates remain closed to prevent the water from flowing back on the outgoing tide (known as ebb tide). Therefore, water is stored in a basin behind the barrage. When the sea level decreases to make a substantial head between the basin and the open water, the basin water is allowed to flow out through low head turbines and to generate electricity. Usually, once the head is nearly half of the tidal range, the basin water will only be allowed to flow through the turbines. Using this method, electricity will be produced around 40% of the tidal range [17]. The flood generation mode utilizes the incoming tide to produce electricity. The basin is filled through the turbine, which generates electricity during flood tide. During flood tide the sluice gates and turbines are kept closed. Thus, the water level can build up outside the barrage. In case of flood generation when a sufficient head is built the turbine and sluice gates are opened, which allow the water to flow into the basin. Therefore, it generates electricity. This scheme is less efficient than the ebb method since keeping a tidal basin at low tide for longer periods can have negative impacts on the environment and on shipping. Further, flood generation mode will give less energy production since the basin area would be larger at high tide than at low tide.

3. Tidal Power Generation

The conventional, one-way generation (ebb mode generation) and two-way generation methods will be applied at Sadarghat. These methods are also suggested for power generation at Sandwip [14]. The detail illustration of these systems can be obtained from [14]. In one way tidal power generation system, a single turbine is used to generate power. For this system, both sea water level and the river basin water level should be considered with equal importance. One of the benefits of this type of plant is that only one turbine is needed and the cost of the turbine, function and maintenance are comparatively less. The type of turbine used in this plant is accessible industrially. Though less electricity is generated from this type of system it is widely used in many parts of the world [14]. Compared to the one way power generation system, two-way generation system generates more electricity. However, there will be no load period (NLP) [14] in this system that refers to "no power production." It is one of the significant problems related to this system. When no load period comes there will be load shedding for some time. While dealing with generating massive amount of power, two way power generation system will face more NLP. Though it has the ability to generate substantial amount of electricity, due to the presence of NLP it is not preferred normally [14].

3.1. Energy Calculation

To estimate the average power output from a tidal barrage, an energy balance model can be applied where the average change of potential energy will be taken into account during the draining process. The mathematical model of tidal power for a day can be calculated as:

$$P_{actual}\ (\text{one-way}) = 0.029\ A\ R^2 \quad (1)$$

$$P_{actual}\ (\text{two-way}) = 0.0568A\ R^2 \quad (2)$$

where, P is the actual power, R is tidal range of tide at a particular place and A is the surface area of the barrage basin

Both equations one and two are efficient and straightforward which results in significant outputs. These equations indicate that actual power generation can be calculated easily if the surface area of the basin and the tidal difference is known. Since the prediction for tidal ranges is reasonably correct it will make these equations more reliable in predicting power generations. These equations can be used to estimate power generation capacity for Sadarghat provided that tidal data and basin surface area are known.

4. Study Area and Data Source

Several coastal areas in Bangladesh have been identified as low head and medium head tidal movements. Stations such as Khulna, Barisal, Bagerhat and Satkhira regions falls under low head tidal movement (2- 5 m) and stations such as Sandwip Mongla Sadarghat Cox's Bazar falls under medium head tidal movement (more than 5 m head). Sadarghat falls under category of medium head tidal movements. The mean spring tidal range at Sadarghat, Gauge No 31 is about 4.72 m [18]. Tidal variation data were collected from Bangladesh Inland Water Transportation Authority (BIWTA). Using the tidal software WXTide32 monthly tidal range data were collected. WXTide32 is a windows tide and current prediction program [19]. It predicts tides from 1970 through 2037. It gives result for more than 9,500 stations globally. It shows result for daily, monthly and incremental tide with tidal graph. WXTide32 software is based on UNIX program XTide version 1.6.2 designed by Dave Flater.

5. Result and Discussion

5.1. Selection of Site

Guohai and Xiaofen [20] list some key guidelines to evaluate whether the area would be appropriate or not for installing a tidal power station. Among them are

a. Preferable tidal range is 3 ~ 5m
b. Preferable power generation technique is one way operation
c. Examine climatic situation of the nearby weather station and hydrologic station, basin traits, rainfall, runoff, flood and sediment concentration of sea bed
d. Research on implementation of mitigation measures, such as fixing time periods when construction should not take place
e. The issue of marine development and exploitation should be considered equally while developing tidal power station
f. Conducting experiments to observe the negative impacts on upstream and downstream marine function utilization

Also a good quality tidal site should have characteristics such as vast amounts of rapidly moving water, bathymetry, seabed characteristics for device setup, no collisions with other portions of sea space and near to a load and grid interconnection [21]. Chittagong, the second largest city in Bangladesh, is located on the edge of the Karnaphuli River. It has a renowned riverside area, Sadarghat, which is famous for its river trade. It contributes greatly to the economic development of Chittagong. Sadarghat is located at the Karnaphuli River estuary of the Bay of Bengal. Karnaphuli is the longest and the most important river in Chittagong and Chittagong hill tracts. It is approximately 667 meters wide and is located in the South-Eastern part of Bangladesh. It originates from the Lusai hills in Mizoram, India and flows 270 km South-West through Chittagong Hill Tracts and Chittagong and finally, falls into the Bay of Bengal. Sadarghat begins from the north bank of the Karnaphuli River port and expands towards northward up to Chawkbazaar. The Latitude of Sadar ghat is $22^{0}20'0''$ North, $91^{0}50'0''$ East. At Sadarghat, during monsoons, salinity differs from 0.1 gms/ 1000c.c at low water to 0.2gms/ 1000c.c at high water; during dry season it varies from 2.5gms/1000c.c at low water to 16.5gms/1000c.c at high water [22]. The tides are semi diurnal with a prominent diurnal effect.

The approximate tidal range at Sadarghat is 1.2 m to 4.2 m, which makes it suitable to set up a tidal power station.

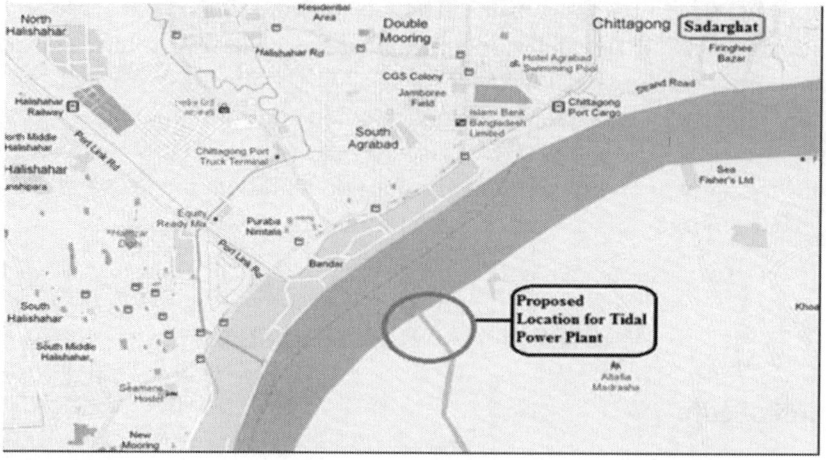

Figure 1. Projected Study Area: Sadarghat (Google Map).

However, Sadarghat is a very busy and crowded area with dense settlings; it will not be feasible to construct a tidal station in that area. A tidal station can be built in the west part of the Karnaphuli River, considering low risk of impact. Some suitable places which are a few kilometers away from Sadarghat possess similar tidal range. A barrage can be constructed around 2.5 to 3 km away from Sadarghat where a narrow channel of Karnaphuli River flows towards the west portion of Chittagong (Figure 1). Further, the absence of resettlement issues and availability of road and sea communication makes the place appropriate to install tidal power plants. Since Sadarghat falls under the central areas of Chittagong that connects to the other important places, it is recommended to install a tidal power station within a place near Sadarghat.

5.2. Suitability of WXTIDE32 Tidal Data

To examine the consistency of tidal software, data received from WXTide32 is compared with the actual tidal data collected from the Bangladesh Inland Water Transportation Authority (BIWTA). Figure 2 shows the monthly average tidal range data for the duration of 5 years (2006-2010) at Sadarghat. Correlation between these two data sets is performed using SPSS software and the results are shown in Table 1. The simulated data from the software is 90.4% resemblance with the data collected from BIWTA. The correlation between these two sources is R = .904 which means over 90% data of observed value can be explained by the data from WXTide32. Thus WXTide32 can be used to reliably predict tidal range.

5.3. Average Power Production

The tidal data used for the power calculation are gathered from the software, WXTide32. In case of basin area, around 6 square kilometers are taken into account. Table 2 shows an analytical analysis of annual power generation for the year from 2013 to 2015 at Sadarghat power station. The results are obtained for both one-way and two-way power generation system. On average 36.52 MW of electricity will be generated from a single turbine and single basin annually.

The amount of power generation can be increased if more turbines are incorporated with it. Tidal range and basin area influence the power generation rate. Figure 3 demonstrates the fact that power generation increases with the increase of basin area.

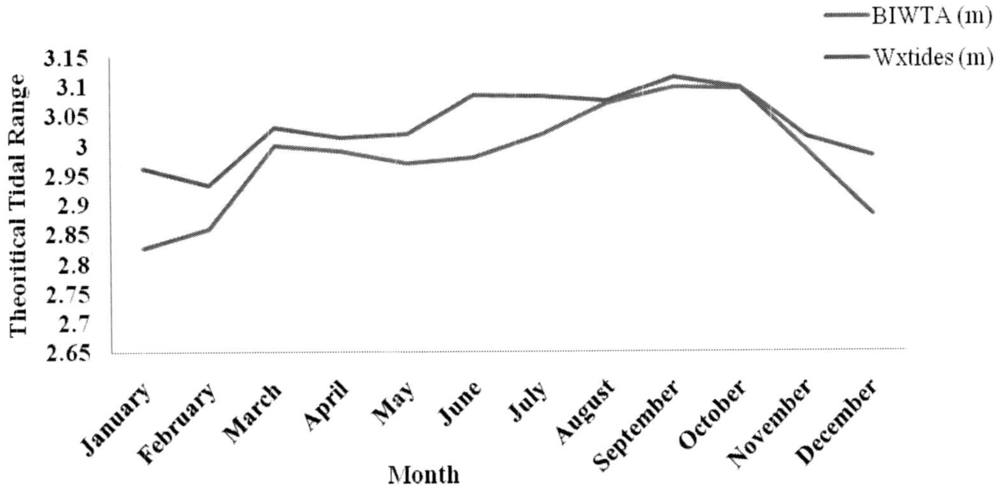

Figure 2. Correlation Between data from BIWTA and WXTide32 Comparison between tidal range data, collected from BIWTA and WXTide32.

Table. 1. Correlation results obtained from SPSS software

		BIWTA			BIWTA
BIWTA	Pearson Correlation	1	BIWTA	Pearson Correlation	1
	Sig. (1-tailed)			Sig. (2-tailed)	
	N	12		N	12
WXTide32	Pearson Correlation	0.904	WXTide32	Pearson Correlation	0.904
	Sig. (1-tailed)	0.000		Sig. (2-tailed)	0.000
	N	12		N	12
**. Correlation is significant at the 0.01 level (1-tailed).			**. Correlation is significant at the 0.01 level (2-tailed).		

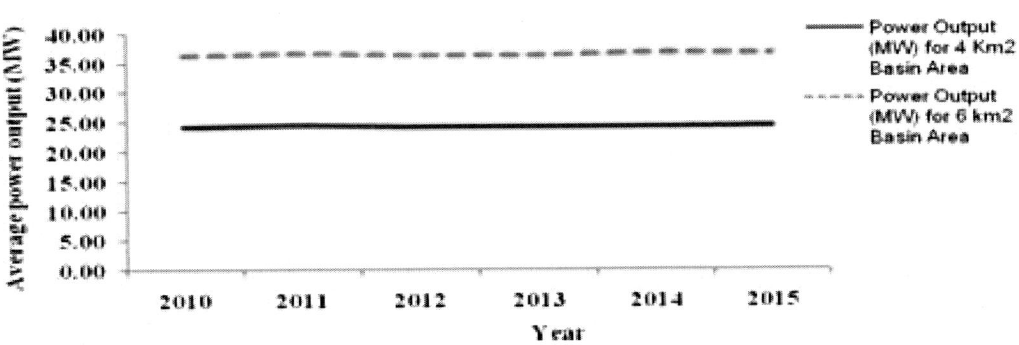

Figure 3. Average Power Output for 6 km^2 basin area and 4 km^2 basin area.

Table 2. Annual power output

Year	Tidal Range	Power Output (One way) (MW)	Power Output (two way) (MW)
2013	4.16	36.26	72.53
2014	4.19	36.86	73.73
2015	4.18	36.62	73.23

Figure 4 shows predicted monthly electricity generation in the year 2013 for both one way and two way system. With a basin area of 6 km2 nearly 36.26 MW of electricity will be generated from one way system and 72.53 MW from two way system in 2013. The reduced tidal head (3.9 m) in March resulted in less power generation. From April, power production increases and remains constant till August (tidal head 4.3 m). In September, there is a sudden reduction in power generation because of the smaller tidal range (4.1 m). In summary, from April to August power production seems to be high in 2013. Power output is larger for the two-way system compared to the one-way. But one of the negative aspects of two way system is that the rate of load shedding will be frequent. Thus, it is recommended to utilize the one way power generation system.

Figure 5 shows monthly power output for one- way system for the year 2014. From the month January to March it shows a stable power output (average 3.08 MW), and in April there is a sudden drop of power generation due to less tidal variation. However, from July to October there is a significant amount of power generation.

The highest power output is 3.54 MW, in August 2014. Similar kind of figure is also produced for the year 2015 (not shown here) where on average 3.05 MW of electricity will be produced monthly.

In summary, power output varies from time to time since it depends on tidal variation. In general, analyzing the above graphs it can be stated that predicted power generation seems to be higher in August. In case of other months the predicted power generation shows significant fluctuations.

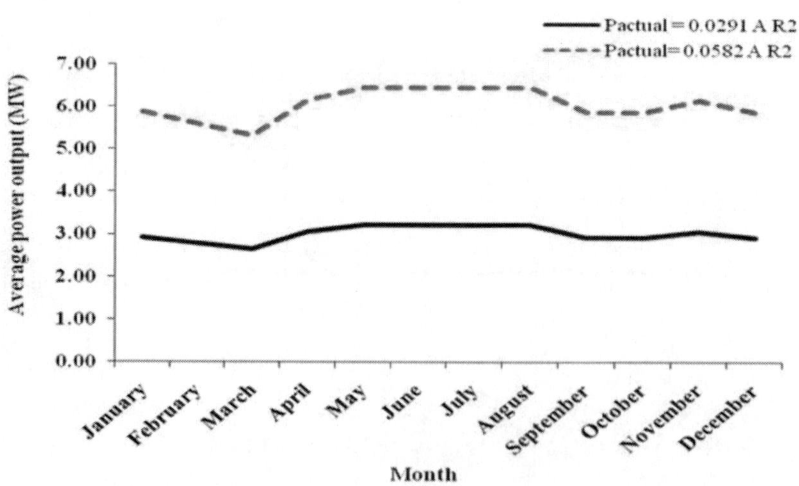

Figure. 4. Average Power Output for both one-way and two-way power generation system for the year 2013.

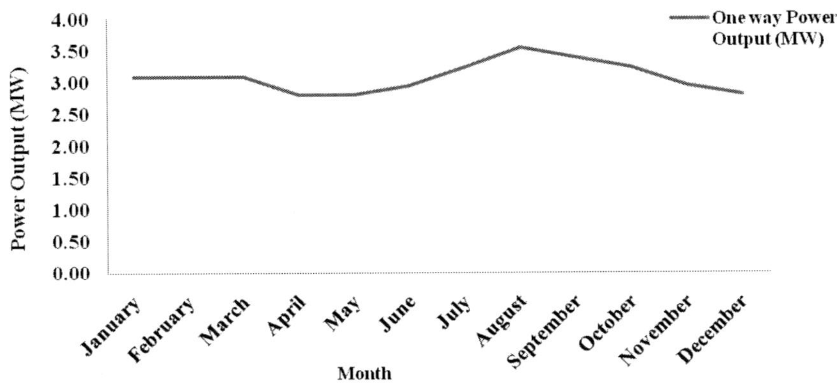

Figure. 5. Monthly Power Output for One way power generation method in the year 2014.

CONCLUSION

Sadarghat station in Chittagong, Bangladesh falls under the medium head tidal movement category. A significant amount of power generation can be expected from this project if a larger basin area is taken into account. Tidal data obtained from WXTide32 can be successfully used when observed data are not available. With a basin area of 6 km2 nearly 36.26 MW of electricity will be produced from one way system and 72.53 MW from two way system in the year 2013. Considering Sadarghat station, the one way generation with ebb generation method is preferred to generate electricity because of its continuous rise and fall of tides. Though the two-way method generates more electricity, because of having No Load Period (NLP) there will be load shedding for some time. The limitation of this study is lack of data from the specific location that has been selected for tidal power generation. The study should help engineers and researchers to continue further work on tidal power in Chittagong as well as other coastal areas of Bangladesh.

ACKNOWLEDGMENTS

The authors would like to acknowledge the financial support made available by the Asian University for Women, Chittagong, Bangladesh.

REFERENCES

[1] Renewable Energy in Bangladesh. 2011. Renewable Energy. Power Division. Ministry of Power, Energy and Mineral Resources. Government of the Peoples Republic of the Bangladesh. Available at http://www.powerdivision.gov.bd/user/brec/49/90 (accessed January 2, 2012).

[2] Energy-The Facts. 2013. The World Bank. Available at http://web.worldbank.org /WBSITE/EXTERNAL/TOPICS/EXTENERGY2/0, contentMDK:22855502~pagePK: 210058~piPK:210062~theSitePK:4114200,00.html (accessed March 7, 2012).

[3] Salequzzaman, M. and Newman, P. (2002) Integration Prospects of Tidal Energy as a Contribution to the Sustainable Development of Coastal Bangladesh. International Review for Environmental Strategies, vol. 3, no. 1.
[4] Chy, N.R., Reza, S.E. et al. 2012. Present scenario of Renewable Energy in Bangladesh and a proposed hybrid system to minimize power crisis in Remote areas. International Journal of Renewable Energy Research. Vol 2. No.2.
[5] Anam, K. & Bustam, H.A. 2011. Power Crisis & Its Solution through Renewable energy in Bangladesh. Journal of Selected Areas in Renewable and Sustainable Energy (JRSE). Available at http://www.cyberjournals.com/Papers/Sep2011/02.pdf (accessed January 20, 2013).
[6] Araquistain, T.M. 2012. Tidal Power: Economical and Technological assessment. Master Thesis. Avaialble at http://upcommons.upc.edu/pfc/bitstream/2099.1/14731/1/Master%20Thesis_%20TIDAL%20POWER_%20Economic%20and%20Technological%20assessment_%20Tatiana%20Montllonch.pdf (accessed January 6, 2013).
[7] Kerr, D., 2007. Marine Energy. Philosophical Transactions: *Mathematical, Physical and Engineering Sciences,* 365, 971-992.
[8] Tidal. 2012. Energy BC. http://www.energybc.ca/profiles/tidal.html (accessed January 2, 2013).
[9] Grafenberger, P., Wurm, E., Lang, M. et al. 2007. Advanced technologies of large bulb units for tidal plants. Available at http://www.drukgreen.bt/library/documents /CPSU /11.02.%20Grafenberger%20P..pdf (accessed February 2013).
[10] Hammons, T.J. 1993. Tidal Power. Available at http://www.inference.phy.cam.ac.uk/ sustainable /refs/WaveTidal.pdf. IEEE. VOL 81.NO.3. (accessed March 19, 2012).
[11] Shelley, L.M. 2012. Bay of Fundy. Tidal Energy: An opportunity for the fundy region and a call for socioeconomic research. http://fern.acadiau.ca/custom/fern/document_archive/repository/documents/143.pdf (accessed January 2, 2013).
[12] Black, Richard. 2011. India plans Asian tidal power first. http://www.bbc.co.uk/news /science-environment-12215065 (accessed February 12, 2013).
[13] Salequzzaman M., Newman. P. Ellery M. and Corry B. 2000. Prospects of Electricity in Coastal Region of Bangladesh: Tidal Power as a Case Study. *Journal of Bangladesh Studies* 2(1): 53-61.
[14] Ali, T. Gupta, S.D., Arnab, I.Z. & Ferdous, S.M. 2012. An Analysis of the Potentiality of Tidal Power in Swandwip by Using One way & Two way power generation Technology: An Ideal Model. Vol.2, No.3. Available at http://www.academia.edu /2349000/An_Analysis_of_the_Potentiality_of_Tidal_Power_in_Swandwip_by_Using _One_Way_and_Two_Way_Power_Generation_Technology_An_Ideal_Model (accessed January 6, 2013).
[15] Salequzzaman, M. 2003. Can Tidal Power Promote Sustainable Integrated Coastal Development in Bangladesh? PhD thesis, Institute for Sustainability and Technology Policy (ISTP), Murdoch University, Western Australia.
[16] Whittington, H.W. 2012. Electricity Generation: Options for Reduction in Carbon Emissions, Philosophical Transactions: Mathematical, Physical and Engineering Sciences, Vol. 360, No. 1797, Carbon, Biodiversity, Conservation and Income: An Analysis of a Free-MarketApproach to Land-Use Change and Forestry in Developing

and Developed Countries (Aug. 15,2002), pp. 1653-1668, The Royal Society, http://www.jstor.org/stable/3066583. Accessed: Jan 10, 2013.

[17] Survey of Energy Resources 2007. n.d. World energy Council. Available at http://www.worldenergy.org/publications/survey_of_energy_resources_2007/tidal_energy/755.asp (accessed February 5, 2013).

[18] Salequzzaman, M. 2001. Sustainability of Shrimp Aquaculture in Coastal Bangladesh. Proceedings of the Fifth International Conference on The Mediterranean Coastal Environment, October 23-27, at Tunisia, 2:863-879, Available at http://www.metu.edu.tr/home/1qwwwmdcst/med01/AcceptedPapers.htm (accessed Jan 20, 2013).

[19] WXTide32. Windows Tide and Current Prediction Program. Chittagong. 91°50.00'E, 22°20.00'N, Region: Indian Ocean, Country: Bangladesh. Available at http://www.wxtide32.com/ (accessed Jan 10, 2013).

[20] Guohai, C., Xiaofen, C. et al. 2009. Planning and Site Selection Report for Preselected Tidal Power Stations in Zhejiang Province. Available at http://www.cresp.org.cn/uploadfiles/114/1670/a2-b12-cs-2008-001-en.pdf(accessed January 26, 2013).

[21] Bedard, R. 2006. North America Tidal IN-Stream Energy Conversion Technology Feasibility Study. Electric Power Research Institute (EPRI). EPRI TP-008-NA. http://oceanenergy.epri.com/attachments/streamenergy/reports/008_Summary_Tidal_Report_06-10-06.pdf (accessed Feb 8, 2013).

[22] Hydrographical Information. 2013. Chittagong Port Authority. Available at http://cpa.gov.bd/portal/home.php?option=article&page=52 (accessed March 7, 2012).

In: Renewable Energy and Sustainable Development
Editors: R. T. Durai Prabhakaran, S. A. Kale et al.
ISBN: 978-1-63463-429-8

Chapter 10

ESTIMATION OF TIDAL POWER POTENTIAL OF NIGERIA COASTAL AREA

Collins N. Nwaokocha[1*], Mufutau A. Waheed[2] and Abayomi T. Layeni[3]
[1,3]Mechanical Engineering Department,
Olabisi Onabanjo University, Ago-Iwoye, Nigeria
[2]Mechanical Engineering Department,
Federal University of Agriculture, Abeokuta, Nigeria

ABSTRACT

It's no hidden fact that the world energy consumption continues to rise and the effects of global warming is getting noticed, this has made countries to place emphasis on renewable energy sources. Nigeria is one of the developing countries and its dependence on fossil fuels can only be offset by the sustained exploitation of its indigenous resources. Tidal energy is renewable, environmentally friendly and can be harnessed for various end-uses. A precise knowledge of ocean energy regime is a pre-requisite for the efficient planning and implementation of any tidal energy project. This chapter presents an investigation on the tidal energy potential of the coastal region of Nigeria. Four sites were purposely chosen along Nigeria's coastal belt, for evaluation viz - Akassa, Bayelsa State (Lat. 4.3167° N, Long. 6.0667° E); Bonny, Rivers State (Lat. 4.4500° N, Long. 7.1667° E); Calabar, Cross-River State (Lat. 4.9667°N, Long. 8.3167° E) and Lagos Bar Beach, Lagos State (Lat. 6.400°N, Long. 3.400°E). For each of the location, tidal height data were obtained from the Nigerian Navy Office of Hydrography, and the Nigerian Institute for Oceanography and Marine Research, Lagos. The tidal height data obtained for a period of ten years (2001 - 2011) was evaluated to determine the tidal energy potential. The tides exhibit a similar trend of flow over the ten years period: the observed frequency rise and fall steadily, reaching a maximum and minimum value of tide height for each location at regular intervals. The result shows that Calabar has the highest tidal height of 3.07 m, Bonny has 2.40 m, Akassa has 1.80 m and Lagos has 1.3 m. Thus, the establishment of tidal power plant for electricity generation should be integrated into the energy masterplanof the Federal government.

[*] E-mails:collinsnwaokocha@gmail.com.

1. Introduction

It's a plain fact that we are in a world where almost 80 percent of the demanding energy is furnished by sources such as natural gas, coal, or oil, which are quickly being depleted as well as being environmentally unfriendly. As the rate of global warming is constantly increasing and world energy consumption continues to rise, countries are placing more emphasis on energy sources that attenuates its effects. Luckily, we have already realized the importance of making an enormous change in our way of life and our way of using the energy, so looking for renewable resources to substitute current ones is much urgent for us. Nigeria is one of the developing countries and its dependence on fossil fuels can only be offset by the sustained exploitation of other energy resources. In addition to solar and wind power, the ocean holds significant promise as a sustainable energy provider. The abundance and consistency of ocean power offer a regular means of energy harvesting unmatched by wind or solar. An average power density of 2-3 kW/m^2 measured below the ocean surface and perpendicular to the wave front is the largest of all renewable energy sources and is sufficient to meet the world's current energy demands. Tidal power is classified as a renewable energy source, because tides are caused by the orbital mechanics of the solar system and are considered inexhaustible within a human time frame. Energy from tidal power is also a form of pollution free energy, which has a lot of potential [1-6].

Tidal energy is renewable and can be harnessed for various end-uses. This chapter presents a means of harnessing this abundant source of energy available in the Nigerian coastal belt. The adoption of this form of energy resource would help in the conservation of our fossil fuels, facilitate nil-carbon emission and also promote the efficient use of energy.

Ocean waves are one of the world's most abundant sources of renewable energy, generated as a result of wind blowing over the ocean surface [1, 3, 4, 7-9]. This chapter investigated tidal power generation by harnessing the kinetic energy in tidal flows to generate electricity with the intention of reducing the amount of electricity generated from nonrenewable fossil fuels.

2. Background

As we all know that, electricity and heat energy are essential in our life. However, most of them come from traditional sources of energy such as coal, oil, gas and nuclear power. The extraction and use of these fuels are causing many environmental problems of the world, such as climate change, the ozone layer destruction and so forth. The combustion of fossil fuels created an excess of carbon dioxide, Scientists across the world agrees that we must reduce our use of fossil fuels in the years to come, or we will create devastating effects in nature. An acceptable power generation technology must be mechanically sound, environmentally acceptable, and economically profitable in order to become a real alternative for builders of new capacity. An energy policy for a sustainable future will be based on a high level of energy efficiency and greater use of renewable energy, preferably in an energy mix. I choose tidal power as my target because it is such a renewable energy source by which we can solve the serious problems: it produces no waste and pollution, what is more important; it is totally free to use. It is absolutely possible that tidal power will be one of the most popular sources of power in the future and become an attractive option for power companies looking for renewable source power, but unwilling to accept the risks involved in experimenting with unproven technologies. China is the country that has built the most tidal power stations in the world, and tidal power energy is no doubt playing a more and more important role in China's energy consumption process [1-4, 6, 8, 10].

3. TIDAL ENERGY ASSESSMENT

This chapter presents the Nigeria's coastal belt as the primary study area. For satisfactory assessment of the tidal energy potential of Nigeria, details of tidal height for ten years were obtained. It is essential to assess tidal energy potential of a site before any tidal energy based system is set up.

Energy is extracted from the movement of the tides; from the differences in tidal height and from the fast-moving tidal streams and currents (tidal speed). Since, movement of the tides is our primary concern, the tidal height data will be used for the assessment, because it is readily available.

The following evaluations were derived from the tidal height data:

Daily mean: This is the average of the tidal height measured per day. Denoted as $\overline{h}$.

$$\overline{h} = \frac{1}{N}\sum_{i=1}^{N} h_i$$

Where,

$\overline{h}$ (called the h-bar) is the mean value, and
N is the total number of tidal heights in a day.

Monthly mean: This is the mean of the daily means in a month. Denoted as $\overline{d}$.

$$\overline{d} = \frac{1}{N}\sum_{i=1}^{N} d_i$$

Where,

$\overline{d}$ (called the d-bar) is the mean value, and
N is the total number of days in a month.

Yearly mean: This is the mean of the monthly means in a year. Denoted as $\overline{m}$.

$$\overline{m} = \frac{1}{N}\sum_{i=1}^{N} m_i$$

Where,

$\overline{m}$ (called the m-bar) is the mean value, and
N is the total number of months in a year.

Tidal power represents the potentially available tidal energy flux per unit area of flow. It is related to the tidal pressure head (water level above mean sea level) and flow. Tidal power, P, per square meter of flow area is given by:

$$P = \rho\, g\, hQ$$

Where

ρ = density of water
g = acceleration due to gravity
h = tidal height, and

This formulation is applicable where tidal power is to be derived from large ranges in tidal elevation. Typical peak tidal variation in Bass Strait is approximately 2.5m [7,12-14].

A second formulation for tidal power relates to current speed where the extraction of power is via the use of tidal turbines[3,13,14]. In this case, tidal power is given by:

$$P = \frac{1}{2}\rho v^2 Q$$

Where

v = tidal current velocity

Historically, the energy flux method has been the primary method for estimating their source potential of a tidal stream [4,13,14]. This method assumes that the extractable power is directly proportional to the kinetic energy flux through the most constricted cross-section of the channel in the undisturbed state. The head difference across the channel is assumed to have a negligible impact.

The instantaneous power density of a flowing current is given by the following equation:

$$\left(\frac{\mathrm{P}}{\mathrm{A}}\right)_{\mathrm{Water}} = \frac{1}{2}\rho \mathrm{U}^3 \text{ (Watts per square meters)}$$

Where A is the unit cross-sectional area (i.e., 1 m^2) of flow intercepted by the device (in square meters), ρ is the density of water (in kilograms per cubic meters; 1,000 kg/m^3 for freshwater and 1,025 kg/m^3 for seawater) and U is the current speed (meters per second). For tidal currents, U varies with time in a predictable manner and also depends on depth beneath the water surface and position in the channel (Figure 1).

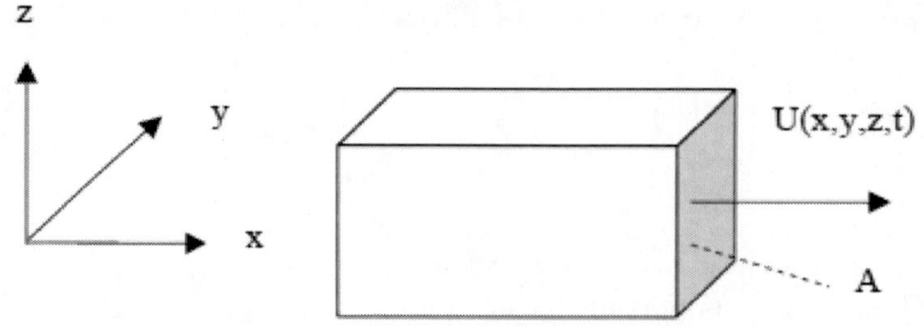

Figure 1. Tidal Velocity over Cross Section of Simplified Channel.

For an isolated turbine, the extractable power is assumed to be some fraction of the instantaneous kinetic energy flux upstream of the turbine, defined as

$$Energyflux = \frac{1}{2}\rho A U^3$$

Where ρ is the density of water, *A* is the cross-sectional area of the turbine, and *u* is the instantaneous depth-averaged upstream current velocity. The challenge here is that the tidal speed data for Nigeria's coastal site is not available for now, it makes evaluating the available energy a bit inconclusive. Tide height data will be used for the evaluation.

In deep water, the wave energy is expressed as [2]:

$$P = \frac{\rho g^2}{64\pi} H_{m0}^2 T \approx \left(0.5 \frac{\mathrm{kW}}{\mathrm{m}^3 \cdot \mathrm{s}}\right) H_{m0}^2\ T,$$

with *P* the wave energy flux per unit of wave-crest length, H_{m0} the significant wave height, *T* the wave period, *ρ* the water and *g* the acceleration of gravity. The above formula states that wave power is proportional to the wave period and to the square of the wave height. When the significant wave height is given in meters, and the wave period in seconds, the result is the wave power in kilowatts (kW) per meter of wave front length.

For example, consider moderate ocean swells, in deep water, a few kilometers off a coastline, with a tidal height of 3 meters and a wave period of 8 seconds (typical of Nigeria's coasts). Using the formula to solve for power, we get

$$P \approx 0.5 \frac{\mathrm{kW}}{\mathrm{m}^3 \cdot \mathrm{s}} (3 \cdot \mathrm{m})^2 (8 \cdot \mathrm{s}) \approx 36 \frac{\mathrm{kW}}{\mathrm{m}},$$

Meaning there are 36 kilowatts of power potential per meter of coastline.

Extractable Tidal Power

Since tidal current energy is intermittent and reaches its maximum output for only a fraction of the year, the average power is often quoted within resource assessments. The *average extractable power* is calculated by averaging the instantaneous power over the tidal cycle.

Based on the energy flux method, it is acknowledged that the extractable power is only loosely related to the kinetic energy flux [10,14]. Also defined is the instantaneous extractable power as [8,10]:

$$P = \frac{1}{2} N \eta \rho E_{turb} U^3$$

Where N is the number of turbines and η is the turbine efficiency. Assumptions for these scaling values vary widely. The number of installed turbines is limited by construction issues related to water depth and inter-unit wake considerations. The addition of more turbines will produce greater amounts of power, but will also impede the flow and hence have a negative feedback on the power generated.

Garrett and Cummins [8] quantify the maximum extractable power from a channel lining a bay to the open ocean as:

$$P_{avg} = 0.24\rho g A\omega a^2$$

Where g is the acceleration due to gravity, *A* is the surface area of the bay, and *a* andω are the amplitude and frequency of the dominant tidal constituent just outside the channel in the open ocean, respectively. This analysis neglects mechanical and electrical inefficiencies, drag on supporting structures, and assume that the drag associated with installed turbines is quadratic in the flow velocity.

Probability of Exceedence

The Probability of Exceedence is a statistical tool which analyzes the distribution of annual mean values for a particular tidal height data, and estimates the probability that the annual mean will exceed certain values. This estimate becomes more accurate with more years of data.

The Probability of Exceedence graph shows both the measured values and the normal distribution that best fits those values. The y-axis indicates the probability that the annual mean value will exceed a particular value.

4. Result and Discussion

The data collected was analyzed to get the daily, monthly and yearly mean tidal height for the locations. Figure 2 presents the monthly means tidal heights for the four (4) locations. Also the annual mean obtained for the ten years duration for the four (4) locations is compared in Figure 3.

The results showed that the tides exhibited similar trends over the ten-year period. The yearly tidal height for Calabar was 3.07 meters, Bonny 2.40meters, Akassa 1.80meters and Lagos 1.3 meters.

In Akassa, there was a large variation of monthly mean tidal height, with the highest tidal height of 1.88 meters occurring in March and the least (1.56 meters) in July. The same wide variation was also observed for the yearly mean tidal height, 1.61 meters at 2008 and highest at 1.85 meters in 2010. The highest mean frequency distribution was 1.71 meters at 18%. The probability of exceedence is 1.75 meters at 45%.

The highest monthly mean tidal height (2.48 meters) for Bonny was observed in February and the lowest value of 2.04 meters in July. It is obvious that for any particular year, the probability that the mean annual tidal height will exceed 1.75 meters is about 45%.

October has the highest monthly mean tidal height of 3.08 meters in Calabar, while June has the minimum value of 2.62 meters. The highest mean frequency distribution was 2.88 meters at 18%. It is obvious that for any particular year, the probability that the mean annual tidal height will exceed 2.60 meters is about 75% and that it will also exceed 2.80 meters is 25%.

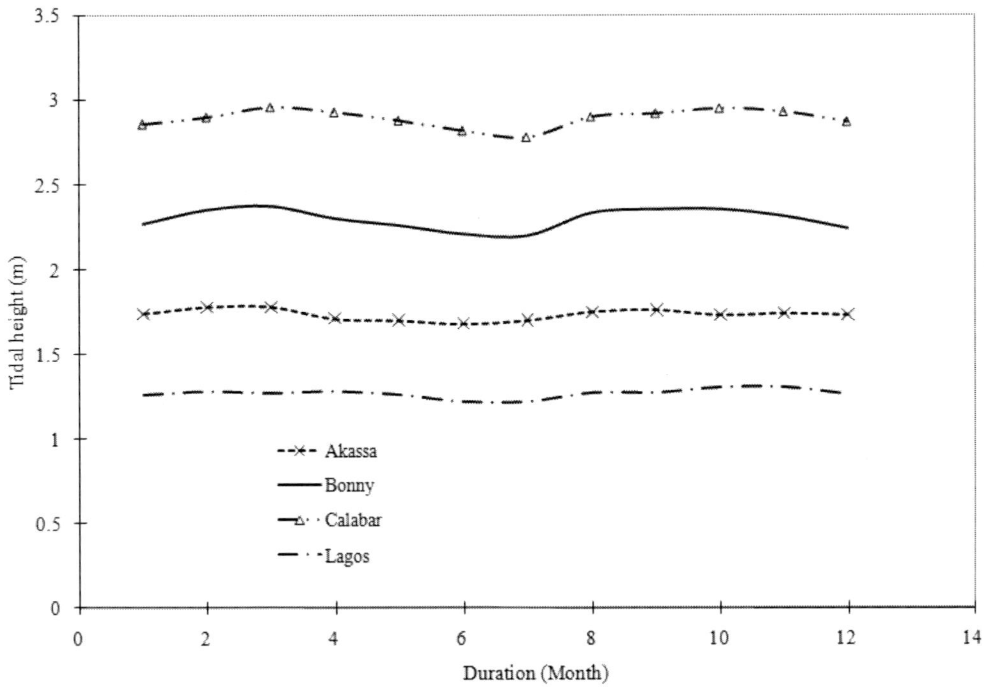

Figure 2. Monthly Mean Tidal Heights for the four locations.

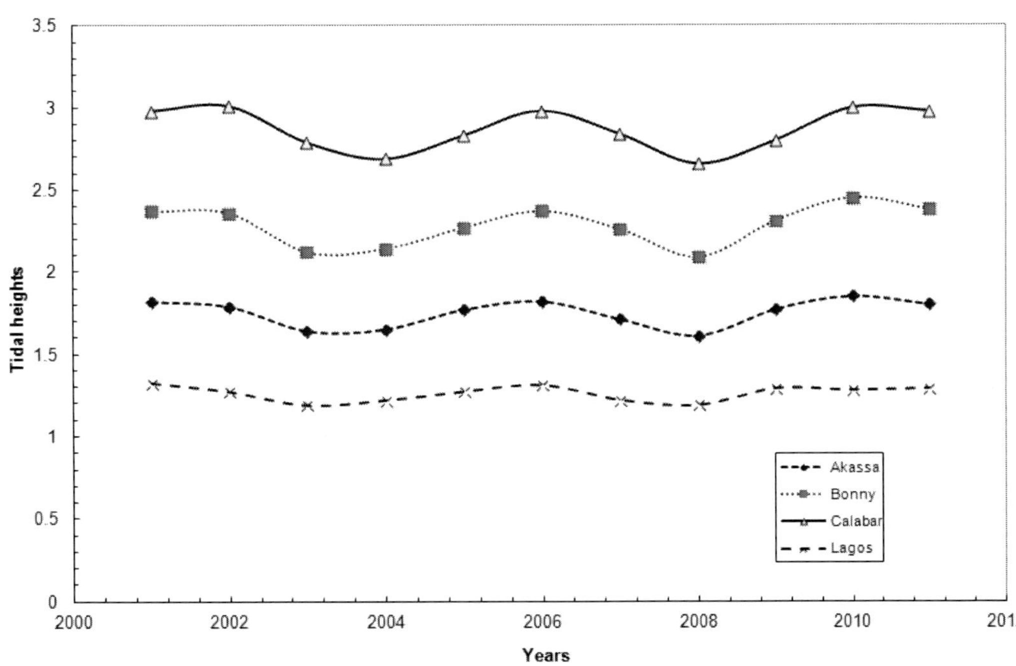

Figure 3. Yearly Mean Tidal Heights for the four locations.

In Lagos, there was a slim variation of monthly mean tidal height, with the highest tidal height of 1.37 meters occurring in November and the least (1.1 meters) in July. The same minute variation was also observed for the yearly mean tidal height, the lowest as 1.19 meters at 2003 and highest at 1.32 meters in 2001. The highest mean frequency distribution was 1.2 meters at 18%. The probability that the mean annual tidal height will exceed 1.25 meters is about 45%.

By way of comparison, the tidal technology in use in the Tacoma region (USA) with average tidal height of 2.5 meters, produces 6.75 MW electric power and has an estimated annual energy capacity of 13 GWh – thus Calabar coastal site with an average tidal height of 3.07 meters will do better. Since the percentage of time the tidal height exceeds 2.50 meters is 92% and that it will also exceed 2.60 meters is 75%.

Conclusion

This chapter showed that the tidal power potential of Nigeria varies from one location to the other. It was discovered that tidal energy availability in Calabar will be the highest with the yearly mean tidal height of 3.07 meters, Bonny 2.40 meters, Akassa 1.80 meters and Lagos 1.3 meters. The chapter therefore establishes that tidal power plant is best established in Calabar. Such electricity generation does not introduce additional heat into the ecosystem, hence keeping entropy and global warming at bay. It is recommended therefore that the establishment of tidal power plant for electricity generation be integrated into the new energy master plan being drafted by the Federal government.

Acknowledgment

The authors appreciate the assistance rendered to him by the Nigerian Navy Office of Hydrography, and the Nigerian Institute for Oceanography and Marine Research, both in Lagos, Nigeria.

References

[1] Young (1995), R.M., Requirements for a Tidal Power Demonstration Scheme, *Journal of Power and Energy*, Vol. 209, pp.215-220.

[2] Bahaj, A S and Myers, L E (2003), Fundamentals applicable to the utilization of marine current turbines for energy production. *Renewable Energy* 28, 2205-2211.

[3] Blunden L. S. and Bahaj A. S. (2007). Tidal energy resource assessment for tidal stream generators. *Proc. IMechE* Vol. 221 *Part A: J. Power and Energy:* 137 – 146.

[4] Chul-Hee Jo, H. Jeong, R.S. Park and W.C. Cho (2008), Application of Floating Tidal Current Power System in Cooling Water Channel, *Proceeding of the Eighteenth International Offshore and Polar Engineering Conference*, Vol. 1, pp.466-468.

[5] Twidell, J., and Weir, T., (2006). *Renewable Energy Resources.* Second Edition. Publisher, Taylor & Francis, 2 Park Square, Milton Park, Abingdon, Oxon OX14 4RN, Canada.

[6] Sambo, A. S. (2007). Demand and Supply Projections for Sustainable Energy Sector Development in Nigeria. Paper Presented at the Annual General Meeting/ Conference of the Nigerian Institution of Mechanical Engineers, 24th October 2007 at Kaduna, Kaduna State.

[7] Blanchfield, J. B. (2007). The Extractable Power from Tidal Streams, including a Case Study for HaidaGwaii. MSc Thesis, University of Victoria.

[8] Garrad Hassan and Partners (2001). *Scotland's Renewable Resource*, Scottish Executive: Edinburgh.

[9] Hagerman G, and Polagye B. (2006). Guidelines for Preliminary Estimation of Power Production by Tidal-In-Stream (Current) Energy Conversion Devices. EPRI Guideline, EPRI– TP – 001 NA Rev 3.

[10] Duckers, L. (1998). Wave Energy. In *Renewable Energy: Power for a Sustainable Future*; Boyle, G., Ed.; Oxford University Press: Oxford, 1998; 315-352.

[11] Zou Fan, (2012). *Tidal Power Energy,* University of Galve.

[12] *Sustainable Energy Authority (SEA),* United Kingdom. Wave and Tidal Power Assessment for the Victorian Coastline (2004).

[13] Herbich, John B. (2000). *Handbook of coastal engineering.* McGraw-Hill Professional. p. A.117, Eq. (12).

[14] Triton Consultants Ltd. (2006). Canada Ocean Energy Atlas (Phase 1) Potential Tidal Current Energy Resources Analysis Background. Prepared for the Canadian Hydraulics Centre.

In: Renewable Energy and Sustainable Development
Editors: R. T. Durai Prabhakaran, S. A. Kale et al.
ISBN: 978-1-63463-429-8

Chapter 11

GEOTHERMAL ENERGY IN INDIA: FROM EXPLORATION TO END USER

Parimal Arjun Patil*
Well Engineering Partners, Hoogeveen, The Netherlands

ABSTRACT

Geothermal energy is getting popular worldwide because it is a renewable source of energy. The importance of geothermal energy is realized long time ago but due to the increasing demand for energy, the need to develop this renewable source is increasing. Renewable electricity energy contributes 12.45% of total electricity production in India. Out of this 12.45%, the share of geothermal energy is almost zero. 8 countries viz. US, Philippines, Indonesia, Mexico, Italy, New Zealand, Iceland, and Japan have made significant progress in harnessing geothermal energy from the earth's crust and contribute 91% of global geothermal production but India is still finding its way difficult in defining policies for exploiting geothermal energy due to certain reasons. The chapter will highlight the worldwide development of geothermal energy, overview of geothermal energy in India, typical stages involved during geothermal project development and technology transfer of oil and gas to the geothermal industry. The chapter also identifies some of the technical barriers, financial barriers and the environmental impact which hinders the development of geothermal energy in India. In order to exploit geothermal energy faster in India, this chapter recommends few thoughts which will help meet India's power deficiencies in the future.

Keywords: Geothermal, electricity, drilling, technical barrier, financial barrier

* Corresponding author: Parimal Arjun Patil. Well Engineering Partners, Hoogeveen, The Netherlands. Tel.: +31 (528) 227710, fax: 31 (528) 227715, e-mail: parimal.patil@gmail.com, parimal.patil@we-p.nl.

INTRODUCTION

India is facing enormous shortage in electrical power and about 40% of the rural India still needs to be electrified. Current massive power shortages in India are forcing industries in parts of the country to operate only 4 days a week. The agricultural sector is also affected and the future prospects of electricity from coal or petroleum products or hydro sources do not appear any too bright. The cost of conventional fuels keeps rising and so does the cost of electricity. Targeting the rapid economic growth and "power for all", renewable energy is seen as the only future source to meet the India's escalating demand. The geothermal power plant can provide constant base load power for more than 95% of the time with very low carbon emission (91 g CO_2/kWh) using heat from hot rock resources or high temperature hydrothermal resources. It has been projected by the International Energy Agency that the geothermal electricity generation could reach 1,400 TWh per year by 2050 which will reduce CO_2 emission by almost 800 Megatons (Mt) per year. It has been estimated in the past that India has around 10^6 GW of geothermal potential and the total estimated hot–rock (granite) suitable for such enhanced geothermal system (EGS) at the depth of 2–3 Kms is greater than 150,000 km^2. It has also been estimated that these granites have the capacity to generate a quantity of energy equivalent to 9.18e24 Wh. It has been estimated that an investment of the order of Rs. 9,000 billion in 2002 – 03 price level would be required to finance generation, transmission, sub–transmission, distribution, and rural electrification projects. Considering the magnitude of the expansion of the sector required, a sizeable part of investments will also need to be brought from the private sector. The act creates a conducive environment for investments of the industry, both for public sector and the private sector, by removing barriers to entry in different segments.

Presenting the worldwide trend of geothermal energy potential and current activities in India, the chapter will elaborate the steps involved and the need of technology transfer from oil and gas industry to the geothermal industry in order to produce electricity from geothermal sources. The chapter will also highlight the parameters which have an influence on the geothermal electricity development. Economics plays a vital role in the development of geothermal projects. The Long debt period before the costs can be recovered and the higher cost for drilling wells are the main barriers. There is need to discuss the barriers and the factors which can hinder in the growth of geothermal projects. Discussing these barriers in the present study, part of the paper also highlights the aspects which are facilitating the geothermal energy development in India as it contributes less than 0.2% of the total primary energy supply.

GEOTHERMAL ENERGY WORLDWIDE

Prince Piero Ginori Contri in Larderello, Italy in 1904 built, tested and commercialized the world's first geothermal power plant. Later, in 1919, Japan drilled the first geothermal well in Beppu followed by wells at The Geysers, California, US in 1921. During late 1960s small facilities producing geothermal energy began operating in New Zealand, Mexico and in the US. In recent years, Australia, France, Germany, Switzerland, the United Kingdom, and the United States of America are developing geothermal projects in order to reduce their dependency on fossil fuels for the production of electricity. The global installed capacity of geothermal power plants is currently around 11,224 MW [1]. The total growth of geothermal energy till the end of 2010 has been shown in Figure 1 [2]. The reasonable amount of geothermal energy was produced. It can also be seen that till 2000 there was a steady growth but after 2000 the growth was reduced till 2004. There might be several reasons for this diminishing growth such as economic slowdown or crises, but it is clear that due to subvention after 2004, the growth suddenly stimulated which is highlighted in the Figure 1. If we expect the similar behavior after 2010, in order to predict the total geothermal energy growth worldwide, we will have the capacity more than 17 GW till 2020 which is the maximum growth we can expect. The optimistic growth will let us have the installed capacity of around 17 GW in 2050. The trend of development of geothermal energy in past decades tells us that the share of geothermal energy usage will increase steeply in future but we need to have several subvention periods in between in order to keep the pace and produce the abundantly available renewable energy.

GEOTHERMAL ENERGY IN INDIA

More than 75% of its primary energy needs are being imported in the form of fossil fuels. India has increased installed power from 1,362 MW to over 112,058 MW after independence and electrified more than 500,000 villages. This is a significant achievement but not impressive because still today India is facing a deficiency in power and many of the rural areas still need to be electrified.

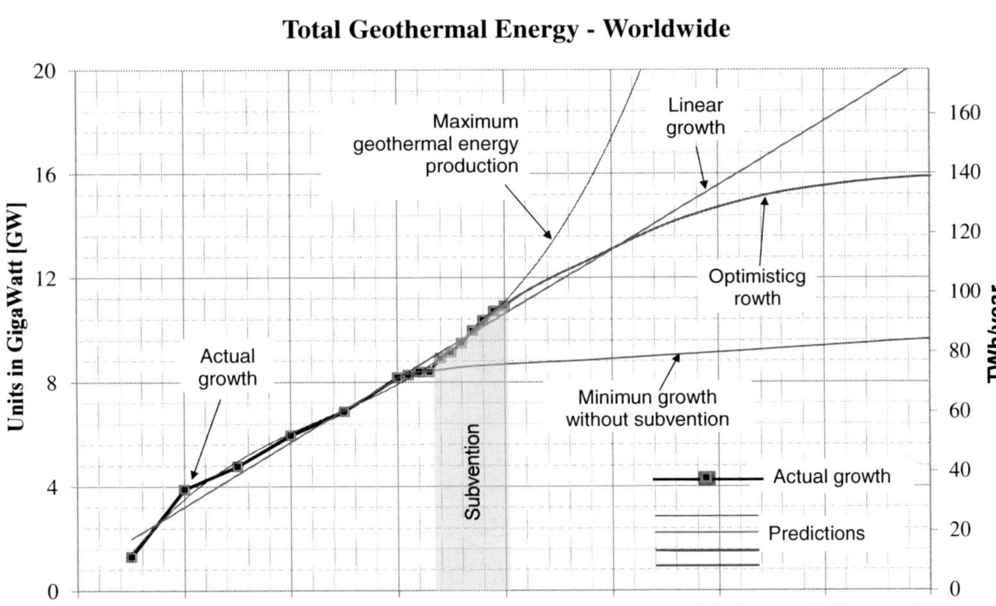

Figure 1. Total Geothermal Energy Production Worldwide and its prognosis till 2050 [2].

To meet the objective of rapid economic growth and "power for all" including household electrification, it is estimated that an investment of the order of Rs. 9,000 billion at 2002 – 03 price level would be required to finance generation, transmission, sub-transmission, distribution and rural electrification projects.

Power being most crucial infrastructure, public sector investments, both at the Central Government and State Governments, will have to be stepped up. Considering the magnitude of the expansion of the sector required, a sizeable part of the investment will also need to be brought in from the private sector.

Renewable grid capacity in India has increased more than five times, from around 2% to around 11% in only eight years, and is contributing about 4.13% to total electricity generation [3]. Government's ambitious target of 'Power to all by 2012' under the National Electricity Policy may remain a distant dream. The power sector is struggling to add about half of its initial planned capacity of 78,577 MW in the 11th five-year plan ending March 31, 2012, owing to shortage of fuel linkages (coal and gas) and environmental concerns. The power ministry is struggling to reach a capacity of even 50,000 MW even after the Planning Commission has reduced the target to 62,000 MW in the midterm review of the current 2007-2012 plan. The current capacity addition has so far been 34,462 MW till FY11.

Figure 2 (left) shows the electric generation using different energy sources in India in 2009. It can be clearly seen from the figure that the coal is the dominant resource, almost 42% of the total primary energy contribution, for the generation of electricity followed by oil and gas. Oil and gas contribute 23.6% and 7.20% of the total primary resources, respectively. Renewable and waste expects geothermal, solar and wind contributes almost a quarter of the portion of total resources but geothermal, solar and wind are contributing just 0.3%. The electric generation from 1971 to 2009 using different fuel sources is also shown in Figure 2 (right). Rapid economic growth has increased the demand over the years.

The current installed capacity in India according to the Central Electricity Authority on May 31, 2012 is 199.627 GW out of which 143.713 GW of the total installed capacity was online. The rest of the capacity of 33.895 GW was under outage due to factors such as maintenance, forced outage, and other reasons.

Stages Involved for Producing Electricity from Geothermal Energy

Geothermal energy which is stored in the earth's crust can be used for many different purposes depending on the temperature of producing water on the surface such as electricity production, as a heating source, food processing/drying, fish farming, bathing and other applications. This wide range of applications makes geothermal unique source of renewable energy compared to other renewables. One of the great advantages of this renewable source is that it provides a real base load capacity for electricity generation of more than 95% and in addition to that is a cleaner alternative to fossil energy.

Historically, electricity production from geothermal energy was only possible from high temperature resources but due to the binary heat exchange technology and various new technologies at various stages of the power generation could provide advantages of producing electricity even from moderate temperature resources.

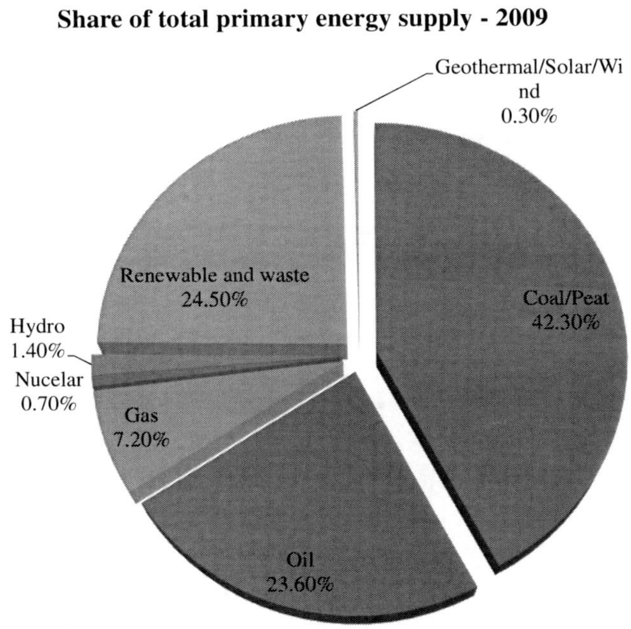

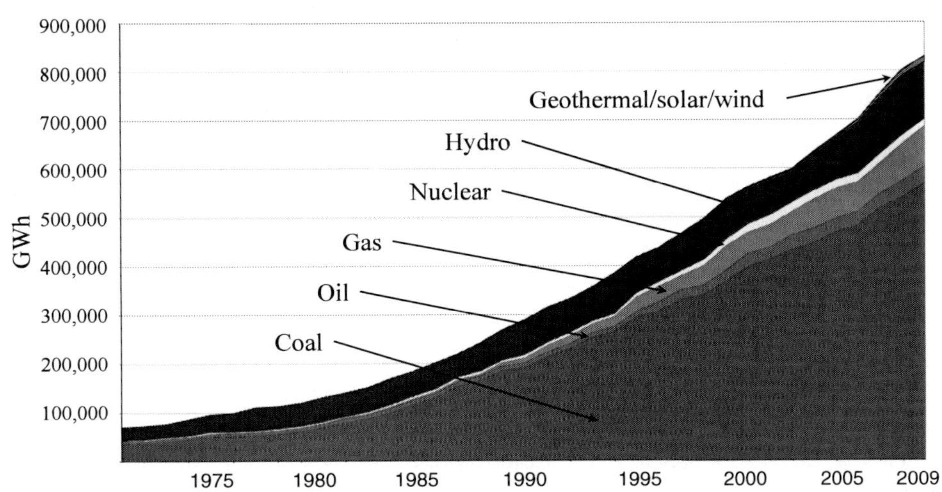

Figure 2. Share of primary energy in % age in India (left) and total electricity generation by fuel in India (right) using 2009 data [4].

Typically, a geothermal project is divided into five different stages; see Figure 3 [5]. The time to be spent in every stage cannot be estimated as there are many external factors affecting the progress of the project. For electricity generation projects using geothermal hot hard dry rock, external factor could be the available technology, financing, issues regarding permits and licensing, etc.

During start-up of the project, legal work, concessions, and permits need to be finalized for the project size which from the economical point of view is feasible at a particular location. During exploration and pre-feasibility stage, work such as geophysical surveys, collection and analysis of geochemical and geological data, and the temperature gradient need to be carried out.

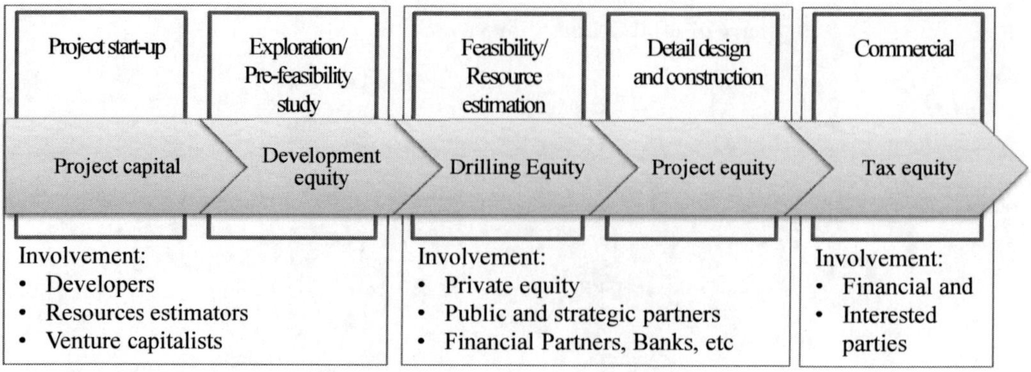

Figure 3. Stages involved for producing electricity from geothermal energy during a power project [after 5].

All this work will directly affect the project cost, project success and project lifetime. The capital invested during this time bears a high element of risk. Based on the pre-feasibility study, look for the prospect where the project could be set up. In this region the temperature gradient must be higher than normal, sufficient geological data available and proper data analysis has been carried out. In the next stage of feasibility and resource verification, drilling of the first successful well must be achieved which will govern the estimation of an underground heat in the prospect. The equipment can be redesigned for the actual scenario. During the detailed design and construction stage, drilling and testing of production and injection wells need to be carried out, the communication of the underground heat exchanger need to be tested. As drilling geothermal doublet wells cost 60% to 80% of the total geothermal project cost, it is very important to use high end technology and if required the novel technologies from oil and gas drilling industry to be applied. Recently, research has been carried out within drilling industry in order to minimize the cost for drilling geothermal wells to make the project economically viable. The funds from private equity, strategic partners, and bank loans need to be used properly in order to make the investment attractive and encouraged. Figure 4 shows the usual aspects which hold us back from such investments. Risk and the investments are inversely proportional to each other. As the investment goes into the project phase with high risk, the level of risk comes down as the assumptions are then the reality. Risks defined in the geothermal projects are as follows:

- Actual temperature and the predicted temperature of the underground formation
- Permeability of the formation
- Price fluctuation in oil and gas industry
- Water chemistry is also important and plays a crucial role
- Underground heat exchanger, i.e., communication between injector and producer well

OBSTACLES TO OVERCOME TO DEVELOP GEOTHERMAL ENERGY

The uneven distribution of geothermal energy worldwide makes it difficult and challenging to say whether it is economical over other energy sources. This is the reason experts have identified the high enthalpy and low enthalpy regions.

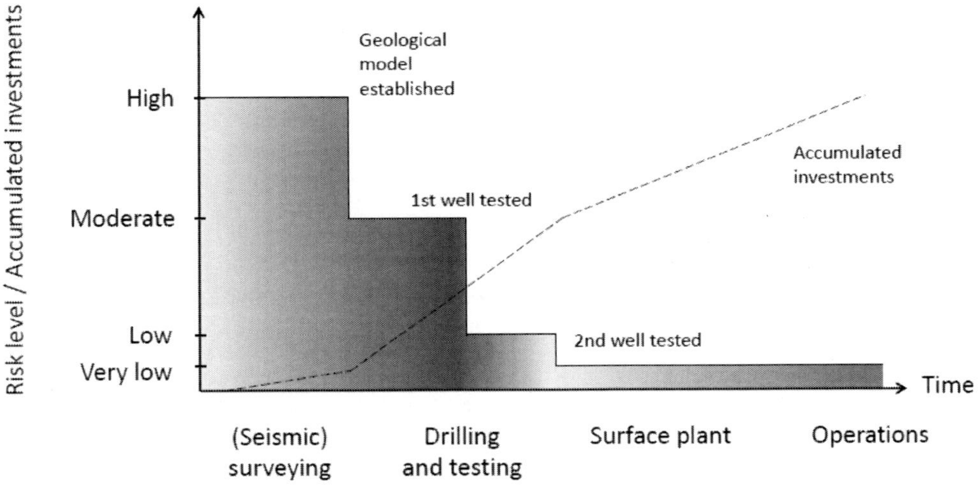

Figure 4. Usual suspect that hold us back [6].

Due to the several benefits of geothermal energy, heat can also be extracted from low enthalpy regions by drilling deep well. In such environments, drilling represents from 30% (dry and flash steam) to 70% (enhanced geothermal system, EGS) of the total cost of the geothermal electricity project. Dry steam plant use hydrothermal fluids that are primarily steam and in flash type plant the fluid at temperatures greater than 182 °C is pumped under high pressure into a tank at surface held at much lower pressure, causing some of the liquid to rapidly vaporize or flash. Geothermal energy when exploited from hot, dry, impermeable rock by creating the underground heat exchanger in this hot dry rock through hydraulic stimulation, it is then called enhanced geothermal system. Geothermal energy has so many advantages over other renewable and nonrenewable energy sources; however, there are several barriers which hinder its wider use. It is very important to identify these obstacles which need to be overcome in order to develop the geothermal energy and make it viable source of energy. Some of the obstacles are mentioned below:

- Technical barriers
  - Lack of information on geothermal energy resources available in Indian subcontinent
  - Inadequate research and development compared to other renewable energy
  - High chances of failure of drilling tools due to sever down hole conditions during exploratory drilling
  - Limited availability of geothermal drilling rigs in the area of interest
  - Lack of skilled professionals for geothermal drilling
  - High geothermal well cost

- Regulatory barriers
  - Lengthy permitting process for getting the licensed area for drilling geothermal wells due to unavailability of standard set of procedures in place
  - Lack of industrial standards to define geothermal resources and associated risks.
- Financial barriers
  - High CAPEX and long payback time
  - Competition to natural gas
  - Relative long lead time from concept to production due to very less information from the exploration wells in the region of interest
  - Uncertainty/Volatile tax incentives on geothermal energy compared to other renewable energies
- Environmental impact and public acceptance
  - Gaseous and solids emission during the drilling and completion activities as well as during life time of the project
  - Water and noise pollution due to drilling and completion activities
  - Induced seismicity due to hydraulic stimulation activities

Geothermal project, as described in the previous section, is usually divided into five different phases, i.e., resource study, exploration, pre-feasibility, feasibility, and design and construction. Before actual drilling starts, legal procedure such as permits needs to be acquired. During this phase, the lack of clear process to gain access to the land and rights to explore and tap the underground heat not only make the process complex but also lengthy and lasting for several years [7, 8].

High chances of failure are associated while drilling the first well in case of inadequate data on potential geothermal resources. It is reported that the success rate for geothermal wildcat wells is only 25 to 40%, so cost reduction in exploratory drilling costs would be a major incentive for promotion of exploitation of this resource. Figure 3 shows the actual and predicted well costs which is a function of well depth. Reduction in drilling costs can have a very positive impact on geothermal development as drilling of geothermal wells is a major factor in project costs, and can consume as much as 40% – 60% of total project cost. High drilling cost is due to the hard drilling rock at high temperatures. From Figure 5, it can be clearly seen that new drilling technologies need to be incorporated to lower the drilling well cost. In order to incorporate and implement the advanced drilling technology, the knowledge and expertise need to be transferred from oil and gas to the geothermal industry.

One possible action to address the above problems is to increase co–operation between government, academia and industry. Government must play a key role by providing the continued financial support with clear sets of legal procedures.

There should be a common platform to discuss critical issues throughout the development of any geothermal project. This will give an advantage to personnel to develop their expertise and accordingly modify rigs available.

One of the main financial hurdles is the high Capital Expenditure (CAPEX) and long payback period. Most of the Capital Expenditure is spent on the wellbore drilling and power plant construction. These expenses occur during the initial stage of the whole project; hence, they are also called up-front cost. The cash flow prediction is based on almost all the uncertain parameters which make the feasible project looks like a non–feasible one. The most important impacting factors with high uncertainty and risk include (a) resource grade, (b) temperature gradient, (c) geological parameters, (d) total well cost, (e) total investment cost, etc. The less important impacting factors yet still important include (a) power plant design and total cost, (b) stimulation cost, (c) well configurations (singlet, duplet, multilateral wells), (d) operating cost, (e) flow rate, etc. To overcome the problem of high CAPEX, many European countries, e.g., France, Spain, Germany, have tried to use the public subsidies and venture loans. On the other hand, the longer payback time, compared to other types of energy, e.g., natural gas makes it risky for a high investment in geothermal projects.

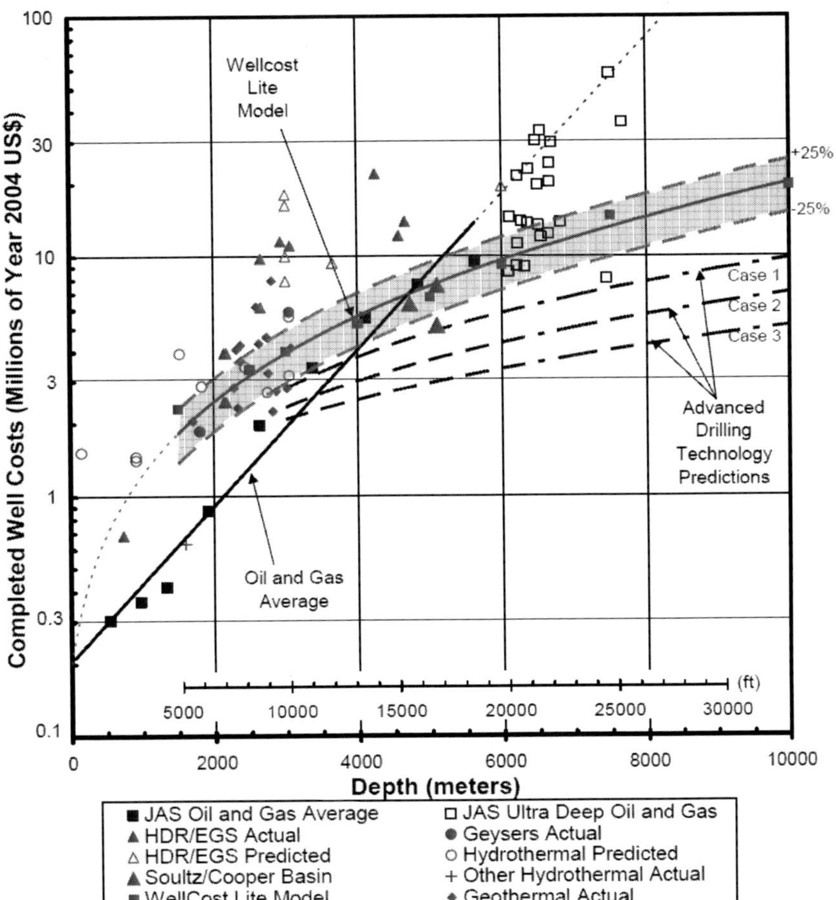

1. JAS = Joint Association Survey on Drilling Costs. 2. Well costs updated to US$ (yr. 2004) using index made from 3-year moving average for each depth interval listed in JAS (1976-2004) for onshore, completed US oil and gas wells. A 17% inflation rate was assumed for years pre-1976. 3. Ultra deep well data points for depth greater than 6 km are either individual wells or averages from a small number of wells listed in JAS (1994-2002). 4. "Geothermal Actual" data include some non-US wells.

Figure 5. Usual suspect that hold us back [9].

Many investors are, therefore, unwilling to invest. As such, the bank facilities will be used to cover the high investment costs. The weakness of bank facilities is that it is difficult to provide loans in the early stage of exploration without insurance mechanism.

Another typical financial barrier is the low outcome of the exploitation. For this, the feed-in tariff policy and tax reduction are commonly used. Tax reduction can on one hand promote increased capital investment, but on the other hand, will affect mainly operational phase when revenues are generated. Likewise, the feed-in tariff is good because it secures income over a long-term period, but it also has disadvantages because it comes into action at the later stage of the project. People still need to be aware of the fact that, the cost of producing electricity over time is lower because the price and availability of the fuel is stable and predictable, which is also advantageous. A more detailed description of the financial instruments can be found in the reference [10].

Environmental impact and the public acceptance need not to be ignored. The typical adverse gases discharged during the production of geothermal energy mainly include CO2, H2S, and methane etc. In many countries, CO2 and H2S are strictly prohibited. During the process of drilling, stimulation and production, the liquid has always high TDS (Total Dissolved Solid), especially for high temperature reservoir. Some of the solids are poisonous, e.g., arsenic and boron. They pose a potential damage to the ground waters and vegetation.

Besides, these operations can also generate high noise which affects the lives of local residence.

Policy Making for Geothermal Energy Development in India

The policy framework and the allotment of acreages over heat are still not clear from the Government of India. The industry is still waiting for the financial and technical support from the government [11]. Avin Energy Systems Ltd seems to be the only firm who has made efforts to exploit geothermal energy to generate power in the State of Gujarat in Western India. They claims to have made a study of the geothermal resources of the region identified most likely spots and prepared a detailed project report.

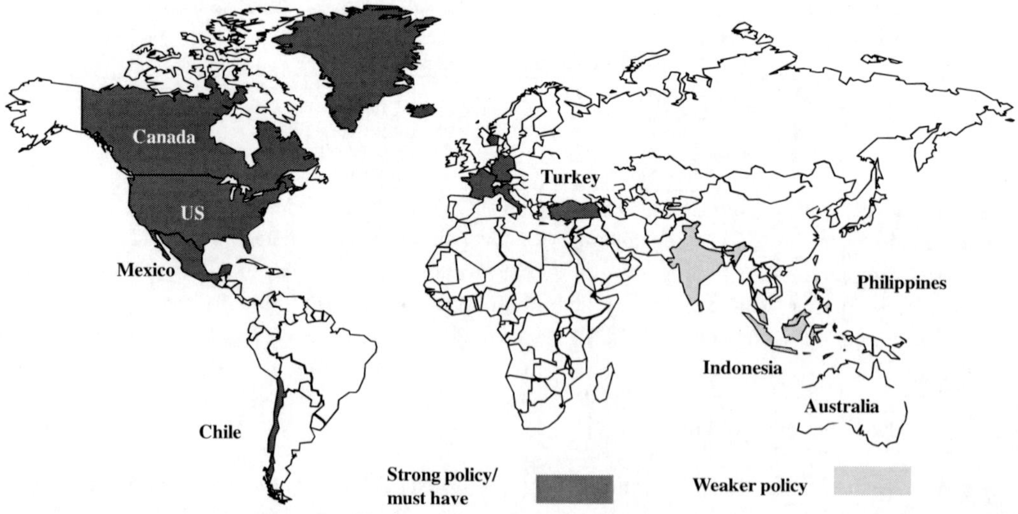

Figure 6. Government Policies for Geothermal Energy Development [12].

However, for some reason or the other Government (State and Central) do not seem excited or even interested in helping (directly or indirectly) to get the project moving. In fact only private investors have, so far, shown concrete interest in promoting this geothermal power project. If in the near or distant future, geothermal power generating projects do come up all over India, it will not be due to any encouragement by the Government.

The government policies for the development of geothermal energy has not changed a lot in the Asian and African countries but stronger policies have been implemented in the US and the European countries. Figure 6 shows regions having stronger and weaker policies and India has weaker government policies which need to be changed as soon as possible in order to boost geothermal energy development.

The latest news is that the Government of Gujarat has framed a new policy and passed a government resolution (No REP-102000-502-B) aimed at formulating an incentive policy for solar photo voltaic, geothermal, waste utilization, biomass, etc. Under this policy,

- Project promoters in any of these categories wishing to set up power generating projects to sell power to Gujarat Electricity Board shall have to enter into a power purchase agreement at Rs. 2.25 per unit (1994-95 as base with 5% escalation in rate every year for a period of 10 years from date of commencement of generation of power)
- Investor shall lay down power evacuating lines to the grid lines of GEB at his own cost
- Metering at site shall be done by Gujarat Energy Development Agency and GEB jointly and GEB shall make payment within 30 days of receipt of invoice. An "Electricity Credit Note" shall be issued by GEB to the power supplier which can be transferred to any HT consumer of GEB who can adjust his electricity bills against the "credit note".
- Wheeling charges of 4% of the generated units shall be deducted.
- Third party sale of power shall be permitted subject to payment of 4% wheeling charges.
- Banking for a period of 12 months shall be permitted by GEB
- GEDA will function as nodal agency for the purpose of implementation of this scheme.
- Investors will submit applications to GEDA in prescribed form and for grid interface to GEB along with a payment of Rs. 200,000 per MW (to cover incidental expenses)
- Investor shall prepare and submit a detailed project report within 6 months from date of "in principle" approval to be got approved from E and PCD department of Government of Gujarat.
- Other approvals/NOC shall be required for the project and have to be arranged/ obtained by the investor.
- Once these procedures are completed investor shall enter into PPA/wheeling agreement/third party sale, etc. in consultation with EPD in Government of Gujarat.
- If investor does not take "effective steps" to implement project within 12 months from date of government approval of the project, government shall terminate approval without any further notice. "Effective steps" is interpreted as incurring at

least 25% of project cost within 12 months from date of government order allocating the project.

FACTORS BOOSTING GEOTHERMAL ENERGY IN INDIA

One of the major driving factors for geothermal energy in India while evaluating the energy options is the land required for production of electricity. According to Islandsbanki Geothermal Energy Team, the land required for producing electricity is much less compared to land required for other energy sources [5], see Figure 7. It gives an idea of land required for different renewable to produce 1GW of electricity.

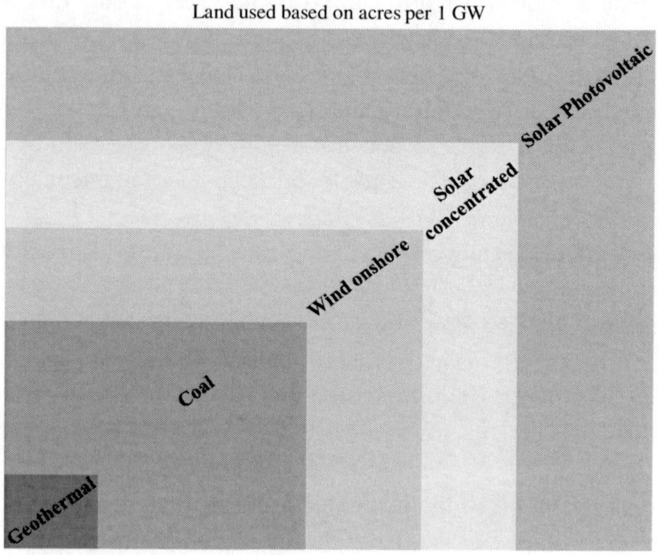

Figure 7. Comparison of land required for different renewable sources to produce electricity [5].

Table 1. National Policies for Geothermal Energy Development

Policy (year)	Description
Geothermal Energy R and D Program (2010)	Ministry of New and Renewable Energy seeking proposal for geothermal energy projects based on results of assessment studies (Geothermal energy projects, 2010)
India – Brazil – South Africa Declaration of clear energy (2007)	Agreement to work together in promotion of nuclear energy, clean energy, technologies, other renewable energies, and climate change mitigation
Integrated Energy Policy (2006)	Address of aspects of energy, including energy security, access and availability, affordability and pricing, efficiency and the environment
Tariff Policy (2006)	Included provision of renewable energy such as percentage for energy purchase made applicable for tariffs
National Electricity Policy (2005)	Released in accordance with the Electricity Act 2003 and promoted non-conventional energy resources
Electricity Act (2003)	Comprehensive frame work for power development with key provisions recognizing the role of renewable energy

Other few boosting factors to mention as

- Geological Survey of India identified more than 300 hot spring location and estimated around 10600 MW of geothermal energy.
- Glitnir Bank of Iceland announced exploration in India in February 2008.
- In January 2010, India and Iceland announced a cooperation on geothermal including technologies used in deep drilling, reservoir assessment, and geothermal demonstration power plants
- National policies (Geothermal-Energy-Association, 2010)

CONCLUSION

Geothermal energy is playing a more and more important role in the energy market. However, there are still some barriers which hinder the faster development of geothermal energy. This paper has discussed the typical barriers including the technical barriers, financial barriers and environmental impacts. To overcome the geothermal development barriers, the experience in Europe has been introduced in many other parts of the world, proposing many instruments. As to the technical barriers, an advanced technology transfer from traditional petroleum sector to the geothermal sector might offer much convenience for a faster development of geothermal technologies. These technologies include the areas of geology and geophysics, drilling and completion, reservoir modeling, formation stimulation.

A trend toward an advanced geothermal production technology includes: new drilling technologies making the usage of smaller and less complex equipment possible, new drilling processes to increase the rate of penetration and to improve productive drilling time, new technologies allowing for a smaller initial diameter of the borehole.

Office of Energy Efficiency and Renewable Energy of USDOE (United States Dept. of Energy) has documented lessons learned for high upfront risks and costs for geothermal environment where the exploration success is low. Following are the lessons learned [13]:

- Commercial size EGS reservoir creation is possible
- Enabling technologies are necessary for cost effectiveness of geothermal energy
- Without government support industry is reluctant to develop an EGS project
- EGS reservoirs are sustainable over time
- Over thrust stress environments are ideal for stimulation
- Pressure drops in the well need to be monitored
- Knowledge of local stress field and geology is essential
- Drilling into the seismic cloud yields the best connection between injector and producer

REFERENCES

[1] Jennejohn, D., Hines, B., Gawell, K., and Blodgett, L. *"Geothermal: International Market Overview Report",* May 2012. Retrieved from http://www.geo-energy.org: http://www.geo-energy.org/pdf/reports/2012-GEA_International_Overview.pdf.

[2] Bai, M., Patil, P., Teodoriu, C. *"Identification of Growth Barriers for Exploitation of Geothermal Energy in China"*, Book section of Clean Energy Systems in the Subsurface: Production, Storage and Conversion, Springer Series in Geomechanics and Geoengineering by Hou Michael Z., Xie Heping, and Were Patrick, pp. 105-116, 2013.

[3] Ministry of New and Renewable Energy I. "*Road map for Up-scaling and Mainstream Renewables. India: Ministry of New and Renewable Energy"*, 2010.

[4] International Energy Agency. *"IEA Energy Statistics", 2011.* Retrieved from http://www.iea.org/stats/pdf_graphs/INTPESPI.pdf and http://www.iea.org/stats/pdf_graphs/INELEC.pdf.

[5] Islandsbanki Geothermal Energy Team. *"United States Geothermal Energy Market Report", 2011.*

[6] Lorenzen, S. B. *"Danish Geothermal District Heating, a case study from Denmark", Dec. 2013.* GeoHeat and Cool Global Congress, Amsterdam.

[7] Ministry of Economic Development, New Zealand. *"Geothermal Energy: Summary of emerging technologies and barries to development"*, 2010, New Zealand.

[8] Falcone, G. and Teodoriu, C. *"Oil and gas expertise for geothermal exploitation: The need for technology transfer", SPE 113852, presented at the SPE Europec /EAGE Annual Conference and Exhibition*, Rome, Italy, 2008.

[9] Thorsteinsson, H., Augustine, C., Anderson, B. J., Moore, M. C., Tester, J. W. "The impacts of drilling and reservoir technology advances on EGS exploitation", January 2008. *Proceeding, 33rd Workshop on Geothermal Reservoir Engineering*, Stanford University, Stanford, California.

[10] Wendel, M. and Hiegl, M., "GEOFAR-Financing geothermal energy in European regions". *Proceedings World Geothermal Congress*, Bali, Indonesia.

[11] Jacob, S. *"Firms warm up to geothermal energy"*, Business Standard, Kolkata, India, March 2011.

[12] Lavigne, C. "Geothermal Energy: India Market Overview, Outlook and Potential for Geothermal Energy", *Geothermal Conference, 2009.*

[13] Stillman, G. *"Presentation to Geothermal Energy and High Performance Drilling Group",* US Dept. of Energy, Geothermal Technologies Program, Office of Energy Efficiency and Renewable Energy, Washington, May 2011.

In: Renewable Energy and Sustainable Development
Editors: R. T. Durai Prabhakaran, S. A. Kale et al.
ISBN: 978-1-63463-429-8

Chapter 12

EFFECT OF ELECTRODE SPACING AND ELECTRODE MATERIALS ON PERFORMANCE OF SEDIMENT MICROBIAL FUEL CELL

D. A. Jadhav[1,*] and M. M. Ghangrekar[2,†]
[1]School of Water Resources, [2]Department of Civil Engineering,
Indian Institute of Technology, Kharagpur, India

ABSTRACT

Organic matter present in the sediment deposited at the bottom of the water body provides a carbon source for microbes. This organic matter in sediment originates due to deposition of degraded parts of aquatic life. Sediment microbial fuel cell (SMFC) is a sustainable approach for bioelectricity production from sediment helping in remediating sediment and overlaying water. In present investigation, performance of SMFCs was evaluated for two different electrode spacings (viz. 8 and 24 cm) and electrode materials (graphite plate and polycrystalline carbon plate). During polarization study, the maximum power density of 1 mW/m^2 was obtained for SMFC with 8 cm spacing between electrodes; which is higher as compared to SMFC with 24 cm spacing (0.89 mW/m^2). The power output was enhanced when electrodes were closely spaced (8 cm) in SMFC. Maximum power density of 2.81 mW/m^2 was achieved while using graphite plates as electrodes, which is 64% higher than polycrystalline carbon plate electrodes (1.72 mW/m^2). Thus, the graphite plate can be effectively used as electrode material in SMFC mainly due to its better electrode kinetics for electron transfer compared with polycrystalline carbon plate electrodes. The lower power output from SMFC than previous studies was noted mainly due to the sandy nature of sediment and higher internal resistance of SMFC. Thus, SMFC with lower electrode spacing and graphite plate electrode can efficiently used to harvest electricity from the organic matter present in the sediment and protect water quality.

---

* D. A. Jadhav: School of Water Resources, Indian Institute of Technology, Kharagpur – 721302. India. E-mail: deepak.jadhav1795@gmail.com.
† Corresponding author: M. M. Ghangrekar. Department of Civil Engineering, Indian Institute of Technology, Kharagpur – 721302. India. Tel.: +91-3222-283440, E-mail: ghangrekar@civil.iitkgp.ernet.in.

Keywords: Electrode spacing; Electrode materials; Power overshoot; Sediment microbial fuel cells; Sediment Organic matter

INTRODUCTION

Microbial Fuel Cell (MFC) is a novel and sustainable technology that recover electrical energy along with wastewater treatment. MFC is considered as renewable energy source for operating the remote monitoring equipment in environmental studies. Sediment microbial fuel cell (SMFC) or benthic MFC (BMFC) could offer a possible solution for in-situ remediation of sediment and harvest energy in the form of electricity. SMFCs are bio-electrochemical devices that generate direct electricity from the natural redox gradients [1-3] that commonly occur across the sediment-water interface (Figure 1).

The generated electricity can be used to operate certain on-site devices such as water quality sensors or other environmental sensors.

The major advantages of this type of MFC over other fuel cells are the lower construction cost, maintenance-free operation and suitability for scale-up operations.

The key limitation of SMFCs is that it generates low amount of power and output potential [4]. However, increase in power and decrease in the internal resistance of SMFC can be possible by increasing the ionic strength and by decreasing the electrode spacing.

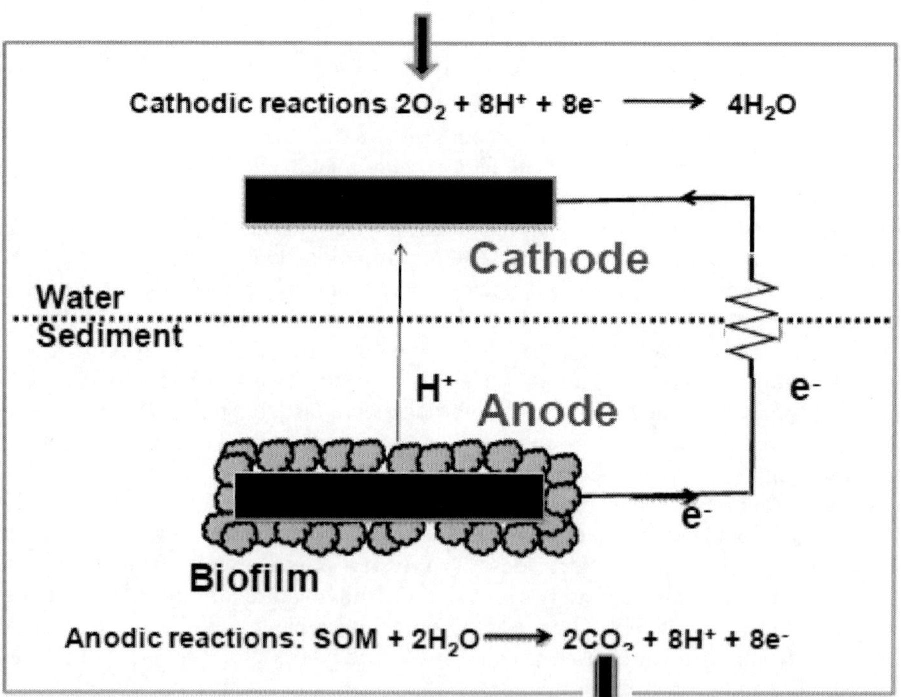

*SOM = Sediment Organic matter.

Figure 1. Diagram of SMFC showing different reactions and electron transfer.

Previously, Hong et al. [5] found that the current density can be expressed as a function of distance between electrodes in SMFCs.

The current densities of 11.5, 7.64, 2.51 and 2.11 mA/m^2 were obtained in SMFC with electrode spacing of 12, 20, 80 and 100 cm, respectively. Cheng and Logan [6] demonstrated that decrease in electrode spacing (ES) resulted into the enhancement of H_2 production rate in microbial electrolysis cell (MEC), but too close electrode spacing (1 cm or less) do not essentially harvest the maximum possible H_2 production rate.

Due to the possibility of oxygen diffusion through interface, lower electrode spacing should not be preferred in SMFCs.

Liu et al. [7] reported that power generation was enhanced from 720 to 1210 mW/m^2 with reduction in ES from 4 to 2 cm, mainly due to decrease in internal resistance (IR).

The SMFC with graphite cathode showed relatively poor performance due to lower reduction reactions [8]. In general, graphite materials with different forms like sheet, plates and rods is widely used as an anode material, because of its higher material strength, durability, inertness, and commercial availability.

In the earlier studies, Reimer and co-workers [9-11] placed graphite plate anode within sediment at subjective depths, and cathode (suspended) in the overlying seawater on frames. For plain graphite electrodes, a marine BMFC showed sustainable and maximum power density of 20 mW/m^2 and 28 mW/m^2 of anode surface area, respectively [11].

Reimer et al. [10] reported similar kind of system with graphite fiber as electrodes generating power density of about 10 mW/m^2; whereas, maximum power density of 34 mW/m^2 was reported for MFC constructed with carbon brush cathode [12].

Scott et al. [13] studied the effect of different electrode materials on performance of SMFC. They compared different anode materials like carbon sponge, carbon fiber, carbon cloth and reticulated vitreous carbon (RVC) with graphite cloth cathode. Power density of 51-54 mW/m^2 and current density of about 170 mA/m^2 (300 mV operating voltage) was reported in MFC with carbon sponge anode. However, in a similar study, anode with RVC material showed low power density (0.18 mW/m^2 at current density of 1.0 mA/m^2) due to higher anodic charge transfer resistivity.

The effect of various cathode electrode materials like graphite and RVC, Co and Fe-Co tetramethoxyphenyl porphyrin (FeCoTMPP), carbon sponge, cloth and paper and Pt-carbon and titanium, etc. were also studied by Scott et al. [14] using Tafel plot analysis. Maximum power density of 60 mW/m^2 was attained for SMFC with FeCoTMMP cathode, which was almost double than that achieved with unmodified carbon material. In freshwater catholyte condition, SMFC with graphite felt [5] and carbon paper [15] as cathode was capable to achieve power output of 4 mW/m^2 and 1 mW/m^2 of effective footprint area, respectively.

The power output of SMFC is mainly governed by microbial activities in anodic chamber and internal resistance of the system, which depends on design parameters such as electrode spacing, electrode material and its properties.

In present investigation, the parameters like electrode spacing (viz. 8 and 24 cm), and different electrode materials (graphite plate and polycrystalline carbon plate) was studied in order to evaluate its effect on performance of SMFC.

The concept of power overshoot in MFC was also explored in this study.

MATERIAL AND METHODS

Sediment As Source of Inoculum

For construction of SMFC, the sediment was collected randomly from downstream of Sandbach catchment area close to the Hordorf and Schandelah, Lower Saxony, Germany [16]. The selected site was located close to the municipal effluent treatment plant (ETP) and during sampling, care was taken that the samples were not influenced by any kind of contamination. The sediment was collected at 5 cm depth with scrapper from the upper stratum, which also served as a source of inoculum for MFC [17]. The collected sediment was then stored in a plastic container and native water (freshwater) was added. To study the properties of sediment, freshwater and sediment was characterized according to the Organisation for Economic Cooperation and Development (OECD) guideline [17-19].

MFC Design and Construction

For this study, four laboratory scale SMFCs were fabricated with an effective anolyte volume of 5 L. To study the effect of electrode spacing (ES), SMFCs were fabricated with anode and cathode made up of graphite plate material (total projected surface area of 2800 mm^2 each) placed with facing parallel to each other (Figure 2; Figure 3). The ES was maintained at 8 cm and 24 cm in MFC-1 and MFC-2, respectively (Figure 2). The anodic site of SMFC was filled with homogeneously mixed sediment after debris and gravel removal and cathodic site was filled with freshwater, which serve as a catholyte.

Both anode and cathode electrodes were hold together in SMFC by polyvinyl chloride (PVC) cylindrical pipe holder and groove arrangement. The connection between wire and electrodes was made conductive and watertight using mixture of carbon powder, resin and hardener to make lower down the contact resistance [16]. All connections were made with watertight using silicon and flexible parafilm. Wet sediment was added in anodic side of SMFCs without any pre-treatment [18]. Within 24 h, sediment settled at the bottom of the MFC and sediment-water interface was formed (Figure 1).

Figure 2. Arrangement of electrode before inserting it into the sediment.

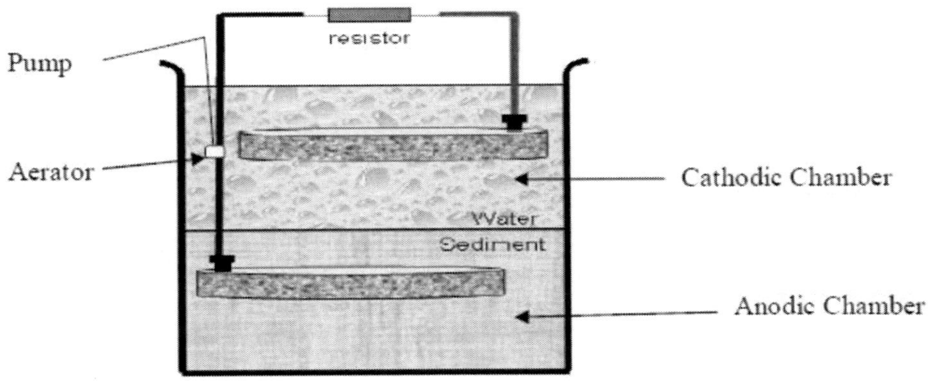

Figure 3. Working of sediment MFC.

In catholyte, oxygen (fine air bubbles) was supplied continuously by using porous diffuser with an air pump (ELITE 799, Hagen, Mansfield, US). The evaporative water loss was maintained by adding either freshwater or distilled water daily and a constant water level in MFC was maintained over the period of experimental run. During operation, 100 Ω external resistor (unless stated otherwise) was connected in the circuit [16, 18].

Measurement and Analysis

The potential difference and current produced by SMFCs were measured with data acquisition unit and digital multimeter (model 2700 and 2701, Integra Up and Running, Keithley Instruments, Data Acquisition, Ohio, US). The data corresponding to each cell were recorded at interval of 5 min on a spreadsheet using ExceLINX™ (Keithley) software. The graphs from data were plotted using Origin Pro 8 software. Power was calculated according to the formula: $P = I^2R$, where 'P' = power, 'I' = current and 'R' = external resistance. The power density (mW/m^2) and current density (mA/ m^2) were determined as the power or current generated per m^2 surface area of an anode. Individual electrode potentials were measured using Ag/AgCl reference electrode (+197 mV vs. SHE, Bioanalytical Systems Inc., US). The open circuit voltage (OCV) for each MFC was measured when circuit remained in the open condition (no current flow) till the potential value stabilized [20].

Conductivity of catholyte solution in SMFC was measured regularly with the help of conductivity probe (GLF 100, Greisinger electronics, Germany). The pH of catholyte was also measured by means of pH meter (GPHR 1400A, Greisinger electronic, Germany).

Study of Different Electrode Materials

To study the effect of two varieties of graphite electrode materials on SMFC performance, two SMFCs were constructed with both electrodes made up of polycrystalline carbon plate material in SMFC-G_1 (CP Graphite GmbH, Germany) (Size: 40*35*5 mm^3) and other SMFC-G_2 with similar size of graphite plates (VEB Elektrokohle Lichtenberg, former GDR, Germany) working as anode and cathode. The prototypes of SMFCs were fabricated as mentioned previously with 8 cm ES (Figure 2).

Polarization

The polarization study was carried out to determine the maximum power output and highest current density of MFC under different operating conditions. Polarization behavior was analyzed by changing the external resistance from 100000 Ω to 1000 Ω in forward (FP) manner and from 1000 to 100000 Ω resistances in backward (BP) manner for each SMFC. The voltage and current were recorded as a function of variable external resistance after allowing the circuit to stabilize condition (nearly to steady state) during polarization. To obtain polarization and power density curve, the voltage and power output were plotted against current. Internal resistance was determined by maximum power point (MPP) method as described by Logan [21].

RESULTS AND DISCUSSION

Energy generation by SMFCs was influenced by particulate present in sediment, sediment properties, microbial interaction and organic contents. It also depends upon the sediment structure and texture. In general, rivers present in Lower Saxony, Braunschweig, Germany contains lower organic matter as also reported by Mohamad [19, 18]. The sediment collected was sandy (53%) in nature which contains about 20% silt and 27% clay particles with water holding capacity of 30% [18, 19]. The sediment was characterized by lower total organic carbon content of about 0.4%. Hence, sufficient growth of bacteria in SMFCs was observed [16] during experiment. The substrate induced respiration (SIR) was 1.90 mg O_2/ 100 g (dry substrate) [19]. The freshwater was used as catholyte with near neutral pH of about 7.2. Temperature of freshwater catholyte was nearly about 18 ± 0.5 °C in winter and 20 ± 0.8 °C in summer season.

Effect of Electrode Spacing

Some scientific studies reveal that there is a possibility of the increase in oxygen diffusion from the cathodic side to the anodic side through sediment water interface due to lower ES. Most of the SMFCs were installed in lower sediment depth with the cathode near to the water-air interface or in deep-ocean environment with comparatively higher O_2 availability [12]. In order to evaluate the effect of ES on SMFC performance, distance between the anode and cathode electrodes was maintained 8 cm in SMFC-1 and 24 cm in SMFC-2, respectively (c/c). Power density of 0.72 mW/m^2 was obtained for larger electrode spacing of 24 cm with the 623 mV of OCV (Table 1); in contrast, low electrode spacing (8 cm) yielded higher power density of 0.968 mW/m^2 with OCV of 699 mV (Figure 4 a). This 35% increase of power was resulted due to reduction in electrode spacing from 24 to 8 cm and reduction in ohmic losses of cell. This increased power density corresponds to a decrease in internal resistance from 45 kΩ to 35 kΩ with decrease in ES from 24 to 8 cm (Figure 4b). The cathode and anode potentials were also enhanced with decrease in ES (from 24 to 8 cm). The power density increased with decrease in distance between the electrodes mainly due to the decrement in ohmic losses reducing internal resistance of the system [22].

As noted earlier, due to the reduction in distance between both electrodes, the protons generated during oxidation process have to travel less distance (24 to 8 cm), which causes decrease in ohmic resistance [22]. Power of MFC with 8 cm spacing was affected by power overshoots in the cell. However, electrode spacing should be kept optimum because too large (or too small) electrode spacing can increase the difficulties in installation and also increases the ohmic losses.

The operating voltage of SMFC with 24 cm ES was lower than SMFC with 8 cm spacing at different external resistances.

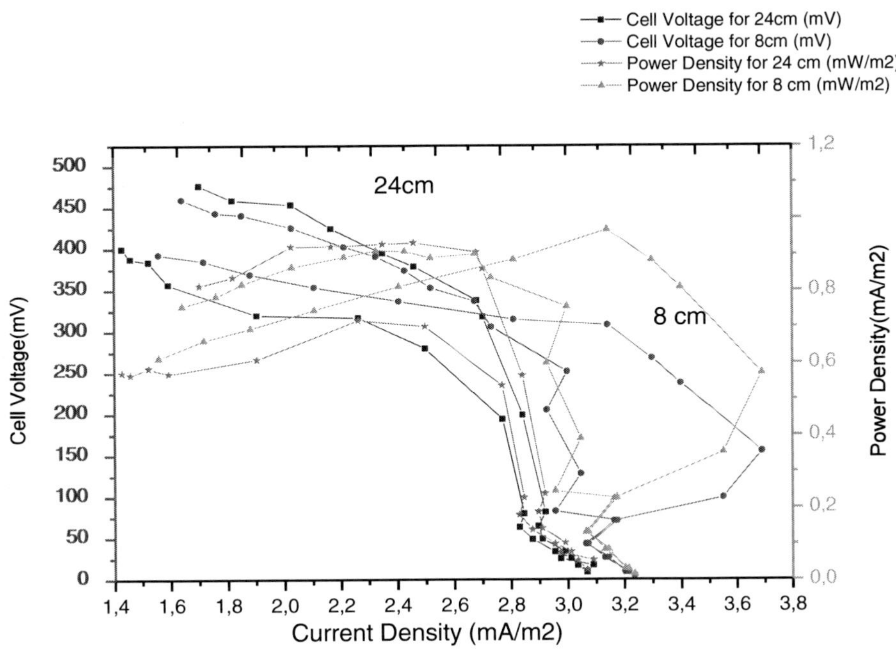

a

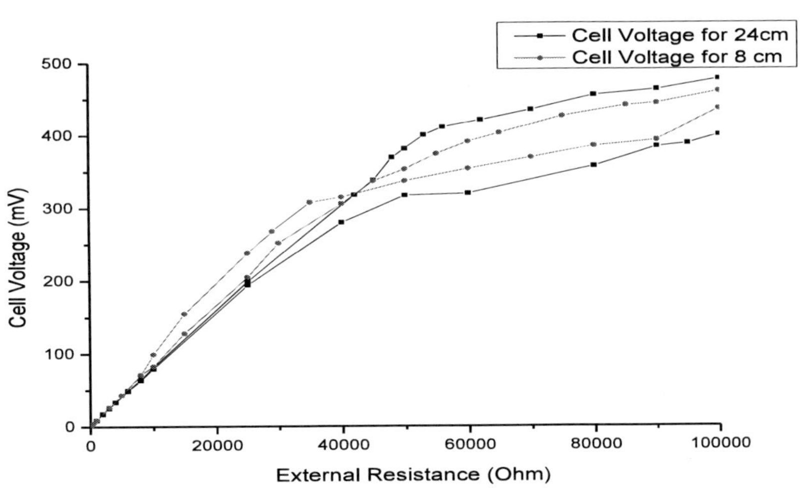

b

Figure 4. (A) Comparisons of power curve for SMFC with 8 and 24 cm electrode spacing; (B) Comparison of cell voltage vs. external resistance for SMFC with 8 and 24 cm ES.

Table 1. Summary of performance of SMFC with different electrode spacing during polarization

Distance between electrodes (cm)		Maximum current density mA/m^2	Maximum power Density mW/m^2	OCV mV	IR ohm
24	FP*	2.69	0.72	623	50000
	BP*	2.69	0.89		45000
8	FP*	3.14	0.968	699	35000
	BP*	3.14	0.9		45000

*FP = Forward polarization, BP = Backward polarization.

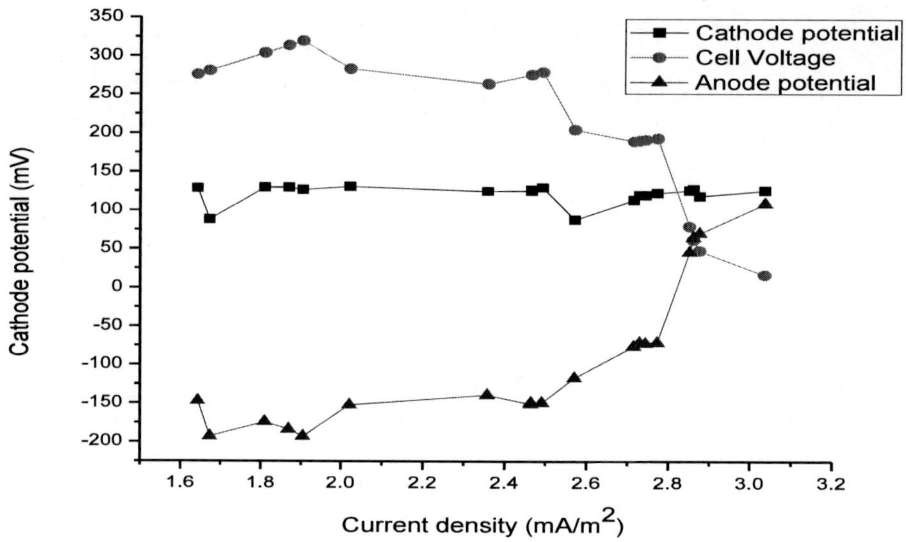

Figure 5. Current density as a function of electrode potential in MFC with 24 cm electrode spacing.

During polarization, it was observed that the cathode potential of MFC with 24 cm ES remained in the same range (Figure 5), which interprets that the difference in their operating voltages could be due to the variation in microbial activities on the anode. Hence, results support that anode is the main limiting factor in these SMFCs. The anode potential of SMFC with 24 cm spacing varied considerably from -200 mV to + 120 mV during polarization.

Polarization curves for SMFC with 8 cm ES show often suddenly large drop in voltage with increased current density. This kind of behavior of SMFC was observed at lower external resistances which leads to a phenomenon called as power overshoot [18]. This is responsible to restrict the power generated at higher current densities by SMFC resulting into over or under-estimation of exact power.

The cause for overshoot phenomenon was reported previously, but still it is not well understood. Winfield et al. [23] suggested that forward (FP) and backward polarization (BP) for single fuel cell can eliminate power overshoot. However, present study shows that there is a possibility of overshoot in power measurements in both the conditions as observed during both forward and backward polarization study. Watson and Logan [24] and Nien et al. [25] reported that power overshoot occurred due to drastic increase in the anode potential primarily due to a reduction in resistance to the flow of current.

The decrease in power density with increasing current at lower resistance also represents electron transfer limitation during anodic reactions. Higher time required (than provided) for the bacteria present in the sediment for adjusting the oxidation rate to the new external resistance or due to the temporary system overload might cause this power overshoot.

During MFC operation, it was noted that the declining phase of the power density curve occurs at the lower resistance, where the magnitude of external resistance step decreased (i.e. external resistance varies from thousands to hundreds Ohm) (Figure 4a).

At these lower external resistance or higher current capturing region, the requirement for electrons exceeds than microbial rate that they can provide by catalyzing oxidation of organic matter [26]. This results in underestimation or overestimation (unsustainable) of actual power in SMFC, leading to unexpected cell failure over the period of time.

Effect of Electrode Materials on SMFC Performance

The electrode material is an essential component that affects power production in MFCs [27]. During polarization, cell voltage can be expressed as the function of external resistance (Figure 6 b). Polarization curve of graphite material G_2 shows right skewed shape (Figure 6 a). The maximum power densities obtained for graphite material G_2 (2.81 mW/m^2) was 64% more than for G_1 (1.72 mW/m^2) (Figure 6 a). Also, utilization of graphite electrodes in SMFC resulted into reduction in internal resistance from 30 to 10 kΩ (Table 2).

Thus, graphite material gave better power performance over polycrystalline carbon electrode material during the experimental period. The enhanced performance with graphite electrode material was observed mainly due to the better electrode kinetics supported by this material for electron transfer reactions.

Therefore, graphite can be effectively used as electrode material in SMFC fabrication.

After 4th day onwards, cell with material G_1 showed lower performance than G_2, probably due to kinetic limitations of that material to microbial activities in sediment. The graphite electrode G_2 gave higher OCV (679 mV) as compared to carbon electrode G_1 (615 mV).

Also, graphite is brittle and durable electrode material and can be utilized for scaling up applications of SMFCs to operate power devices for practical use.

Due to cost constraints and durability, graphite material is more suitable for electrodes in SMFC for long term use in sediment.

Table 2. Polarization characteristics of SMFC with different electrode materials

Graphite material	Max. Current density mA/m^2	Max. Power Density mW/m^2	Max. Voltage (OCV) mV	IR Ω
G_2	14.17	2.81	679	10000
G_1	10	1.73	615	30000

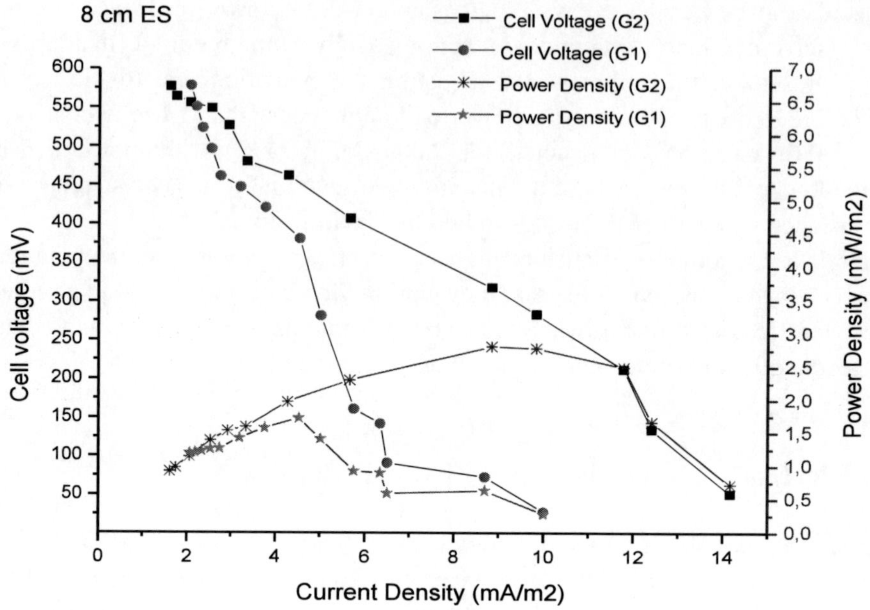

a

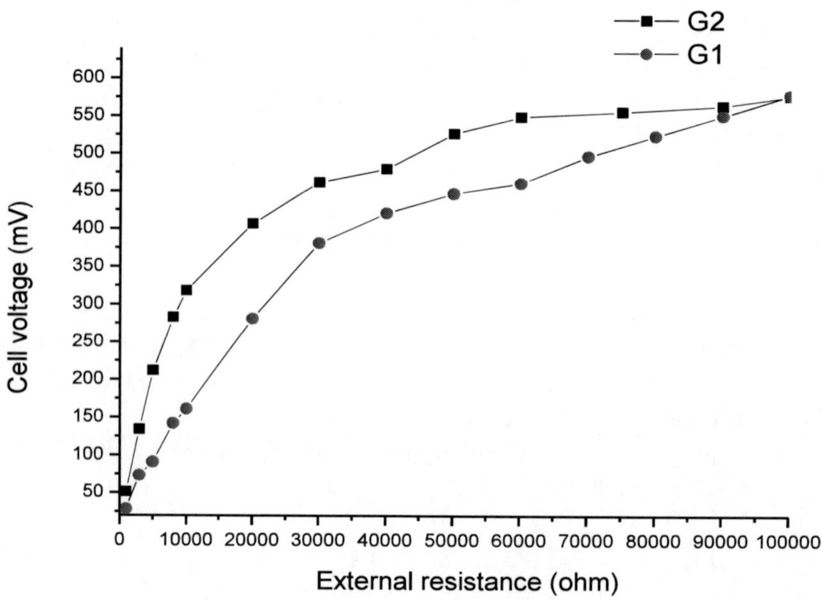

b

Figure 6. (A) Polarization curve for SMFC G_1 and G_2 with 8 cm spacing between the electrodes; (B) Plot of cell voltage vs. external resistance for 8 cm spacing between the electrodes.

Conclusion

As the performance of SMFC was stable over extended time period of around 7 to 8 months, SMFC can be used as energy and power source along with simultaneous remediation of sediments. The lower power output obtained from SMFCs was mainly due to the sandy nature of sediment, lower organics content and higher internal resistance of SMFC. The result supports that the decrease in distance between anode and cathode (i.e. electrode spacing) results into enhancement in power generation from SMFC due to faster proton exchange. SMFCs with 8 cm spacing attained higher power output than SMFC with 24 cm spacing due to reduction in internal resistance of the cell. However, power overshoot was observed due to electron limitations when SMFC were operated at lower resistances. SMFC with graphite electrode material achieved higher power production than polycrystalline carbon electrodes.

Therefore, graphite can be effectively used as electrode material in SMFCs due to better electrode kinetics, lower cost, durability, and suitability of this material for field applications. Thus, lower electrode spacing and graphite plate electrode can be efficiently used to improve the electricity generation from SMFC. In future, more efforts are required to reduce or eliminate power overshoot in SMFC. Also, optimization and modeling studies on SMFC domain required to maximize power generation as well as reduction in overall operation cost.

ACKNOWLEDGMENTS

The both authors gratefully acknowledge the scholarships provided by 'German Academic Exchange Service (DAAD)' during research stay in TU Braunschweig, Germany. The authors also thank Prof. Uwe Schröder, Falk Harnisch, Alessandro, Pol and team-mates of Institute of Environmental and Sustainable Chemistry, Technical University of Braunschweig, Germany for their guidance during experimental and research work.

REFERENCES

[1] Lowy, D. A., Tender, L. M., Zeikus, J. G., Park, D. H., Lovley, D. R., "Harvesting energy from the marine sediment-water interface II. Kinetic activity of anode materials", *Biosensors and Bioelectronics*, vol. 21(11), pp. 2058-2063, 2006.

[2] Rezaei, F., Richard, T. L., Brennan, R. A., Logan, B. E., "Substrate-enhanced microbial fuel cells for improved remote power generation from sediment-based systems", *Environmental Science and Technology*, vol. 41(11), pp. 4053-4058, 2007.

[3] Lile, J. R. D., "Energy harvest by benthic microbial fuel cell: Influence of salinity", *M.Sc. thesis*, University of Applied Science, Cologne, Germany, 2008.

[4] Holmes, D. E., Bond, D. R., O'Neil, R. A., Reimers, C. E., Tender, L. R., Lovley, D. R., "Microbial communities associated with electrodes harvesting electricity from a variety of aquatic sediments", *Microbial Ecology*, vol. 48(2), pp. 178-190, 2004.

[5] Hong, S. W., Chang, I. S., Choi, Y. S., Chung, T. H., "Experimental evaluation of influential factors for electricity harvesting from sediment using microbial fuel cell", *Bioresource Technology*, vol. 100(12), pp. 3029-3035, 2009.

[6] Cheng, S., Logan, B. E., "High hydrogen production rate of microbial electrolysis cell (MEC) with reduced electrode spacing", *Bioresource Technology,* vol. 102, pp. 3571-3574, 2011.

[7] Liu, H., Cheng, S., Logan, B. E., "Power generation in fed-batch microbial fuel cells as a function of ionic strength, temperature, and reactor configuration", *Environmental science and Technology*, vol. 39(14), pp. 5488-5493, 2005.

[8] Zhao, F., Harnisch, F., Schröder, U., Scholz, F., Bogdanoff, P., Herrmann, I., "Challenges and constrains of using oxygen cathodes in microbial fuel cells", *Environmental Science and Technology,* vol. 40 (17), pp. 5193-5199, 2006.

[9] Bond, D. R., Holmes, D. E., Tender, L. M., and Lovley, D. R., "Electrode-reducing microorganisms that harvest energy from marine sediments", *Science*, vol. 295(5554), pp. 483-485, 2002.

[10] Reimers, C. E., Tender, L. M., Fertig, S., Wang, W., "Harvesting energy from the marine sediment - Water interface", *Environmental Science and Technology*, vol. 35(1), pp. 192-195, 2001.

[11] Tender, L. M., Reimers, C. E., Stecher, H. A., Holmes, D. E., Bond, D. R., Lowy, D. A., Pilobello, K., Fertig, S. J., Lovely, D. R., "Harnessing microbially generated power on the seafloor", *Nature Biotechnology,* vol. 20, pp. 821-825, 2002.

[12] Reimers, C. E., Girguis, P., Stecher, H. A., Tender, L. M., Ryckelynck, N., Whaling, P. "Microbial fuel cell energy from an ocean cold seep.", *Geobiology*, vol. 4(2), pp. 123-136, 2006.

[13] Scott, K., Cotlarciuc, I., Hall, D., Lakeman, J. B., Browning, D., "Power from marine sediment fuel cells: the influence of anode material", *Journal of Applied Electrochemistry*, vol. 38: pp. 1313-1319, 2008a.

[14] Scott, K., Cotlarciuc, I., Head, I., Katuri, K. P., Hall, D., Lakeman, J. B., Browning, D., "Fuel cell power generation from marine sediments: Investigation of cathode Materials", *Journal of Chemical Technology and Biotechnology,* vol. 83, pp. 1244-1254, 2008 b.

[15] Martins, G., Peixoto, L., Ribeiro, D. C., Parpot, P., Brito, A. G., Nogueira, R. "Towards implementation of a benthic microbial fuel cell in lake Furnas (Azores): Phylogenetic affiliation and electrochemical activity of sediment bacteria", *Bioelectrochemistry*, vol. 78(1), pp. 67-71, 2010.

[16] Jadhav, D. A., Ghangrekar, M. M., "Effect of salinity, acetate addition and alteration of sediment on performance of benthic microbial fuel cells", *Recent Advances in Bio-energy Research*, Vol. III, pp. 404-418, Sardar Swaran Singh National Institute of Research, Kapurthala, Punjab, India, 2014.

[17] Organization for Economic Cooperation and Development (OECD), *Guideline for testing of chemicals No. 308.* Aerobic and anaerobic transformation in aquatic sediment systems, 2002.

[18] Jadhav, D. A., Ghangrekar, M. M., "Effect of biofilm transfer on power generation in sediment microbial fuel cell", *International Journal of Environmental Sciences*, Vol. 2 (1), pp. 53-60, 2013.

[19] Mohamad A., "*Environmental chemical analysis related to drug industries: fate monitoring of Diclofenac in water/sediment systems*", Ph.D. Thesis, Technical University of Braunschweig, Germany, 2010.

[20] Jadhav, D. A., Ghadge, A. N., Mondal, D., Ghangrekar, M. M., "Comparison of oxygen and hypochlorite as cathodic electron acceptor in microbial fuel cells", *Bioresource Technology*, vol. 154, pp.330-335.

[21] Logan, B. E., "*Microbial Fuel Cells*", Wiley and Sons, Inc., Hoboken, NJ, 2008.

[22] Ghangrekar, M. M., Shinde, V. B., "Performance of membrane-less microbial fuel cell treating wastewater and effect of electrode distance and area on electricity production", *Bioresource Technology*, vol. 98(15), pp. 2879-2885, 2007.

[23] Winfield, J., Ieropoulos, I., Greenman, J., Dennis, J., "The overshoot phenomenon as a function of internal resistance in microbial fuel cells", *Bioelectrochemistry*, vol. 81(1), pp. 22-27, 2011.

[24] Watson, V. J., Logan, B. E., "Analysis of polarization methods for elimination of power overshoot in microbial fuel cells", *Electrochemistry Communications,* vol. 13(1), pp. 54-56, 2011.

[25] Nien, P. C., Lee, C. Y., Ho, K. C., Adav, S. S., Liu, L., Wang, A., Ren, N., Lee, D. J., "Power overshoot in two-chambered microbial fuel cell (MFC)". *Bioresource Technology*, vol. 102(7), pp. 4742-4746, 2011.

[26] Ieropoulos, I., Winfield, J., Greenman, J., "Effects of flow-rate, inoculum and time on the internal resistance of microbial fuel cells", *Bioresource Technology*, vol. 101(10), pp. 3520-3525, 2010.

[27] Song, T. S., Yan, Z. S., Zhao, Z. W., Jiang, H. L., "Construction and operation of freshwater sediment microbial fuel cell for electricity generation". *Bioprocess and Biosystems Engineering*, vol. 34(5), pp. 621-627, 2011.

In: Renewable Energy and Sustainable Development
Editors: R. T. Durai Prabhakaran, S. A. Kale et al.
ISBN: 978-1-63463-429-8

Chapter 13

ROLE OF ENERGY STORAGE IN THE POWER SYSTEM NETWORK

V. Arangarajan**, Amanullah M. T. O., Jaideep Chandran, G. M. Shafiullah and Alex Stojcevski

Deakin University, School of Engineering, Faculty of Science, Engineering and Built Environment, Geelong, Waurn Ponds, Victoria, Australia

ABSTRACT

Today's power system network is more complex with enhanced responsibility to maintain reliable, stable and quality supply of power at transmission and distribution level. Maintaining grid balance is a bigger issue, in case of any unexpected generation shortage or grid disturbance or any participation of an intermittent nature of renewable energy sources like wind and solar power in the energy mix. In order to compensate such imbalance and improve reliability, and stability of power system, an energy storage system (ESS) can be considered as a vital solution. Also ESS can be used to mitigate associated issues of renewable energy sources while integration into the power system network. Thus ESS supports to get a reduction in greenhouse gas (GHG) emissions by means of integrating more renewable energy sources to the grid effectively. There are various types of Energy Storage (ES) technologies which are being used in power systems network for large scale (MW) to small scale (KW) level. Based on the type and characteristics, each storage technology is suitable for a particular role of applications. This paper presents an extensive review study on various types of ES technologies in characteristics and applications point of view. It also demonstrates various applications of ESS in detail. Finally, with the aid of ES-select™ tool software, a feasibility analysis has been carried out to identify a suitable ES technology for appropriate applications at different grid locations and also helps to develop a smart hybrid storage system for grid applications in future.

* Corresponding author: Tel:+61 410006730, E-mail:avinayag@deakin.edu.au.

1. Introduction

Energy storage system (ESS) stores excess energy in any form and delivers stored energy in the same or different form according to the application requirements. It supports to provide a reliable, stable, and quality power supply along with the reduction of GHG emissions by integrating more RE sources effectively in the power system network. ESS is available from small scale to large scale level and it fulfills several electrical application requirements on transmission and distribution level of power system network. It can be categorized into several types, based on the type of mechanisms used for storing energy. Each ES technology has its own merits and de-merits based on their characteristics. Their role and feasibility level also varies for particular applications. ESS can be utilized for several applications like energy time shift, peak shaving, load following, RE integration (solar/wind energy time shift, solar/wind energy smoothing out, RE firming capacity) and also provides support for improving voltage regulation and power quality at transmission and distribution level [1-3]. In the upcoming section, different types of ES technologies are discussed in detail.

2. Types of Energy Storage Technologies

Depending on the type of mechanisms used for storing energy, ESS can be classified into (i) Mechanical; Pumped Hydro Energy Storage (PHES), Compressed Air Energy Storage (CAES), Flywheel Energy Storage (FES), (ii) Electrical; Super Capacitor (SC), Super Magnetic Energy Storage (SMES), (iii) Electro chemical; Secondary batteries, redox flow batteries, (iv) Chemical; Hydrogen and other gas storage, (v) Thermal; Thermal Energy Storage (TES) [2, 3, 5].

2.1. PHES Technology

PHES system is a matured and large scale energy storage technology which is same as the conventional hydroelectric power system. During off peak demand, water is transferred from a lower reservoir to an upper reservoir, while at peak, stored water from upper reservoir flows to the turbine and produce electricity through the connected generator. Storage capacity of PHES system depends on the reservoir capacity and hydraulic head. It is suitable for bulk storage applications with long duration of power discharge from hours to days. It is a long life storage system with an efficiency of around 70% to 85%. PHES system highly depends on specific geological formation with a requirement of large land area [1, 2].

2.2. CAES Technology

CAES is also a type of large scale energy storage technology, which consists of a compressor, gas turbine and high pressure air storage tank. During off peak demand, with the help of a compressor, high pressure air is being stored in a storage tank. The storage tank may be either underground caverns or on ground pressure vessel tank. During on peak demand, stored high pressure air is being released into gas turbine combustor, where combustion takes place to run the gas turbine and produce electricity through a connected generator. Its efficiency is around 70% to 80% and it needs a suitable large space for underground air storage which makes potential impact to the environment [1, 2].

2.3. FES Technology

It is a mechanical type storage system which stores energy in the form of kinetic energy. It has a flywheel which is a rotating mass fixed in a rotating cylinder. The flywheel storage system draws power from primary sources and stores energy by means of spinning a high density cylinder at high speed. Whenever there is a power loss from the primary source, the connected motor act as a generator along with flywheel rotation that produces electricity. The FES system can be used in power system for compensating power quality issues. Self-discharge loss of FES is high during an idle condition as compared to the other storage systems [2, 6].

2.4. Super Capacitor

A Super capacitor or double layer capacitor stores energy in the form of electric field between a pair of electrode charge plates. SC electrodes are made from a high porous carbon or high surface material as conductor and used in combination with aqueous or non-aqueous electrolyte. SC has the capability of rapid charge and discharge behavior with higher lifetime. Efficiency of SC is around 85% to 98% with a lifetime more than 500000 cycles [2, 7].

2.5. SMES Technology

SMES stores energy in the form of a magnetic field when direct current flows through super conductive coil. It mainly consists of three subsystems; Super conducting coil unit, cryogenic refrigeration system, power conditioning equipment. To maintain a high conductive condition of super conductive coil, it is required to maintain the temperature beyond -269°C by means of cryogenic refrigeration system. SMES system efficiency is around 90% to 95% and it can be used for power quality applications in the power system network [2, 6, 7].

2.6. Secondary Batteries

It works on the principle of electro chemical technology. Secondary batteries like lead acid (Pb acid), nickel cadmium (Ni-Cd), nickel metal hydride (Ni-MH), sodium sulfur (NaS), sodium nickel chloride (Na-Nicl), lithium-ion (Li-Fe) and metal air belongs to electro chemical storage technology. Secondary batteries can be used for the following applications; energy and power management, voltage regulation, backup power support, and energy time shift [2, 6].

2.7. Redox Flow Batteries

Flow battery storage system stores and release energy through a reversible electro chemical reaction between two electrolytes. Electrolyte used for that battery contains dissolved electro active species, stored in a separate tank. The stored electrolyte flows through an electro chemical membrane where it is converted into electrical energy and vice versa. Zinc bromine and vanadium redox are the most common types of flow batteries, mostly used in power system applications. Lifetime of flow batteries is more with a high efficiency around 70% to 75% [1].

2.8. Hydrogen (H2) Storage

H_2 is available in water, organic materials, biomass and hydrocarbons. It is an energy efficient and environment friendly gas. H_2 gas can be used in the fuel cell to produce electricity and it can be stored in different forms; they are gaseous form at high pressure, liquid form, in chemical compounds and metal hydrides. The hydrogen storage energy density is higher as compared to the other storage technologies. However the storage efficiency is low, which is only around 29% to 49% due to the energy conversion process [2].

2.9. TES Technology

TES is a commercially available storage system, available at an affordable cost. It is a mature and proven technology with higher round trip efficiency as compared to other storage technologies. TES stores thermal energy as sensible heat, latent heat, thermo chemical heat with features of reversible thermo chemical reaction. Intermittent nature of solar resources can be effectively utilized by thermal energy storage system [1].

In the upcoming section, different characteristics of ES technologies are discussed in detail.

3. Characteristics of Energy Storage (ES) Technologies

The characteristics of energy storage technologies are determined by the following factors: energy density, power density, round trip efficiency, response time, discharge duration, lifetime, self-discharge, technical maturity, cost, recharge time, operating temperature, memory effect, footprint (space and weight), environment impacts [3, 5, 7, 8].

3.1. Power Density versus Energy Density

Specific power or gravimetric power density is a relation between powers to weight ratio, which is the maximum amount of power that can be supplied per unit mass (watts/kg). Specific energy or gravimetric energy density is a relation between energy to weight ratio which is the amount of energy that can be stored per unit mass (watt- hour/kg). Volumetric power density is expressed in power per unit volume (watts/liter or watts/m^3) and volumetric energy density is expressed in energy per unit volume (watt-hour/liter or watt-hour/m^3). Normally, most of the storage technology with high power density have low energy density and vice versa. Higher the power and energy density, lighter in weight and smaller in size will be a storage technology. PHES, CAES and Redox Flow Battery (RFB) belong to the low energy density type as compared to other types. SMES, FES and DLC belong to the high power density type system. A lithium-ion battery is a high power and high energy density type storage, which is more suitable for mobile storage applications. H_2 and Synthetic Natural Gas (SNG) storage technologies belong to the high energy density type, suitable for long term storage applications. Figure 1 shows the relationship between power and energy densities of ES technologies [7, 8].

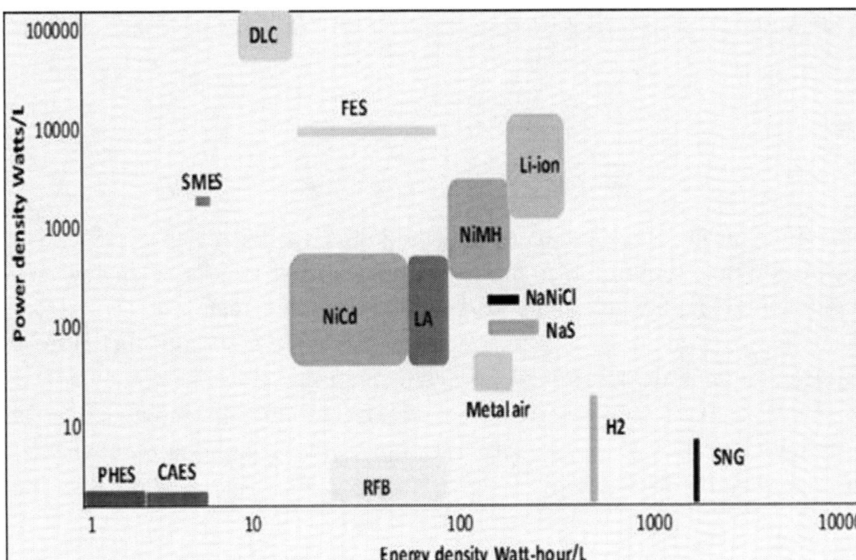

Figure 1. Power density versus energy density.

3.2. Round Trip Efficiency

Round trip efficiency (RTE) also called as cycle efficiency is the ratio between energy outputs to energy input during one full (charge/discharge) cycle. Round trip efficiency (RTE) is based on the amount of losses that occurs during charge, discharge and storing process. The loss of energy is expressed as in percentage (%). Equation-1 shows the efficiency relationship between energy in and energy out and Figure 2 shows the RTE of various storage technologies [7].

$$\eta = E_{out} / E_{in} \quad (1)$$

(Where E_{out} = Energy output, E_{in} = Energy input)

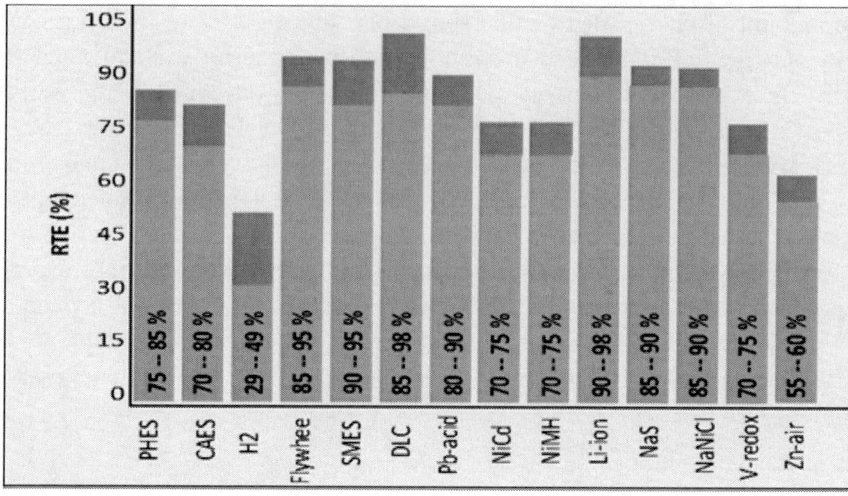

Figure 2. Round trip efficiency of ESS.

3.3. Response Time

Response time means the amount of time required to deliver the power by a storage system, which is the time required from no discharge to full discharge. Conventional batteries like lead acid (pb-acid), nickel cadmium (Ni-Cd), nickel metal hydride (Ni-MH) and advanced batteries; lithium ion (Li-ion), sodium sulfur (NaS), sodium nickel chloride (Na-NiCl) belongs to the quick response type storage technology. Response time of flow batteries is less than a millisecond, similarly flywheel, SMES, double layer capacitor (DLC) also belongs to the quick response type storage technology as compared to CAES, PHES and hydrogen storage technologies. Figure 3 shows the response time details of various storage technologies [7].

Energy Storage Technologies	Response Time
CAES	1--15 minutes
PHES	Sec--minute
Hydrogen	ms--minute
Flywheel	ms--sec
Li-ion	ms--sec
SMES	ms
DLC	ms
Pb-acid	ms
NiCd	ms
NiMH	ms
NaS	ms
NaNiCl	ms
Zn-air	ms
V-Redox	< 1ms
Zn-Br	< 1ms

Figure 3. Response time of ESS.

3.4. Discharge Duration

Discharge duration of a storage technology depends on the power and energy density characteristics. High power density storage technologies such as SMES, DLC and Flywheel can be used for power related applications, which can deliver high power in a short duration. Discharge duration of PHES, CAES and Hydrogen storage technologies are in the range of hours to days whereas for flow batteries (V-Redox and Zn-Br), it is around seconds to hours. For conventional and advanced batteries, discharge duration is around minutes to a few hours.

3.5. Self-Discharge

If the energy storage systems are kept in charge for extended period without use, it tends to lose some of its stored energy due to its self-discharge characteristics. The self-discharge level of each storage technology varies according to the type of storing mechanism. For electro chemical technology type of storage system, self-discharge losses due to a chemical reaction occurs when the battery is in idle condition. In thermal storage system, there is a loss of thermal energy due to heat transfer between the storage area and surroundings. In addition to self-discharge, some of the storage technologies need additional power for the functioning of auxiliaries like control, heating, cooling. This additional power requirement is considered as another loss of energy known as parasitic losses. Both self-discharge and parasitic losses are collectively called as standby losses [7].

The self-discharge rate for a flywheel storage system is more which is around 20 to 100% per day, whereas for DLC, it is around 2 to 40%. For lead acid battery storage system, self-discharge is around 0.1 to 0.3%, whereas for Ni-Cd, it is around 0.2 to 0.6%. For NaS, NaNiCl and SMES, self-discharge is around 10 to 20%, while flow batteries have only 1% per hour. PHES and CAES technologies have an almost negligible rate of self-discharge. Figure 4 shows the self-discharge per day for various storage technologies [7].

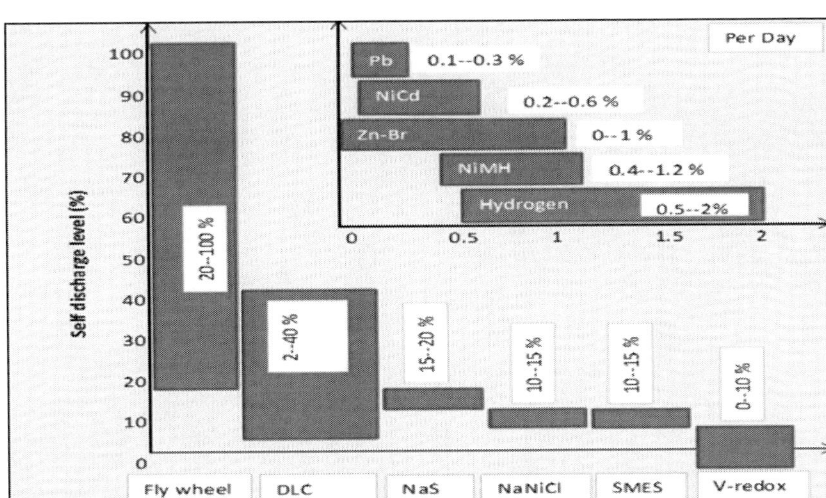

Figure 4. Self-discharge level of ESS.

3.6. Lifetime

All energy storage systems downgrades over a certain period of usage. The life time of the energy storage system depends on the following two criteria's: calendar life (storage system in service for a number of years) and cycle life (number of charge and discharge cycles before failure of the storage system). The lifetime of the storage system depends on the type of storage technology, operational parameters and other environmental conditions. From the Table 1 below, life time related to both calendar years and the number of cycles for PHES, CAES, SMES, Flywheel, DLC storage technologies are high as compared to other storage technologies [7].

Table 1. Lifetime of ES technologies

Energy Storage Technologies	Calendar years	Number of cycles
PHES	40--60	12000—35000
CAES	25--40	9000--20000
SMES	> 20	10000
Flywheel	> 20	> 100000
DLC	> 20	> 500000
NiCd	15--20	1500
Li-ion	8--15	> 4000
Pb-acid	3--15	2000
NiMH	5--10	300--500
NaS	12--20	2000--4500
NaNiCl	12--20	1000--2500
Zn-air	30	> 2000
V-Redox	15--20	> 13000
Zn-Br	5--10	> 2000
Hydrogen	5--15	> 1000

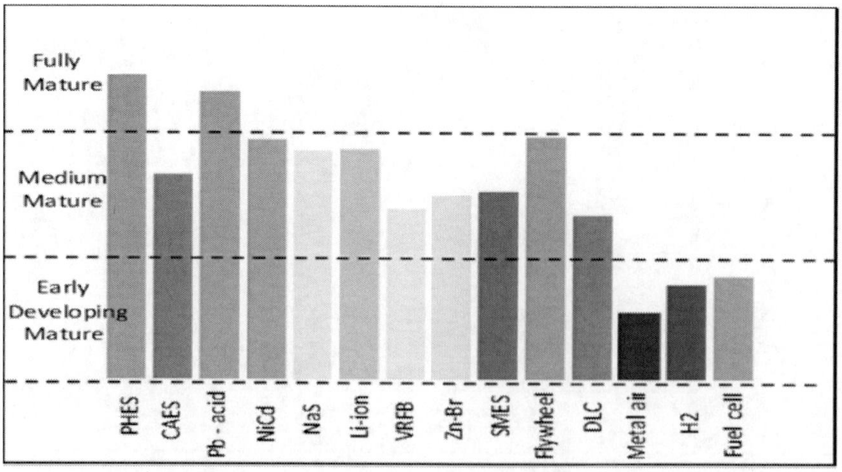

Figure 5. Recharge time of ES technologies.

3.7. Recharge Time

Recharge time or charge rate is a rate at which the storage system can be charged. Most of the storage system's charge rate is same as the discharge rate. In some cases, charge time varies as rapid or slow which depends on the capacity of a power conditioning unit, chemical reaction and the type of storage medium. If a storage system is not getting charged quickly, it cannot provide power for the next cycle of service. Figure 5 shows the recharge time of various storage technologies [7].

3.8. Operating Temperature

Operating temperature varies according to the characteristics of energy storage technologies. DLC can operate in a wide range of temperature from -45° C to 85° C. Operating temperature of NaS battery is highly around 300° C, while SMES temperature is extremely very low. Similarly, NaNiCl and Ni-Cd has operating temperature in the range of -40° C to 70° C, while for Pb-acid, it is around 27° C. If the operating temperature of Pb-acid battery rises in excess of 5° C (from 27° C), it can reduce its life time around 50%. CAES and PHES system can operate at normal ambient temperature [7].

3.9. Cost

Cost of energy storage technology depends mainly on the capital, and life cycle cost. Total capital cost includes energy cost ($/KWh) of storage units, and power cost ($/KW) of the sub system which involves the power conversion system (PCS) and other accessories. Life cycle cost includes operation and maintenance cost, electricity cost for recharging, and cost for component replacement. As per the energy capacity cost of concern, investment cost is high for SMES, Flywheel and DLC storage, whereas operation and maintenance cost is high for NaS and Pb-acid batteries [6]. As in equation-2, the cost of energy ($/KWh) of an energy storage system can be calculated from the known total annual cost, and annual energy discharge values [5, 9-12].

$$\text{Cost of Energy (COE)} = (AC + OMC + ARC)/AED \quad (2)$$

$$\text{Annual energy discharge (AED)} = P \times n \times H \times D \quad (3)$$

(Where AC = annual capital cost, OMC = annual O&M cost, ARC = annual replacement cost, P= rated power capacity of storage, n = charge/discharge cycles per day, H=length of each discharge cycle duration, D = annual operating days of storage).

3.10. Technical Maturity

Technical maturity of the storage technologies can be classified into three categories: fully mature, medium mature, early and develop mature technology. PHES and lead acid battery are fully mature technologies, available for high and medium power range of applications. CAES, flywheel, SMES, DLC, Ni-Cd, Li-ion, NaS, Na-NiCl and flow batteries belong to the medium or developed mature technologies. Ni-Cd, Ni-MH, Li-ion and Na-NiCl batteries can be considered as fully mature technology, in case of portable or mobile applications. Metal air batteries, hydrogen storage with fuel cell are an early and develop mature technologies. Figure 6 shows the technical maturity level of various storage technologies [7, 8].

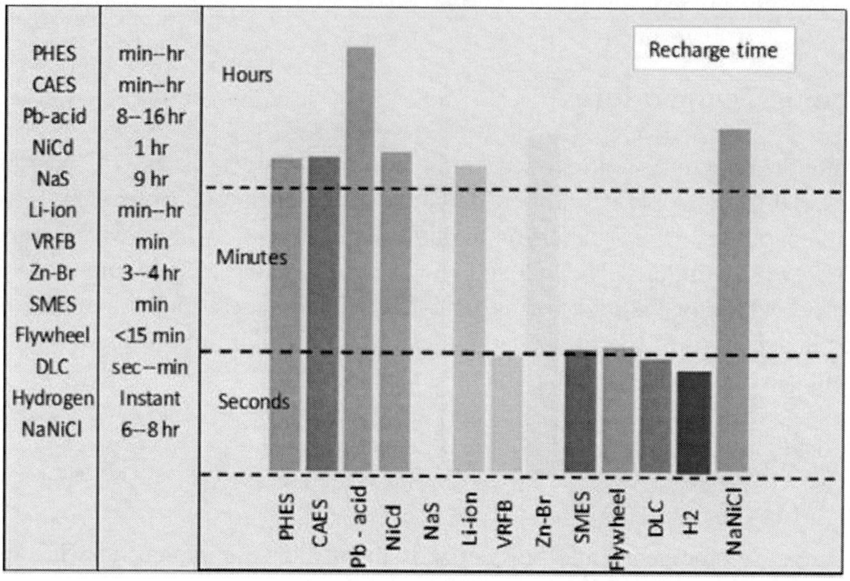

Figure 6. Maturity level of ES technologies.

3.11. Memory Effect

It means that a battery will be repeatedly recharged after being only partially discharged. This effect occurs mostly in Ni-Cd and NiMH batteries. Maximum energy storage capacity of battery loses gradually due to this effect. Full charge and full discharge of a battery for a regular period can help to mitigate this effect [7, 13].

3.12. Foot Print

An area of space required per unit energy capacity of a storage device is called as foot print or it is the spatial requirement which is expressed in m^2/KWh. It is closely related with energy and power density of a storage system. The hydrogen storage system requires less space per unit of energy storage which is around 0.005 to 0.05 m2/KWh. Similarly space requirement for Zn-air battery storage is around 0.005 m^2/KWh. PHES and CAES system are occupying a large area and volume as compared to other storage technologies [7].

3.13. Environment Impacts

Energy storage system becomes toxic and corrosive due to the usage of certain metals as electrodes and acid as an electrolyte. In Lead acid battery, lead is a toxic metal and sulphuric acid creates corrosive atmosphere, whereas for nickel cadmium batteries, cadmium is a highly toxic material. Lithium-ion batteries have less environmental impact as compared to the other batteries, since lithium oxide and salt can be recycled easily. The CAES type storage system needs large underground space with high pressure air storage which can be a hazard for the environment, whereas PHES technology needs specific geological formations. For flywheel storage, there is no chemical impact on environment. However stringent safety guards required while in operation [2, 7, 13].

In the upcoming section, various applications of ES technologies are discussed in detail.

4. Applications of Energy Storage Technologies

ESS can be used to support electrical power system from large scale to small scale applications. ESS can be used to move or shift electricity through the time and can be made to deliver whenever required. ESS helps to balance and smooth out fluctuations of renewable energy (RE) generation with proper deployment and integration into the grid. Also it can be used to support grid reliability and asset utilization. Figure 7 shows the details of different ES Technologies usage in various locations of the power system [4].

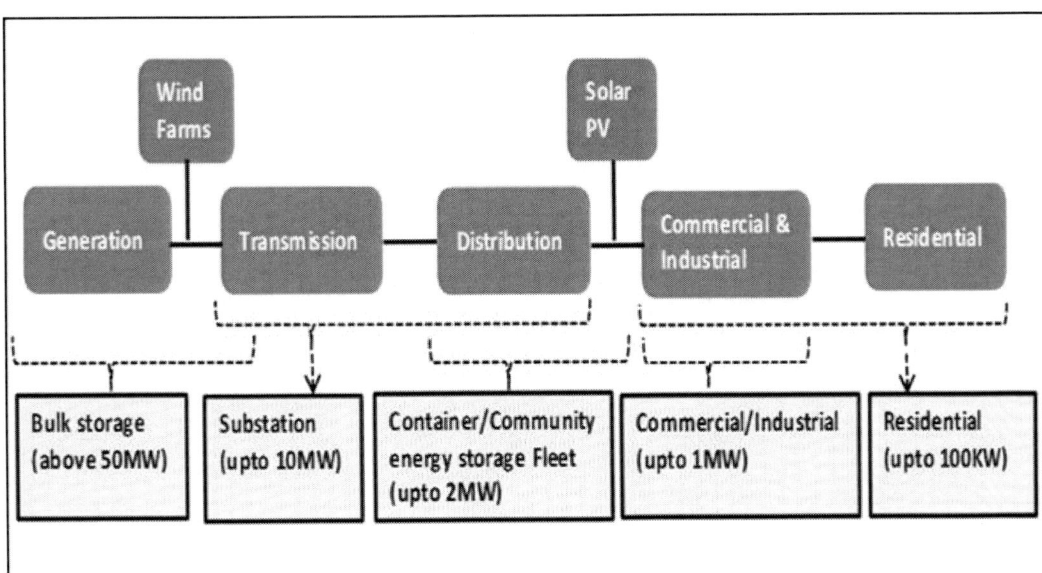

Figure 7. Usage of ES technologies in various grid locations.

4.1. Energy Time Shift (Arbitrage)

By moving or shifting electricity through time, there is a benefit in terms of price difference during on peak and off peak period. purchase of electricity during low demand at low price is stored in the energy storage system and delivered during high demand at high prices in the wholesale network marketplace [4, 5, 9].

4.2. Load Following and Voltage Support

ESS can be used for load following services with varying its output energy to balance generation and load at particular areas of power system network. ESS maintains voltage level in power system network by means of injecting or absorbing reactive power (VAR) on the basis of voltage condition [4, 5, 9].

4.3. Supply Reserve Capacity

ESS can be used as a supply reserve capacity or standby generation capacity that can be ready to call up in case of any sudden or unexpected generation loss. Supply reserve capacity is classified into three subcategories: spinning reserve, supplemental spinning reserve and back up supply. Spinning reserve capacity is an online reserve capacity responds in terms of power system frequency. It responds within 10 minutes of period to compensate for any generation or transmission outages. Supplemental reserve capacity is an off line reserve type, effective after spinning reserve in service. Back up reserve picks up load with a period of 1hour, back up to spinning and supplemental reserve capacity [4, 5, 9].

4.4. Transmission Support

ESS can be used in this ancillary service to support grid stability. Mostly high power density type ES technologies are suitable for this application. ESS provides both real and reactive power with a standard discharge duration of around 2 to 5 seconds in the range for this service [4, 5].

4.5. Transmission Congestion Relief

Transmission congestion occurs when there is any shortage of transmission line capacity due to transmission of excess power at peak demand. ESS can store electricity during off peak with no congestion hours and release during peak congestion hours. To get congestion relief and congestion related charges, ESS can be situated at the downstream of transmission line congestion area [5, 9].

4.6. Retail Demand Charge

Demand charge is based on the maximum load that occurs at the time of peak demand. Electricity end users can be able to reduce overall electricity demand charge by means of using suitable energy storage during the peak demand period [5, 9].

4.7. Electric Service Power Quality

ESS provides a highly reliable service in power systems at industrial and commercial level, in case of any power outage for extended periods. ESS can be used to reduce power quality related financial losses at utility end user level. The suitable storage system can be used to mitigate power quality issues related to voltage or frequency variation during the case of any network disturbance [4, 9].

4.8. Wind and Solar Energy Time Shift

The ESS can be used for moving or shifting excess wind and solar energy between off peak to on peak periods. The ESS reduces overall cost of electricity, enables transmission relief for wind and solar farm, improves wind and solar farm capacity factor by storing excess wind and solar energy during high wind and peak sunshine conditions [5, 9].

4.9. Wind and Solar Energy Smoothing

Wind power generation greatly depends on the wind speed, which varies more frequently in different time scales. Solar power generation mainly depends on the sun position and sunshine hours. Solar power variation also occurs due to partial shading of PV cell by the passing of clouds and weather condition. ESS can be used to smooth out such variation of solar and wind power output and also ensures more reliable solar and wind farm operation [5, 9].

4.10. Renewable Energy Capacity Firming

Solar and wind power output intermittency can be classified into short duration (randomly varies from seconds to minutes) and diurnal (regular variation during 24 hours). Such output variation of solar and wind power is firmed into constant by using suitable storage system. RE capacity firming is effective when there is peak demand in power system. RE firmed capacity offsets a need for additional generation capacity and higher capacity of T&D equipment's [5, 9].

The feasibility levels and analysis results of various ES technologies are discussed in the following section.

5. Feasibility Level of Energy Storage Technologies

In this section, feasibility level of various storage technologies has been identified with the help of ES-Select™ tool. This tool was developed by DNV-KEMA in collaboration with the Sandia National Laboratories. ES-Select™ tool software is useful to study feasibility level of various ES technologies, based on the type of applications at various grid locations. The output feasibility level of each ES technology can be presented in simple and visual form with the help of this software [4]. The Feasibility level of each ES technology can be expressed in percentage (%) of feasibility scores. The feasibility scores of each storage option is estimated on the basis of four criteria's: (i) maturity level: depends on the commercial usage and availability of ES technology, (ii) appropriate grid locations: depends on the availability, size, weight, scalability, mobility of ES technologies, (iii) application requirements: depends on the discharge duration, cycle life, efficiency of ES technologies, (iv) Cost: depends on the installed cost of ES technologies, either in $/KW or $/KWh [4].

The feasibility level of each ES technology can be varied according to the type of applications and grid locations. In ES-select™ tool, grid locations are categorized in five different levels such as Generation and transmission level (Bulk storage of above 50MW), sub transmission level (up to 10MW), sub transmission and distribution level (up to 2MW), first distribution level for commercial/industrial applications (up to 1MW) and second distribution level for residential applications (up to 100KW) [4]. In this study, three grid locations (Generation/Transmission level, distribution at commercial/industrial level, and distribution at residential/small commercial level) with selected applications (energy time shift, load following, Transmission support, wind energy time shift, solar energy time shift, wind energy smoothing, and solar energy smoothing) are considered for identifying feasibility level of each ES technology.

5.1. Feasibility Study at Generation and Transmission Level

5.1.1. Feasibility analysis (Energy time shift/load following/RE time shift)

With the aid of ES-select™ tool, the feasibility analysis of ES technologies has been carried out for the following applications; energy time shift (arbitrage), load following, wind energy time shift, solar energy time shift. The feasibility scores of each ES technology have been estimated on the basis of four criteria: application requirements, maturity level, installed cost, and location requirements. The radar chart in Figure 8 [4] shows the combinational feasibility score level of all four criteria, which determines the overall feasibility level of each ES technology. Figure 9 [4] shows the final results of overall feasibility level of each ES technology. The results clearly show that, in bulk storage level (above 50MW), the Pumped hydro (PHES) and compressed air energy storage (CAES) are highly feasible with 79% score level, and sodium sulfur (NaS) storage technology is the next feasible with score level of 68%, whereas valve regulated lead acid, and advanced lead acid battery storage technologies are less feasible with score levels of 37% and 35% respectively [4].

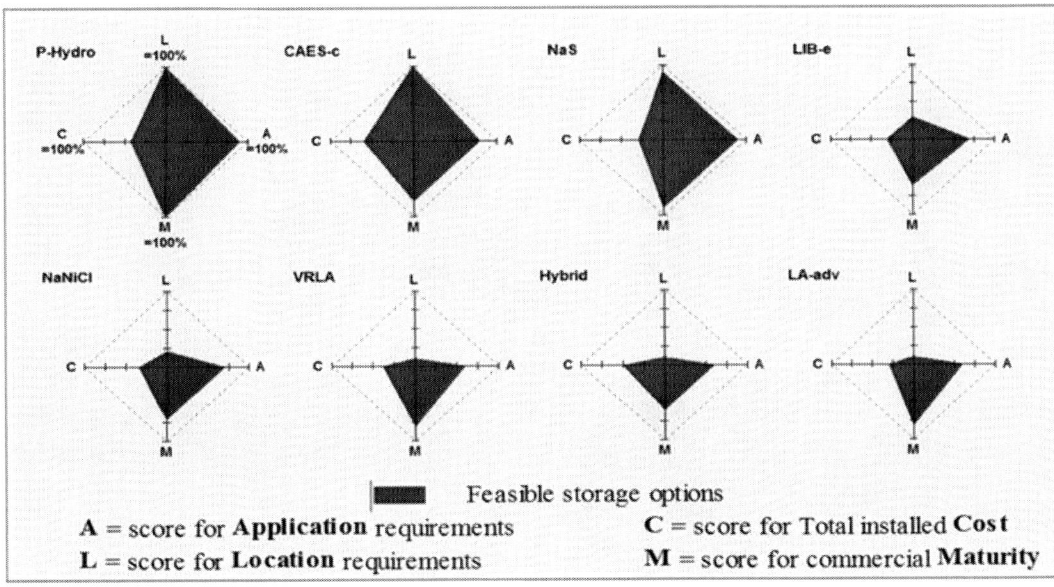

Figure 8. Radar chart for Feasibility score level of ES technologies.

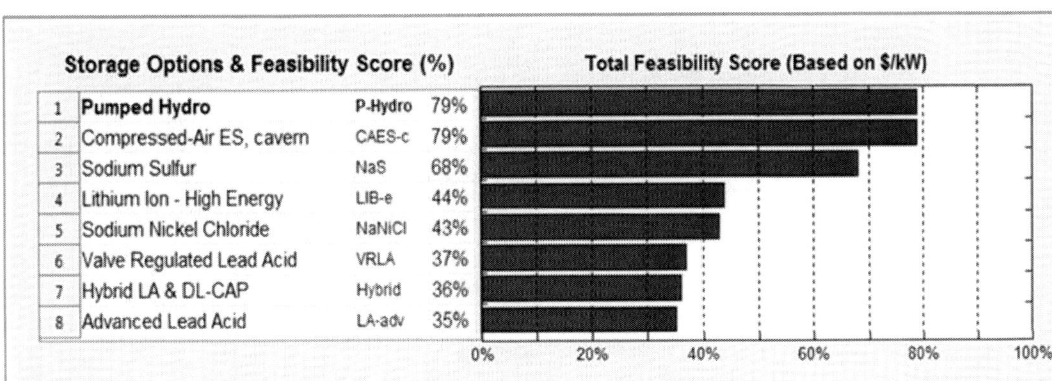

Figure 9. Overall feasibility level of ES technologies.

5.1.2. Feasibility analysis (Transmission support)

Outcomes from the feasibility analysis study with ES-select^{TM} to identify the highly feasible ES technologies for transmission support application in generation and transmission level (bulk storage) is presented in this section. It is clearly identified that high power density type ES technologies are the most suitable to fulfil the application of transmission support. The radar chart Figure 10 [4] shows the combinational feasibility score level of all four criteria for each ES technology. Figure 11 [4] shows the output results of overall feasibility level of suitable ES technologies for this particular application. It is evident from the results that flywheel, Li-ion (high power), and sodium sulfur battery storage technologies are highly feasible with score level of around 66%, 62%, and 61% respectively, whereas advanced lead acid, and hybrid storage (lead acid + double layer capacitor) is less feasible with the score level of around 39% and 36% respectively [4].

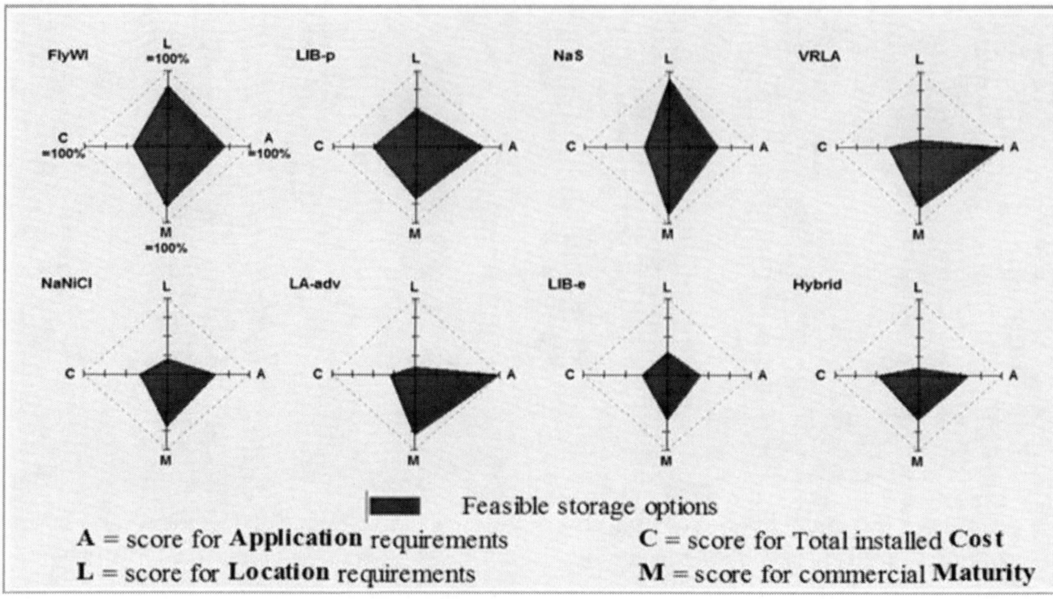

Figure 10. Radar chart for Feasibility score level of ES technologies.

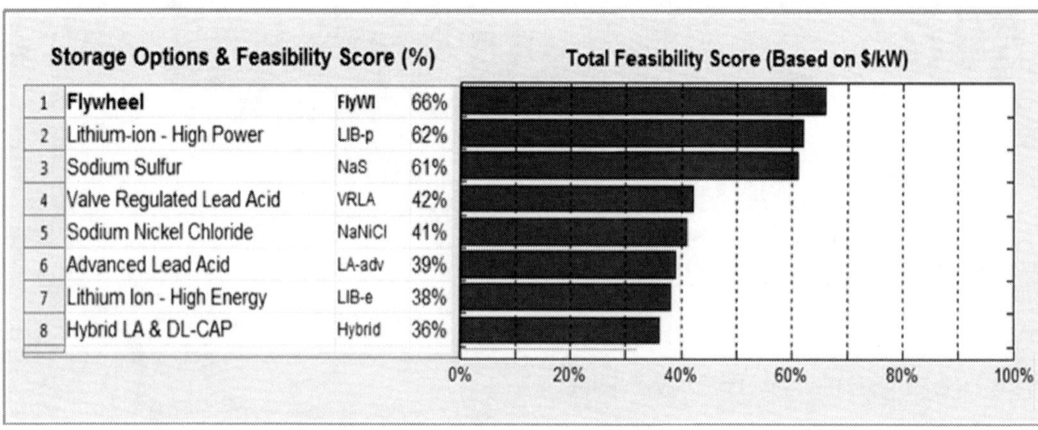

Figure 11. Overall feasibility level of ES technologies.

5.1.3. Feasibility analysis (Wind/solar energy smoothing)

This section highlights the highly feasible ES technologies for wind and solar energy smoothing applications. The radar chart in Figure 12 [4] clearly shows the combinational feasibility score level of all four criteria for each ES technology. As in Figure 13 [4], output results of overall feasibility level, clearly shows that the flywheel, lithium-ion (high power), and sodium sulfur storage technologies are more feasible with feasible scores of around 72%, 62%, and 61% respectively, whereas sodium nickel chloride, hybrid storage (lead acid + double layer capacitor) technologies are less feasible with feasible scores of around 41% and 36% respectively [4].

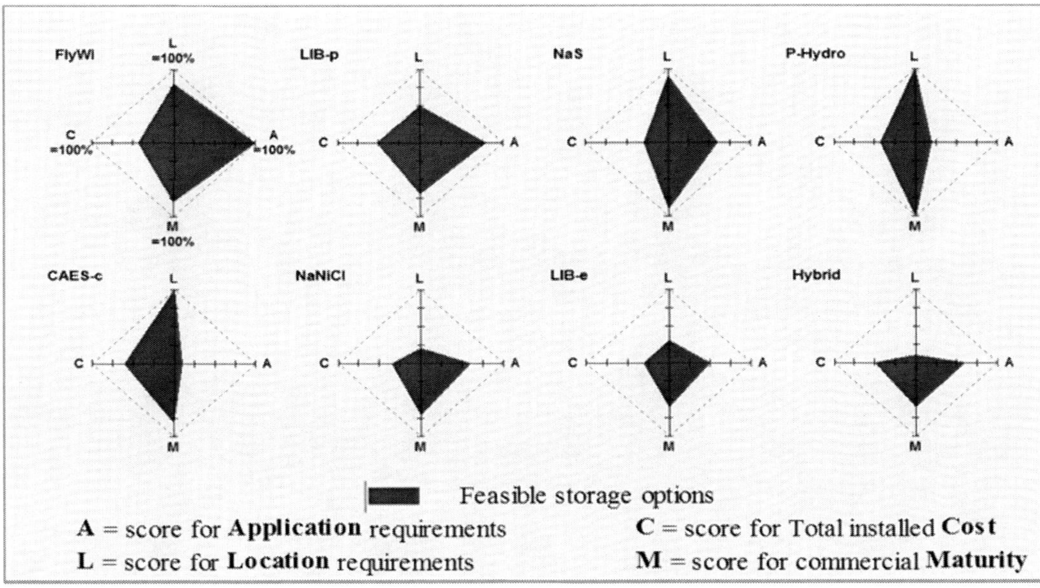

Figure 12. Radar chart for Feasibility score level of ES technologies.

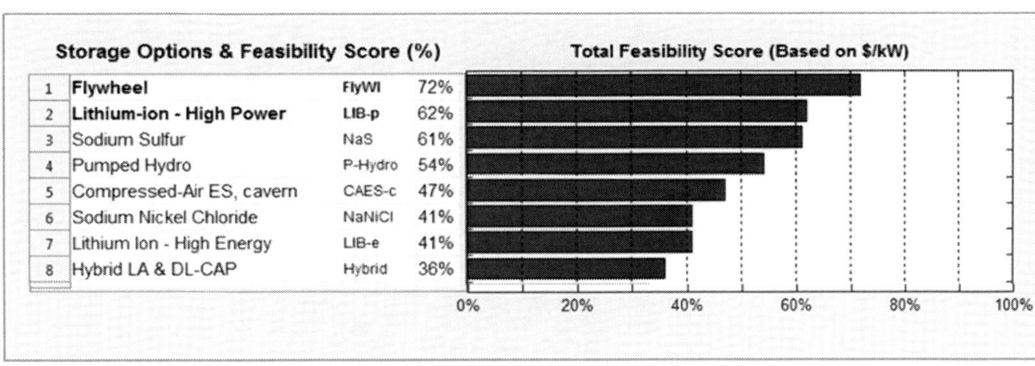

Figure 13. Overall feasibility level of ES technologies.

5.2. Feasibility Study at Distribution Level (Commercial/Industrial)

5.2.1. Feasibility analysis (Energy time shift/load following/RE time shift)

The commercial/industrial power system at distribution level (storage up to 1MW) is now considered for feasibility analysis of ES technologies and used for the following applications: energy time-shift (Arbitrage), load following, wind energy time shift, solar energy time shift. The radar chart in Figure 14 [4] clearly shows the combinational feasibility score level of all four criteria for each ES technology. As in Figure 15 [4], output results of overall feasibility level, clearly shows that the thermal storage (hot), and sodium sulfur battery storage technologies are more feasible with score level of around 70%, similarly thermal storage (cold), and sodium nickel chloride battery storage technologies are the next feasible technologies with the feasible score level of around 64%, whereas nickel base batteries, and vanadium redox flow battery storage technologies are less feasible with feasible score level of around 46% (both) [4].

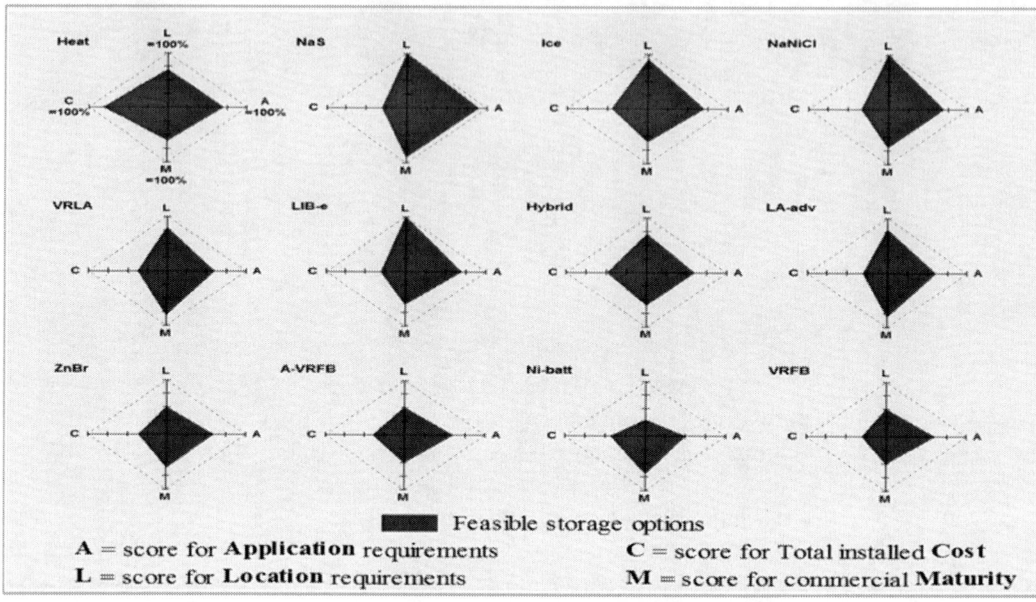

Figure 14. Radar chart for Feasibility score level of ES technologies.

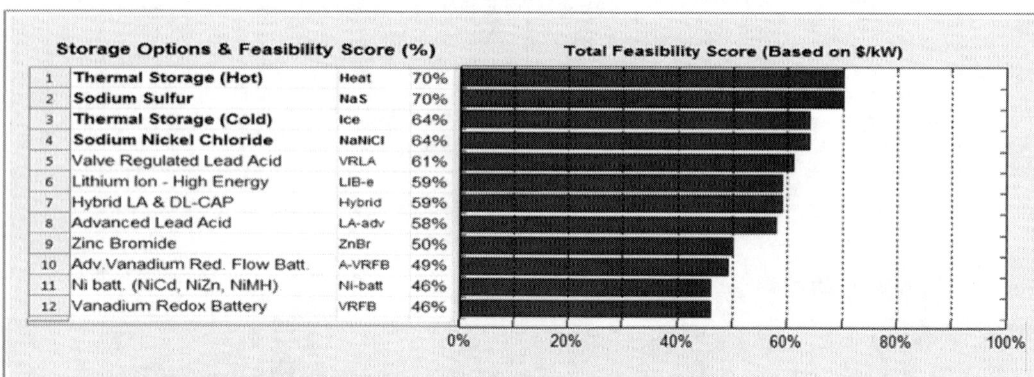

Figure 15. Overall feasibility level of ES technologies.

5.2.2. Feasibility analysis (Wind/solar energy smoothing)

Results from the feasibility analysis study with ES-select™ tool for wind and solar energy smoothing applications has been presented in this section. The radar chart in Figure 16 [4] clearly shows that the combinational feasibility score level of all four criteria for each ES technology. As in Figure 17 [4], output results of overall feasibility level, the Li-ion (high power), sodium sulfur, and sodium nickel chloride battery storage technologies are the highly feasible technologies with feasible scores of around 65%, 63%, and 61% respectively, whereas zinc bromide, advanced vanadium redox flow battery storage technologies are less feasible with the feasible scores of around 42%, and 41% respectively [4].

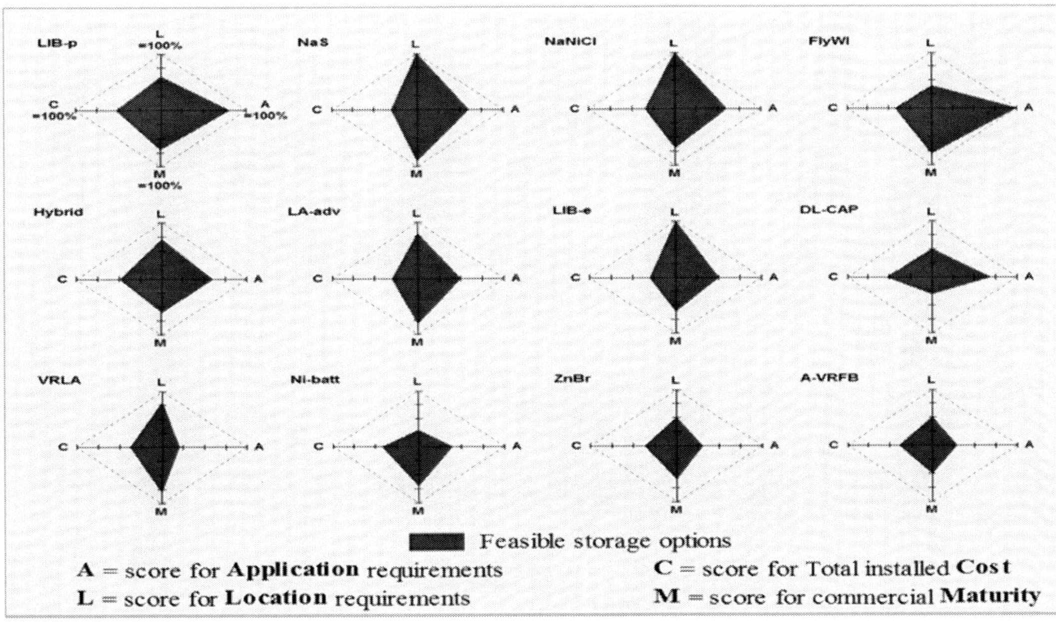

Figure 16. Radar chart for Feasibility score level of ES technologies.

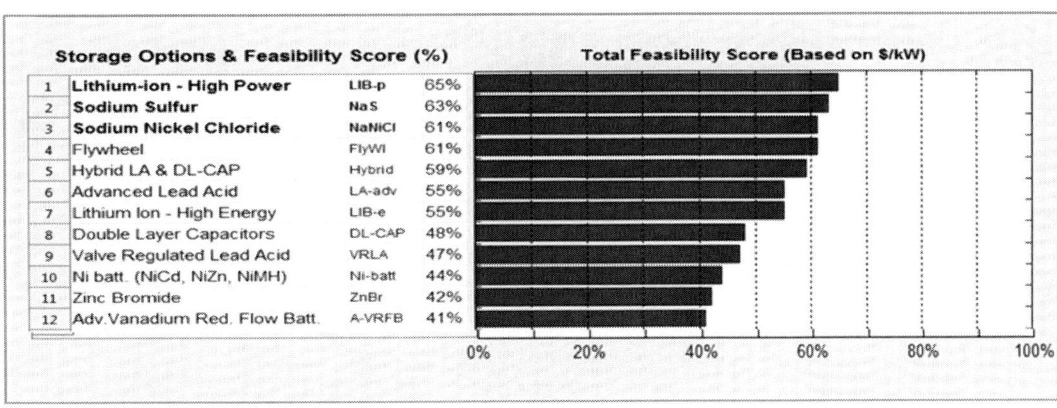

Figure 17. Overall feasibility level of ES technologies.

5.3. Feasibility Study at Low Voltage Distribution Level (Residential/Small Commercial)

5.3.1. Feasibility analysis (Energy time shift/load following/RE time shift)

Results from the feasibility analysis of ES technologies for the energy times shift (Arbitrage), load following, wind energy time shift, solar energy time shift is presented here. The radar chart in Figure 18 [4] clearly shows the combinational feasibility score level of all four criteria for each ES technology. As in Figure 19 [4], output results of overall feasibility level, clearly shows that the thermal storage (hot), valve regulated lead acid battery storage, and Li-ion battery storage technologies are more feasible with feasible scores of around 76%, 61%, and 60% respectively, while nickel base batteries, and zinc bromide battery storage technologies are less feasible with feasible score level of around 46%, and 39% respectively [4].

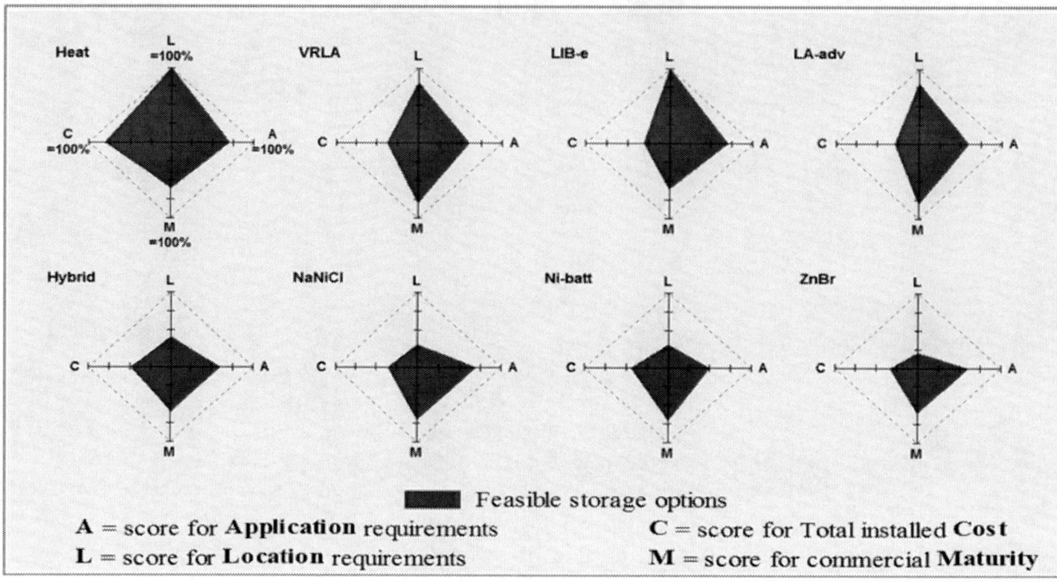

Figure 18. Radar chart for Feasibility score level of ES technologies.

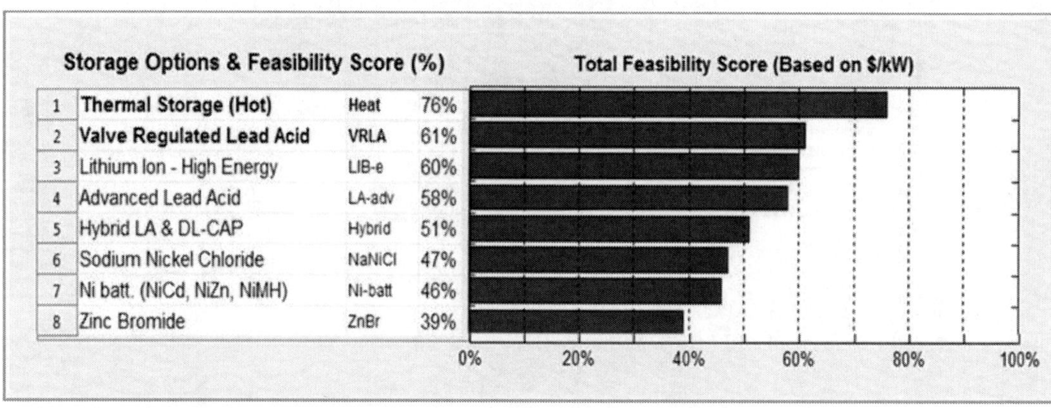

Figure 19. Overall feasibility level of ES technologies.

5.3.2. Feasibility analysis (Wind/solar energy smoothing)

In this section, feasibility analysis results on wind and solar energy smoothing applications is discussed. The radar chart in Figure 20 [4] clearly shows the combinational feasibility score level of all four criteria for each ES technology. As in Figure 21 [4], output results of overall feasibility level, clearly shows that the advanced lead acid battery, Li-ion (high power/high energy) battery storage technologies are more feasible with feasible scores of around 55%, 55%, and 54% respectively, while sodium nickel chloride battery, and nickel base batteries storage technologies are less feasible with score level of around 46%, and 44% respectively [4].

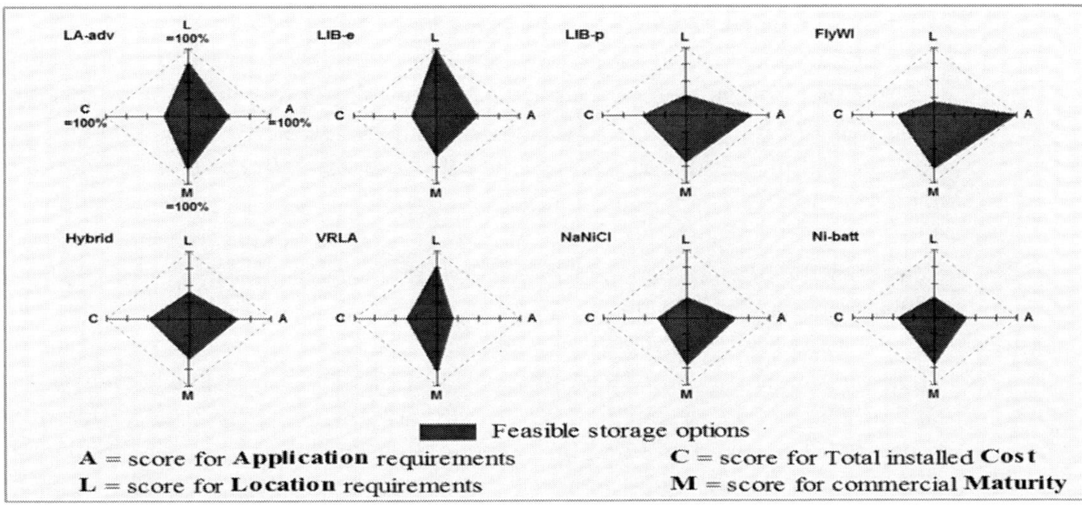

Figure 20. Radar chart for Feasibility score level of ES technologies.

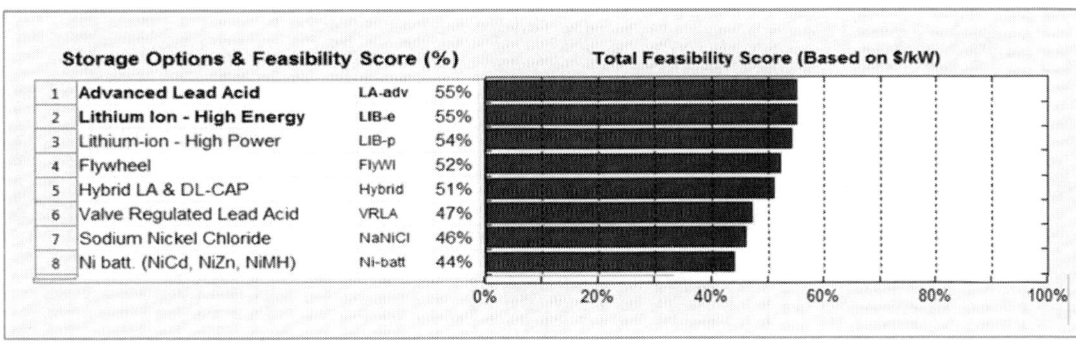

Figure 21. Overall feasibility level of ES technologies.

5.4. Feasibility Analysis Results and Discussion

Tables 2, 3, and 4 shows the summarized overall feasibility results of various ES technologies.

The table-2 shows the overall feasibility level of various storage technologies, used in generation and transmission level. From these results, the following points are concluded.

(i) For the applications: energy time shift, load following, and wind/solar time shift
- Pumped hydro (feasible score: 79%), Compressed air (feasible score: 79%), and sodium sulfur battery (feasible score: 68%) storage technologies are more feasible. This is due to higher feasible scores in the following criteria, like; maturity level, grid location, and application requirements.
- Lead acid battery storage technologies like valve regulated (feasible score: 37%), hybrid storage (feasible score: 36%), and advanced (feasible score: 35%) are less feasible.

(ii) For Transmission support application;

- High power density storage technologies like; flywheel (feasible score: 66%), lithium-ion battery (feasible score: 62%), and sodium sulfur battery (feasible score: 61%) storage technologies are more feasible. This is due to high feasible scores in the following criteria such as maturity level (flywheel, lithium-ion, sodium sulfur) grid location (flywheel, sodium sulfur), and application requirements (flywheel, lithium-ion).
- Lead acid advanced battery (feasible score: 39%), lithium-ion (high energy density with feasible score: 38%), hybrid storage (feasible score: 36%) technologies are less feasible.

(iii) For wind/solar energy smoothing applications;

- High power density storage technologies like; flywheel (feasible score: 72%), lithium-ion battery (feasible score: 62%), and sodium sulfur battery (feasible score: 61%) is more feasible. This is due to high feasible scores to the following criteria's such as maturity level (flywheel, lithium-ion, sodium sulfur), application requirements (flywheel, lithium-ion), and grid location (flywheel, sodium sulfur).
- Sodium nickel chloride (feasible score: 41%), lithium-ion (high energy density with feasible score: 41%), and hybrid storage (feasible score: 36%) is less feasible.

Table 2. Feasibility scores of ES technologies (Generation/Transmission level)

Feasibility level	**High**			**Low**		
ES technologies	**P. Hydro**	**Compress. air**	**Sodium sulfur**	**VRLA**	**Hybrid**	**LA Adv.**
Energy Time shift	79%	79%	68%	37%	36%	35%
Load following	79%	79%	68%	37%	36%	35%
wind time shift	79%	79%	68%	37%	36%	35%
solar Time shift	79%	79%	68%	37%	36%	35%
ES technologies	**Flywheel**	**Li-ion (Hp)**	**Sodium sulfur**	**LA Adv.**	**Li-ion (He)**	**Hybrid**
Transmission support	66%	62%	61%	39%	38%	36%
ES technologies	**Flywheel**	**Li-ion (Hp)**	**Sodium sulfur**	**NaNicl**	**Li-ion (He)**	**Hybrid**
Wind energy smoothing	72%	62%	61%	41%	41%	36%
Solar energy smoothing	72%	62%	61%	41%	41%	36%

The table-3 shows the overall feasibility level of various storage technologies, used in distribution level (commercial/industrial). From these results, the following points are concluded.

(i) For the applications; energy time shift, load following, and wind/solar time shift,

- Thermal energy-hot (feasible score: 70%), sodium sulfur (feasible score: 70%), and thermal energy-cold (feasible score: 64%) storage technologies are more feasible. This is due to higher feasible scores in the following criteria, like; maturity level (thermal energy-hot, sodium sulfur, thermal energy-cold), grid location requirements (thermal energy-hot, sodium sulfur, thermal energy-cold), installed cost (thermal energy-hot), and application requirements (thermal energy-hot, sodium sulfur, thermal energy-cold).

- Advanced vanadium redox flow battery (feasible score: 49%), nickel base batteries (feasible score: 46%), and vanadium redox flow battery (feasible score: 46%) is less feasible.

(ii) For wind/solar energy smoothing applications;

- Lithium-ion (high power density with feasible score: 65%), sodium sulfur battery (feasible score: 63%), and sodium nickel chloride battery (feasible score: 61%) technologies are more feasible. This is due to high feasible scores in the following criteria's like maturity level (lithium-ion, sodium sulfur, sodium nickel chloride), application requirements (lithium-ion), and grid location (sodium sulfur, sodium nickel chloride).
- Nickel base batteries (feasible score: 44%), zinc bromide battery (feasible score: 42%), and vanadium redox flow battery (feasible score: 41%) is less feasible.

Table 3. Feasibility scores of ES technologies (Commercial/Industrial: up to 1MW)

Feasibility level	**High**			**Low**		
ES technologies	**TES (hot)**	**Sodium sulfur**	**TES (cold)**	**VRFB. Adv.**	**Ni-batt**	**VRFB**
Energy Time shift	70%	70%	64%	49%	46%	46%
Load following	70%	70%	64%	49%	46%	46%
wind time shift	70%	70%	64%	49%	46%	46%
solar Time shift	70%	70%	64%	49%	46%	46%
ES technologies	**Li-ion (Hp)**	**Sodium sulfur**	**NaNicl**	**Ni. batt**	**Zn-Br**	**VRFB Adv.**
Wind energy smoothing	65%	63%	61%	44%	42%	41%
Solar energy smoothing	65%	63%	61%	44%	42%	41%

The table-4 shows the overall feasibility level of various storage technologies, used in distribution level (residential/small commercial). From these results, the following points are concluded.

(i) For the applications; energy time shift, load following, and wind/solar time shift,

- Thermal energy-hot (feasible score: 76%), valve regulated lead acid battery (feasible score: 61%), lithium-ion (high energy density with feasible score: 60%) are more feasible. This is due to higher feasible scores in the following criteria's like maturity level (valve regulated lead acid), grid location requirements (thermal energy-hot, valve regulated lead acid, lithium-ion), installed cost (thermal energy-hot), and application requirements (thermal energy-hot, lithium-ion).
- Sodium nickel chloride battery (feasible score: 47%), nickel base batteries (feasible score: 46%), and zinc bromide battery (feasible score: 39%) are less feasible.

(ii) For wind/solar energy smoothing applications;

- Advanced lead acid battery (feasible score: 55%), Lithium-ion battery (high energy density with feasible score: 55%), and lithium-ion battery (high power density with feasible score: 54%) storage technologies are more feasible. This is due to high feasible scores in the following criteria, like; maturity level (advanced lead acid), application requirements (lithium-ion high power), and grid location requirements (lithium-ion high energy, advanced lead acid).
- The valve regulated lead acid battery (feasible score: 47%), sodium nickel chloride battery (feasible score: 46%), and nickel base batteries (feasible score: 44%) are less feasible.

Table 4. Feasibility scores of ES technologies (Residential level: up to 100KW)

Feasibility level	**High**			**Low**		
ES technologies	**TES (hot)**	**VRLA**	**Li-ion (He)**	**NaNicl**	**Ni-batt**	**Zn-Br**
Energy Time shift	76%	61%	60%	47%	46%	39%
Load following	76%	61%	60%	47%	46%	39%
wind time shift	76%	61%	60%	47%	46%	39%
solar Time shift	76%	61%	60%	47%	46%	39%
ES technologies	**LA Adv**	**Li-ion (He)**	**Li-ion (Hp)**	**VRLA**	**NaNicl**	**Ni-batt**
Wind energy smoothing	55%	55%	54%	47%	46%	44%
Solar energy smoothing	55%	55%	54%	47%	46%	44%

CONCLUSION

An extensive review study of various ES technologies was performed at the characteristics and applications point of view. ES-select^{TM} helps to identify the merits and demerits of each ES technology and also gives an idea to identify the suitable storage system for particular applications. According to the different characteristic condition, each ES technology is suitable for a particular application. For example, flywheel, super capacitor and SMES belong to the high power density type and suitable for the power quality applications.

At feasibility point of view, due to higher cycle efficiency, and long discharge duration characteristics that the Pumped hydro, and compressed air storage technologies are most suitable for energy time shift, and load following applications at generation and transmission level. Because of high energy density, high efficiency and long cycle life characteristics, the sodium sulfur battery is more feasible and applicable storage technology for energy time shift, load following, transmission support, renewable energy smoothing applications in transmission and distribution level (generation/transmission, commercial/industrial). Similarly, lithium-ion battery storage is suitable for transmission support application at a large level, and also suitable for load following, renewable energy integration applications at the distribution level, because of its high power/energy density, higher efficiency, and low self-discharge characteristics. Due to higher maturity and affordable cost, the thermal energy storage, and lead acid storage technologies are more suitable for distribution level (commercial/industrial, and residential/small commercial) applications. Finally, the feasibility analysis results clearly show that some of the storage technologies are more feasible for a particular application while some other storage technologies are suitable for other applications.

This study requires further research and development in the following directions:

- Research and study in depth analysis is required to develop a smart hybrid energy storage system for power system applications.
- Further research study is required to develop feasible, smart ESS for smart grid and micro grid applications in future.

REFERENCES

[1] Barnes, FS; Levine, JG. *Large energy storage systems handbook*: CRC Press, 2011.

[2] Ter-Gazarian, AG. "*Energy Storage for Power Systems* (2nd Edition)," ed: Institution of Engineering and Technology.

[3] Carnegie, R; Gotham, D; Nderitu, D; Preckel, PV. "Utility Scale Energy Storage Systems," 2013.

[4] Laborotories, SN. (2012). *ES select*. Available: http://www.sandia.gov/ess/ esselect. html

[5] Rastler, D. *Electricity energy storage technology options: a white paper primer on applications, costs and benefits:* Electric Power Research Institute, 2010.

[6] Hall, PJ; Bain, EJ. "Energy-storage technologies and electricity generation," *Energy policy*, vol. 36, 4352-4355, 2008.

[7] Chatzivasileiadi, A; Ampatzi, E; Knight, I. "Characteristics of electrical energy storage technologies and their applications in buildings," *Renewable and Sustainable Energy Reviews*, vol. 25, 814-830, 2013.

[8] Ibrahim, H; Ilinca, A; Perron, J. "Energy storage systems—characteristics and comparisons," *Renewable and Sustainable Energy Reviews*, vol. 12, 1221-1250, 2008.

[9] Eyer, J; Corey, G. "Energy storage for the electricity grid: Benefits and market potential assessment guide," *Sandia National Laboratories Report*, *SAND2010-0815*, *Albuquerque*, New Mexico, 2010.

[10] Ibrahim, H; Ilinca, A. *"Techno-Economic Analysis of Different Energy Storage Technologies,"* 2013.

[11] Poonpun, P; Jewell, WT. "Analysis of the cost per kilowatt hour to store electricity," *IEEE Transactions on energy conversion*, vol. 23, 529-534, 2008.

[12] Akhil, AA; Huff, G; Currier, AB; Kaun, BC; Rastler, DM; Chen, SB; et al., "DOE/EPRI 2013 electricity storage handbook in collaboration with NRECA," *Report SAND2013-5131*, *Sandia National Laboratories*, 2013.

[13] Zhang, J; Zhang, L; Liu, H; Sun, A; Liu, R.-S. *Electrochemical technologies for energy storage and conversion*, vol. 1: John Wiley & Sons, 2011.

In: Renewable Energy and Sustainable Development
Editors: R. T. Durai Prabhakaran, S. A. Kale et al.
ISBN: 978-1-63463-429-8

Chapter 14

APPLICATIONS OF CHITOSAN NANOPARTICLES IN ACTIVE BIODEGRADABLE AND SUSTAINABLE FOOD PACKAGING

***Farnaz Assa*[1]*, Hoda Jafarizadeh-Malmiri*[2*]*, Navideh Anarjan*[3]*, Aydin Berenjian*[4] *and Younes Ghasemi*[5]**

[1,2]Department of Chemical Engineering, Sahand University of Technology, Tabriz, Iran
[3]Department of Engineering, Collage of Chemical Engineering, Tabriz Branch, Islamic Azad University, Tabriz, Iran
[4]School of Engineering, Faculty of Science & Engineering, The University of Waikato, Hamilton, New Zealand
[5]Department of Pharmaceutical Biotechnology, Faculty of Pharmacy and Pharmaceutical Sciences Research Centre, Shiraz University of Medical Sciences, Shiraz, Iran

ABSTRACT

Chitosan due to its unique properties such as no toxicity, biodegradability, antimicrobial and antioxidant effects have been regarded as a promising alternative for use in sustainable food packaging systems. Utilization of nanotechnology in food packaging significantly enhances the active agent's properties for antimicrobial applications. In this chapter, the influence of nanotechnology especially biodegradable chitosan nanoparticles in food packaging was investigated in order to gain a better understanding about the current trends and future prospects of this protocol.

1. INTRODUCTION

Chitosan (poly-(b-1/4)-2-amino-2-deoxy-D-glucopyranose) is a natural polymer and is the N-deactylated derivative of chitin which is naturally obtained from crabs and shrimp shells or microorganisms such as fungi and bacteria [1, 2]. Chitosan properties can be mentioned due to its chemical and biochemical characteristics. It is chemically inert,

[*] Corresponding author: Email: h_jafarizadeh@sut.ac.ir, h_jafarizadeh@yahoo.com.

hydrophilic and insoluble in water, alkali and organic solvents [3]. Furthermore, chitosan has some unique biological characteristics such as biodegradability, nontoxicity and antimicrobial activity [4]. These special properties of chitosan have led to some variety of applications. It can be used in food and food packaging, waste treatment, cosmetics, agriculture, textile, pharmaceutical and even fuel cell industries. These properties can be improved by applying some modification such as utilization of chitosan nanoparticles.

Nanotechnology means manipulation of material at a very small scale usually between 1 to 100 nanometers. This miniaturization leads to very impressive properties and functions. Nonomaterials can be found in many forms like: nanoparticles, nano rods, nano tubes, nano sheets, nanofibers, etc. It must be noted that when at least one dimension of a discrete particle is in order of 100 nm or less, it can be defined as a nanoparticle [5-7]. In this regard, recently many researchers have investigated the production of chitosan nanoparticles and have successfully modified its different characteristics for specific applications [8, 9].

As previously mentioned, chitosan has remarkable antimicrobial properties. This characteristic has been investigated against many bacteria, yeasts and fungi. Investigations about antimicrobial activity of chitosan demonstrate that it is dependent on some parameters such as microorganism species, pKa, pH, molecular weight, metal cations existence and its degree of deacetylation [10-13].

The main purpose of food packaging is to keep the quality and safety of foods during the process of storage as well as transportation and delivery to customers. Thus, it is used to prolong their shelf life by protecting from undesirable factors and conditions like chemical contaminants, moisture, microorganisms, etc. This role, therefore, can be undertaken when packaging materials can provide physical protection and good physicochemical conditions for food products. Recently, petroleum based plastic materials which are widely used in packaging materials, have resulted to some crucial environmental problems. So biodegradable property of food packaging related to petroleum based one attracts scientists to investigate more about it. Various food grade antimicrobial agents are suggested to utilize in edible food packaging such as sorbic acid, potassium sorbate, benzoic acid, citric acid, polysaccharides, etc. Among these, chitosan offers real potential for applying in these kinds of packaging systems [14-17].

2. Active Packaging

Active packaging is a novel packaging method which is incorporated with materials such as antioxidant, antimicrobial and/or nutrients to make an interaction between packaging and food to improve food quality and safety. This method can employ some technology to delay in oxidation and control the respiration rate, moisture migration and microbial growth to prolong the shelf life of food products [18-20]. Therefore, active packaging does not only provide a barrier against harmful external factors, but also releases or absorbs substances into food and its environment to maintain or even improve the condition of food products [21]. The substances which have the main role in active packaging can be a component of the packaging system or directly exist in whole packaging materials. Due to their different operations, they usually consist of various layers or portions [22].

Active packaging contains some additives depending on their application and packaged food. These additives have some properties such as releasing substances like oxygen, ethanol and carbon dioxide, antimicrobial activities, etc. Table 1 and 2 shows some of these additives and their applications in different industries [23].

Recently two types of active packaging have attracted the attention of researchers and scientist. These include antimicrobial and antioxidant packaging systems. Antioxidants are known as additives which are added to polymers to prevent their degradation during thermal processes. They are also added to food packaging to prevent oxidation, one of the major problems which affects food quality. In the past, synthetic antioxidants incorporate to food packaging systems which created risk of toxicity by migration into food. However, currently food additive antioxidants such as butylated hydroxyl anisole (BHA) and butylated hydroxyl toluene (BHT) have been used more frequently [24, 25]. Though these antioxidants have an effective influence on food packaging industries due to their low cost and high stability, there are still some concerns about their permeation into food products and creation of health risk. Therefore, research is in progress to introduce some natural antioxidants such as phenolic compounds instead of synthetic antioxidants [20, 26-28].

Antioxidant packages prevent oxidation of food product by two methods: firstly, releasing of antioxidant compound to food, and secondly scavenging the unfavorable substance like oxygen. Application of antioxidant packaging which release antioxidant materials to food has benefits related to direct addition of antioxidant to food that in this method these active materials release to food by controlled concentration rates and its determined concentrations before storage maintain constant. In later method, the scavengers which are compounds that influence on substances involved in the oxidation process should be designed properly to allow their access to pro-oxidant compounds. In addition to the selection of suitable antioxidant compound, its interaction with packaging materials is a critical point which should be considered [29-31].

Another kind of active packaging which has recently attracted many attentions is the antimicrobial active packaging. The main function of this packaging is to assist the preservation and to prolong the shelf life of packaged food by suppressing the growth of spoilage microorganisms [32]. The active agents for production of antimicrobial packaging include organic acids, fungicides, bacteriocins, enzymes, ceramic materials with Ag ions, inorganic gases and planet extract. The usage of these agents relates to the kind of microorganisms which causes the spoilage of packaged food. Organic acids, bacteriocins and spice extracts show good influence on control of meat spoilage [33, 34]. Potassium sorbate and nisin have antimicrobial activity against yeast and gram-positive bacteria, respectively [35, 36].

The antimicrobial agents can be incorporated to packaging systems by different methods such as separate coating with volatile antimicrobial agents, which are added to main packaging, coating onto polymeric surfaces, application of antimicrobial polymers such as chitosan and poly-L-lysine, immobilization of agents to polymers, incorporation of volatile or non-volatile antimicrobial agents directly to polymeric packaging [37]. Figure 1 shows the typical structure of antimicrobial multi layer film.

Nowadays, there is an increasing interest in natural polymers which have inherently various properties for active packaging. Due to their biodegradable characteristics, they can be nominated as an alternative to commodity polymers. In addition, since they have intrinsic antimicrobial and antioxidant properties there is no need for incorporating active agents. Most of these polymers are made from proteins and polysaccharides which can be replaced by some of conventional synthetic packaging materials exerted to preserve food products. Among active packaging sources, chitosan is a biopolymer with special properties which are discussed in the upcoming sections [12, 21, 38].

Table 1. Applications of active packaging technologies

Type of application	Foods
Oxygen scavengers	Ground coffee, tea, roasted nuts, potato chips, chocolate, fat powdered milk, powdered drinks, bread, tortillas, pizza, pizza crust, refrigerated fresh pasta, fruit tortes, cakes, cookies, beer, deli meats, smoke and cured meats, fish, cheese
Carbon dioxide absorbers	Ground coffee
Carbon dioxide emitters	Meat, fish
Moisture absorbers	Dry and dehydrated products, meat, poultry, fish
Ethylene scavengers	Kiwifruit, banana, avocados, persimmons
Ethanol emitters	Bread, cakes, fish
Antimicrobial releasing films	Dry apricots
Antioxidant releasing films	Cereals
Flavor absorbing films	Navel orange juice
Flavor releasing films	Ground coffee
Color containing films	Surimi
Anti-fogging films	Some fresh fruit and vegetable packages
Anti-sticking films	Soft candies, cheese slices
Light absorbers	Pizza, milk
Time-temperature indicators	Microwaveable pancake syrup, refrigerated pasta, deli items
Gas permeable/breathable films	Ready-to-eat salads
Microwave susceptors	Ready-to-eat meals

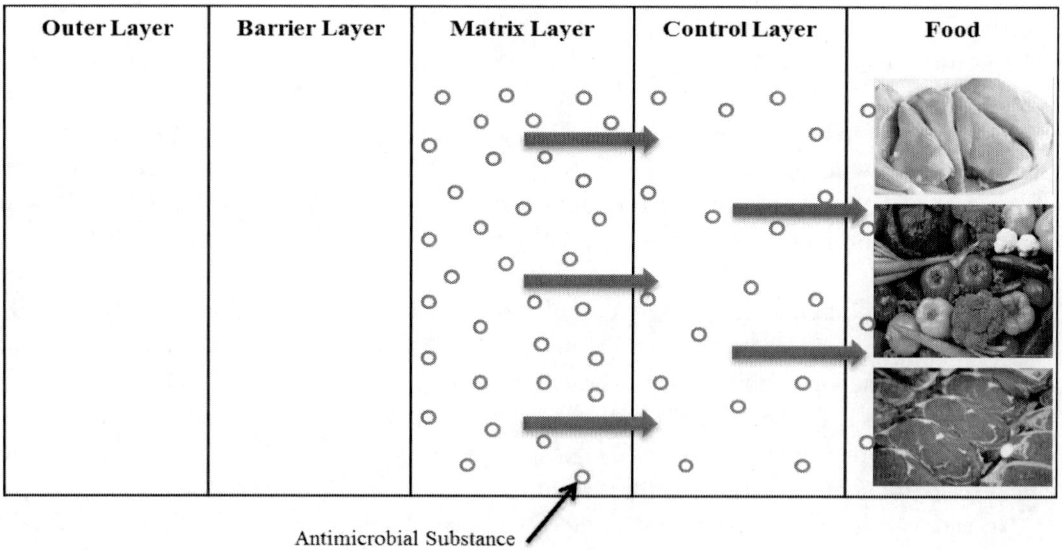

Figure 1. A structure of typical antimicrobial multi-layer active film.

Table 2. Examples of some active packaging systems

Type of active packaging system	Substances used and mode of action
Oxygen absorbing	Enzymatic systems (glucose oxidase-glucose, alcohol oxidase-ethanol vapor) Chemical systems (powdered iron oxide, catechol, ferrous carbonate, iron-sulfur, sulfite salt-copper sulfate, photosensitive dye oxidation, ascorbic acid oxidation, catalytic conversion of oxygen by platinum catalyst)
Carbon dioxide absorbing/ emitting	Iron powder-calcium hydroxide, ferrous carbonate-metal halide
Moisture absorbing	Silica gel, propylene glycol, polyvinyl alcohol, diatomaceous earth
Ethylene absorbing	Activated charcoal, silica gel-potassium permanganate, Kieselguhr, bentonite, Fuller's earth, silicon dioxide powder, powdered Oya stone, zeolite, ozone
Ethanol emitting	Encapsulated ethanol
Antimicrobial releasing	Sorbates, benzoates, propionates, ethanol, ozone, peroxide, sulfur dioxide, antibiotics, silver-zeolite, quaternary ammonium salts
Antioxidant releasing	BHA, BHT, TBHQ, ascorbic acid, tocopherol
Flavor absorbing	Baking soda, active charcoal
Flavor releasing	Many food flavors
Color containing	Various food colors
Anti-fogging and anti-sticking	Biaxially oriented vinylon, compression rolled oriented HDPE
Light absorbing/ regulating	UV blocking agents, hydroxybenzophenone
Monitoring	Time-temperature indicators
Temperature controlling	Non-woven microperforated plastic
Gas permeable/ breathable	Surface treated, perforated or microporous films
Microwave susceptors	Metallized thermoplastics
Insect repellant	Low toxicity fumigants (pyrethrins, permethrin)

3. Vital Nanotechnology Benefits: Application Scopes in Food Packaging

Nanotechnology is the science of manufacturing and application of materials and structures which are at the scale of nanometer. Nowadays, it has been applied in many science and technology such as electronics, energy, catalysts, agriculture and food industry. In recent years, significant investments in the field of nanotechnology have been made by food and agricultural industries. These efforts have improved the quality of products and have reduced the total costs [39, 40].

There are two known categories for application of nanotechnology especially in food industries. These include "top down" and "bottom up" systems (see Figure 2). The top down approach means breaking down of large particles to materials that are in range of nanometric dimensions. This method of manufacturing of nanoparticles is basically included physical and mechanical processing such as grinding and milling [41]. For example dry milling of wheat to flour increases the water binding capacity of flour [42, 43]. Moreover, size reduction technology has been used to improve the antioxidant effect of green tea. When the particle size of green tea is decreased to about 1000 nm, the high ratio of nutrient digestion and absorption leads to high activity of the oxygen eliminating enzyme [44]. The homogenization process, which is widely used in diary industries, is another example of the top down size reduction mechanism. In this process the size of fat globules are reduced by applying pressure [45].

In the other hand, the self-organization and self-assembly of biological compound is categorized in bottom up approach. Crystallization, layer by layer deposition, microbial synthesis, biomass reaction and solvent extraction-evaporation are the methods which can be used in bottom up way of manufacturing of nanomaterials. In this method, molecules arrange step by step with specific features [46]. For example, the casein micelles can result to a stable nanomaterial by their self-assembly. A balance between various non-covalent forces can lead to self-organized biological entities on the nanometer scale [47].

Reduction of particle sizes in nanotechnology develop some properties of nano scale biomaterials like surface area, solubility, delivery properties, absorption by cells and the residence time in the body. Some of these properties such as high surface area increase the efficiency of biomaterials to reduce risk of certain diseases such as cancer [48].

Organic materials which naturally exist in foods such as carbohydrate, fat, vitamins and proteins can be in different sizes such as large macromolecules or a simple mono molecule in the nano range. Due to special properties of biological nano materials, they can be used for various purposes to improve taste, texture and sustainability. Efficiency of these nano materials depends on different parameters such as their biological and chemical properties and composition of the food [40, 46].

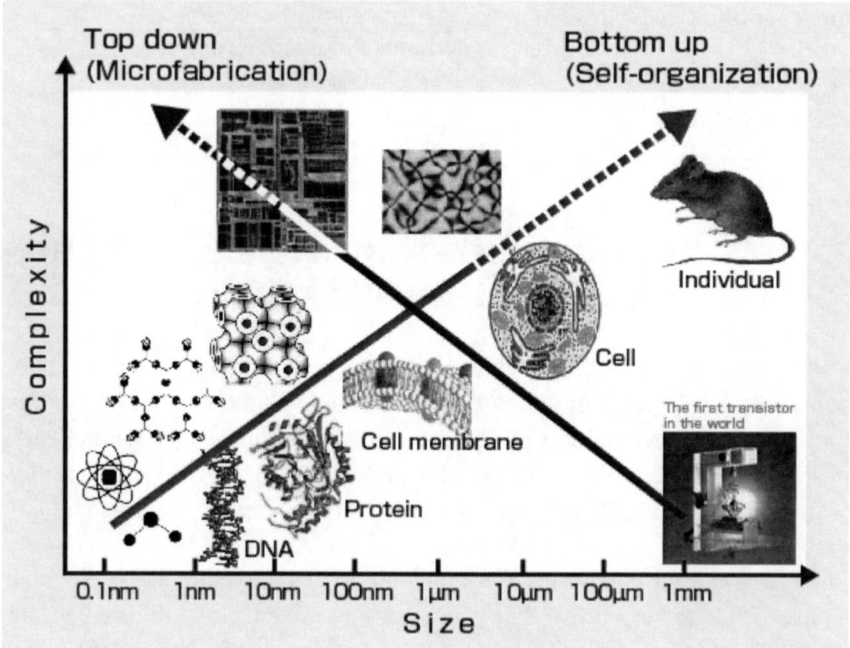

Figure 2. Schematic of “top down” and “bottom up” approaches for nanomaterials manufacturing.

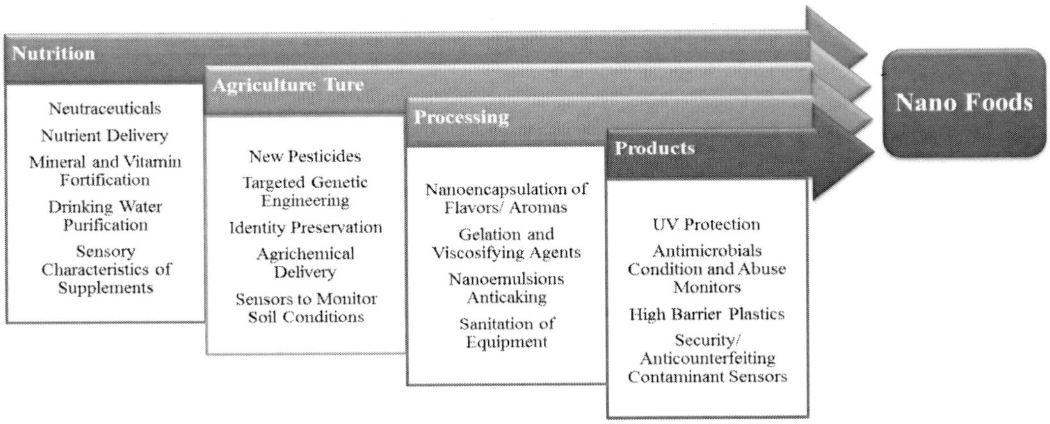

Figure 3. Nanotechnology rule in food industry.

Due to several advantages which were counted above, nanotechnology can play an effective role in food industry (see Figure 3). Bioactive proteins which are used in functional food for their health benefits is one of these applications. Their reduced size helps them to improve their availability and solubility and as a result their ability to be transferred across intestinal membranes [49]. In addition, nanoemulsions which have significant rheological and textural properties can be used in food products to reduce their fat content with no change in their creaminess.

The main application of nanotechnology that has received much attention recently is in food packaging as shown in Figure 4. In this mode of packaging, it is expected that packaging to be more than just a physical barrier that protects the packaged food from surrounding environment. Therefore, the nano materials which are incorporated in food packaging must be able to control the passage of gases and moisture through packaging to prolong the shelf life and preserve its initial freshness and quality. Furthermore, they improve the mechanical and barrier properties of packaging and their biodegradability and could generate less waste [50, 51].

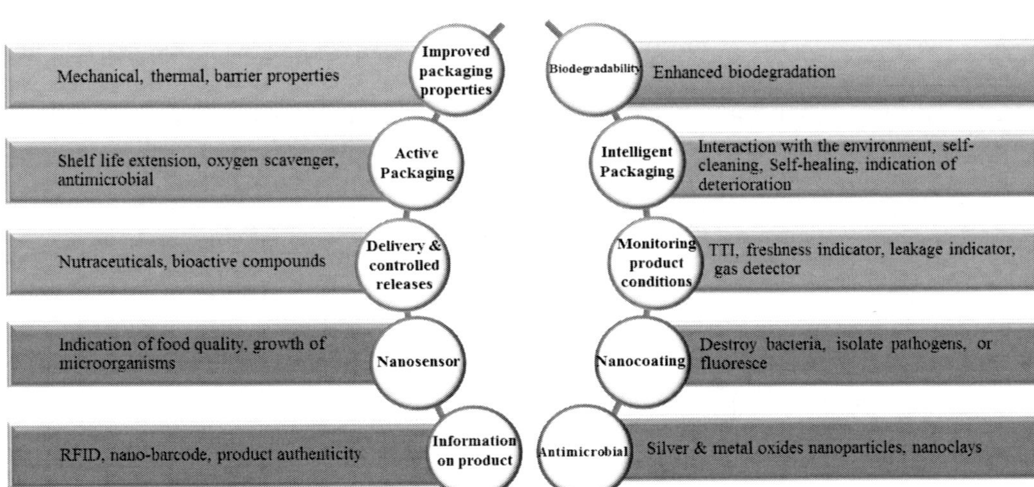

Figure 4. Potential applications of nanotechnology in the food packaging sector.

Bio-nanocomposites are a new class of nanomaterials which are considered as a promising alternative for improvement of food packaging materials. They are hybrid nanostructured materials which are consisted of a biopolymer matrix strengthened by inorganic nanoparticles. Two kinds of bio polymer are known: natural polymers such as agar, starch, chitosan, whey, collagen, gluten, etc., synthetic biodegradable polymers like poly vinyl alcohol, poly glycolic acid, poly butylene succinate, etc. and microbial fermentation based polymers like poly hydroxyl alkanoates and pullulan (see Figure 5) [52, 53].

Biopolymers alone cannot be used specially for food packaging which needs mechanical and barrier properties. Therefore, mechanical and thermal properties of biopolymers can be improved by addition of nano scale fillers. Application of layered silicate fillers such as montmorillonite will improve the mechanical and thermal as well as gas barrier properties of bio polymers [6].

4. Active Agents Incorporated at Nano-Scale

Nanotechnology application in food packaging is not limited to preparation of biopolymers. In addition, active agents which were mentioned above can be incorporated to food packaging at nano scale. Due to high surface area to volume ratio of nano sized active agents their activity will significantly increase. For example, nano antimicrobial agents such as silver, metal oxides like titanium dioxide, zinc oxide and magnesium oxide and clays are able to inactive more microbial cells.

Silver antimicrobial activities against gram negative and gram positive bacteria, fungi and protozoa have been well known since ancient times. Due to its significant antimicrobial characteristics, by recognition of nanotechnology many researches were performed to investigate the effect of its nanoparticles on its antimicrobial activity. Because of their high surface area to volume ratio, silver nanoparticles can suitably interact with microbial cells. Therefore, silver nanoparticles widely were incorporated at nanocomposites used in food packaging and it was found out that although they were used in small amount but they eliminated most or in some cases (such as *E.coli*) all of microbial cells [54, 55]. In other research silver nanoparticles incorporated to low density polyethylene films as orange juicepackage. It was found that by incorporating of 1% silver nanoparticles in film packaging, total amount of yeast and molds were significantly decreased compared to the unloaded polyethylene film and the shelf life was extended up to 28 days at 4 °C [56].

Metal oxides are another type of antimicrobial agents which are studied for their nano scale form in food packaging application. The effect of nanoparticles and nanotubes of TiO_2 against *E. coli* were investigated by Jing et al. It was reported that TiO_2 nanoparticles with high surface area of 587.7 m^2/g displayed better antibacterial activity than TiO_2 nanotubes [57]. Jin et al. demonstrated that ZnO nanoparticles have dramatic effects against gram positive than gram negative bacteria. Furthermore, their antimicrobial activity against *E.coli*, *Salmonella enteritidis* and *Listeria monocytogenes* were reported [58].

Nanoclays can be applied as antimicrobial agents in food packaging in addition of enhancement of their mechanical and thermal properties. It is considerable that nanoclays should be modified due to their hydrophilic characteristics made them ineffective in hydrophobic polymers. This modification method is a series of cation exchange reactions. For example organically modified montmorillonites can destroy the cell walls of gram positive and gram negative bacteria [59].

5. CHITOSAN ROLE

Chitosan as a polysaccharide which derived by deacetylation of chitin, is obtained from crab and shrimp shells. Chitosan is a bio polymer which consists of β-(1-4)-2-acetamido-D-glucose and β-(1-4)-2-amino-D-glucose units as shown in Figure 6. It can be applied in such scientific and industrial fields as biotechnology, pharmaceutics, cosmetics, wastewater treatment, drug delivery, food science and packaging. Its significant characteristics such as biofunctional, biodegradability, nontoxic, film forming and antimicrobial activity have made chitosan a promising packaging material. Beside these characteristics, chitosan films are able to postpone the rate of mass transfer, moisture and shelf life [2, 8, 60].

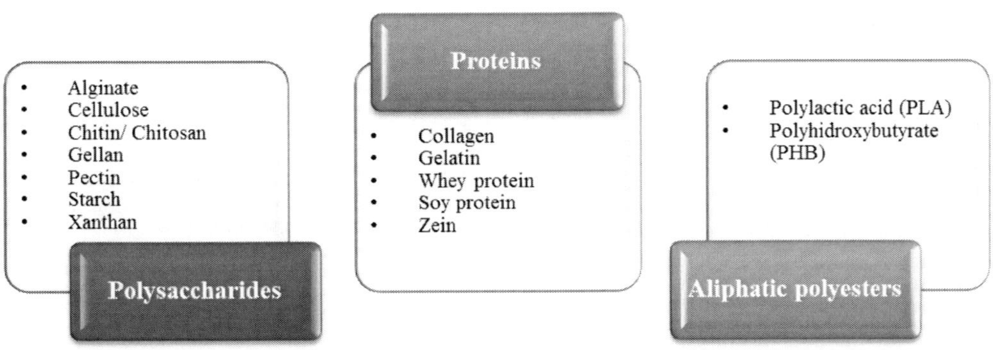

Figure 5. Classification of biopolymers depending on the general chemical composition used in food packaging.

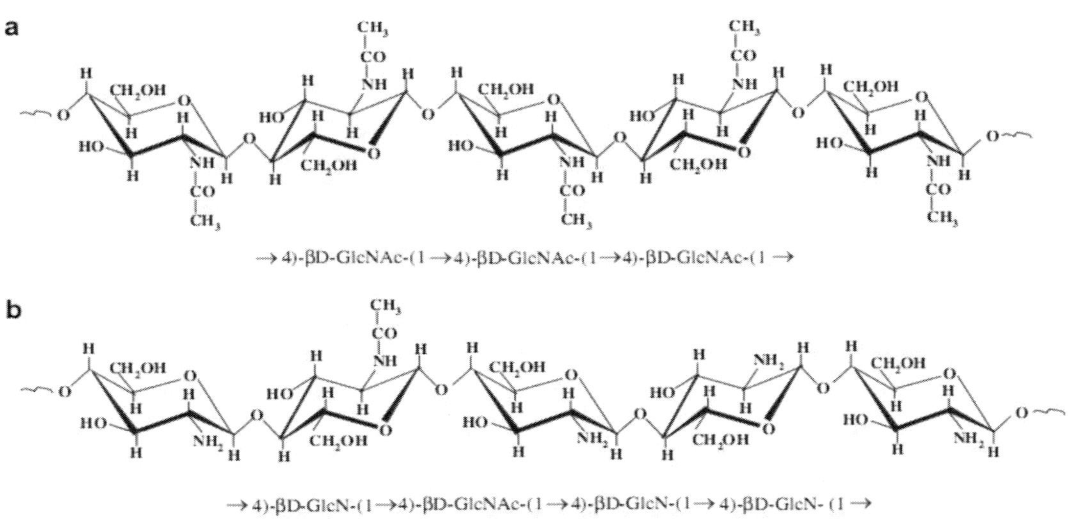

Figure 6. Primary structure of (a) chitin and (b) chitosan [61].

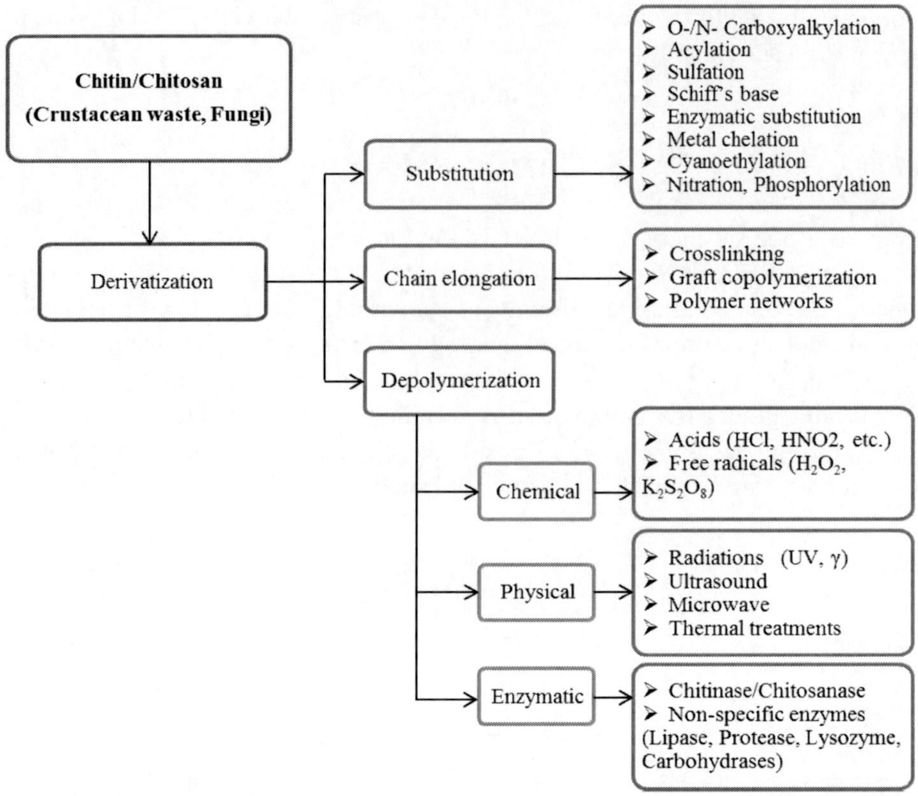

Figure 7. Multifaceted derivatization potential of chitin/chitosan.

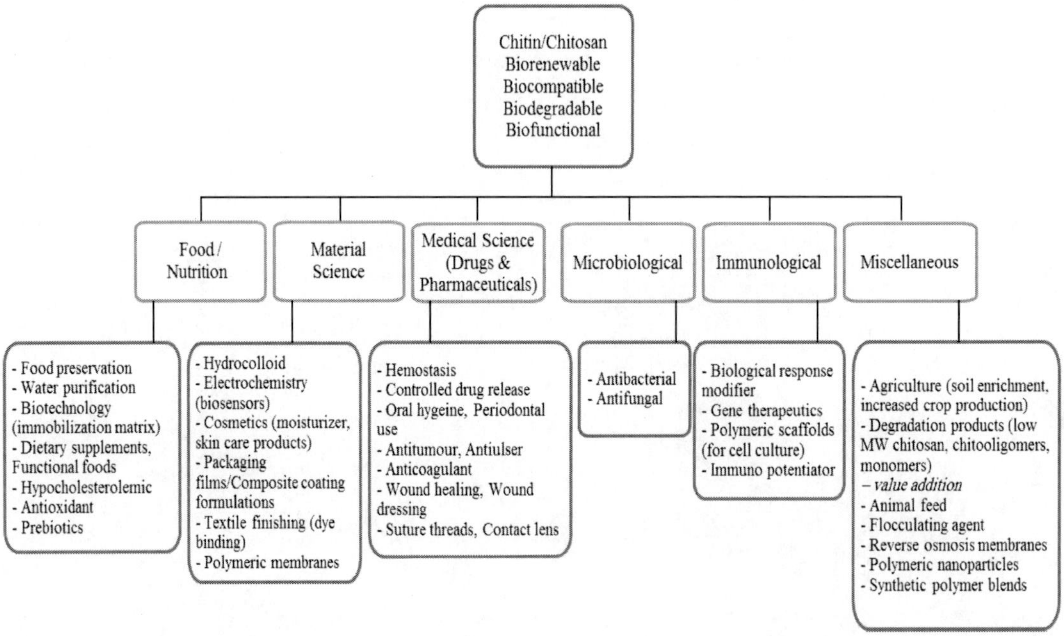

Figure 8. Application potential of chitin/chitosan.

Despite its different advantages, chitosan films have low mechanical strength which can be reinforced by additive like nanoclayparticles [62]. Also its films are highly permeable to water vapor which effects on its application since most of food packaging must control moisture effectively (Figure 7). This problem can be solved by addition of neutral lipids and fatty acid waxes to its films [63].

Chitosan antimicrobial activity has been investigated on many bacteria such as *Staphylococcus aureus, Listeria monocytogenes, Pseudomonasaeruginosa, Escherichia coli, Shigelladysenteriae and Vibrio cholera* [64-67] and fungi like *Sclerotiniasclerotium, Botrytis cinerea ,Moniliniafructicola, Rhizopusstolonifer and Aspergillusniger* [68-71]. The exact mechanism of the antimicrobial action of chitosan is still unknown. However different mechanisms have been proposed such as: a) the charge interaction between positively charged amine group in polymeric chain of chitosan and negatively charged lipopolysaccharides and proteins in the membranes of microbial cells that results release of intracellular materials, b) acting as a chelating agent and binding trace metal to inhibit the production of toxins and microbial growth and c) disruption of cells function by penetrating into cells and interfering the RNA synthesis by binding to DNA[72]. The antimicrobial activity of chitosan depends on some factors like molecular weight, pH, degree of deacetylation and temperature [73].

Chitosan nanoparticles present more effective properties than bulk chitosan due to their higher surface area to volume ratio. As mentioned, the antimicrobial property of chitosan is one of its major characteristics which make it suitable for application in food packaging. By applying nanoparticles instead of large amount of chitosan in food packaging the charge density will increase which provides a higher interaction by the surface of microbes' cell [74].

Furthermore, the size reduction of chitosan to its nanoparticles improves their dispersion in biopolymers which improves their efficiency as an antimicrobial agent in food packaging. Chitosan nanoparticles can be fabricated by various methods such as emulsion droplet coalescence, emulsion cross linking, ionotropic gelation, template polymerization and molecular self-assembly which results in changes in shape and particle size as shown in Figure 8. Nanoparticles are usually prepared by the ionotropic gelation method. In this method, the dissolved chitosan in water or weak acidic medium is added drop wise under constant stirring to the solution of negatively charged groups of polyanion [3, 75].

6. Using of Combinations with Chitosan in Food Packaging

As mentioned, despite of unique characteristics of chitosan for applying in food packaging, its poor mechanical strength and high water vapor permeability forced researchers to use it in a composite system. Furthermore, other properties of chitosan such as its antimicrobial activity and biodegradability can be improved by addition of some additives such as garlic oil, potassium sorbate and nisin [76].

The effect of uniform dispersion of chitosan nanoparticles in a glycerol plasticized starch matrix was investigated and the obvious effect of this combination on tensile strength, water vapor barrier and thermal stability of resulted films was reported [77]. The strength property of chitosan can be improved by applying its combination with cellulose. Because of their structural similarity, their composite benefits from the excellent physicochemical properties of chitosan and the strong mechanical properties of cellulose. Several investigations were performed and were resulted to excellent optical transparency together with enhancement of mechanical properties and thermal stability [78]. Layered silicates nanclays such as sodium montmorillonite (MMT) are also good alternative for improvement of mechanical properties of chitosan films. The chitosan/MMT composite can be fabricated by an ion exchange reaction between soluble oligomeric chitosan and Na MMT [79].

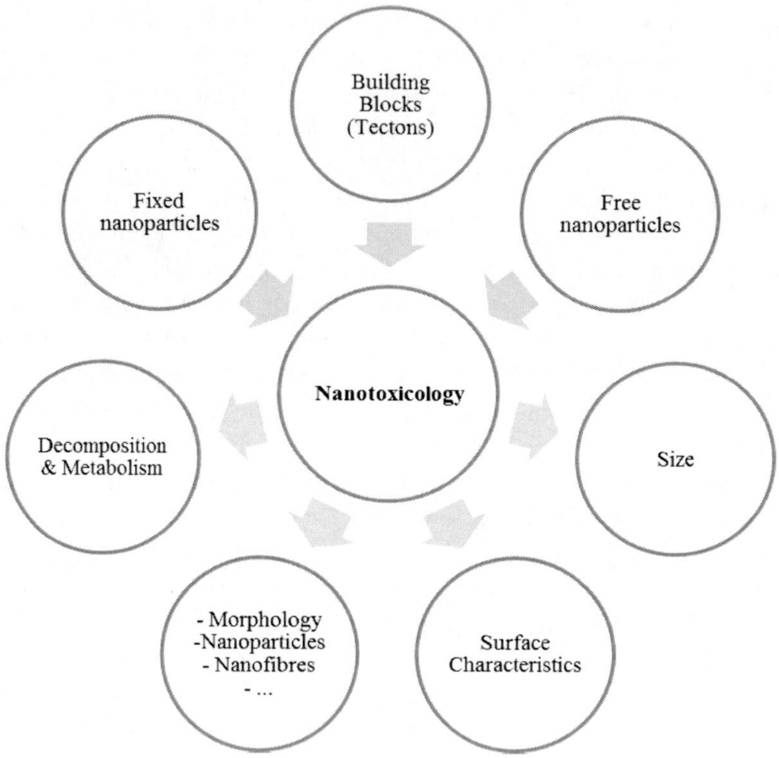

Figure 9. Key nanotoxicology issues relating to the nature of the particles.

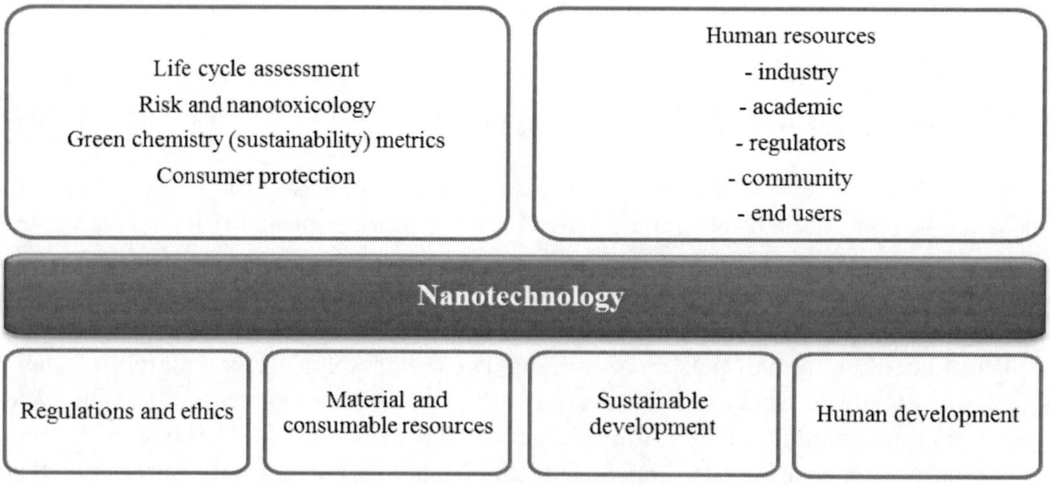

Figure 10. Key issues in dealing with the potential impact of nanotechnology on human health.

The antimicrobial action of chitosan can be improved by addition of some additives to chitosan films. Addition of silver to chitosan was probed against *S. aureus* and it was concluded that their minimum inhibitory concentrations were 500 fold lower than bulk chitosan [80]. ZnO nanofibers which are inherently antimicrobial were mixed with chitosan and the resulted composite was used against *E. coli* and *Candida albicans*. It was observed that *E. coli* and *C. albicans* growth were completely prevented at low minimum inhibitory concentrations (160 μg/ml) [81].

7. Social, Ethical, Regulatory and Toxicological Issues with Nanotechnology

As was already discussed, nanocomposites can be applied in food packaging due to their excellent properties to protect the quality of food. However, it should be considered that properties of particles in nano scale can change. This behavior can significantly influence on health of humans, animals and their environment (see Figure 9). There is no obvious information about the effect of nanoparticles incorporated to food packaging into food or even human bodies by touching [82]. Few investigations were performed to analyze the risk of presence of these extremely small particles in food, human bodies and the environment. Some nanoparticles can penetrate in biological barriers such as blood and brain barriers and potentially enters organs and cells. Their various properties as microbicides and activation catalysts pose a risk if they enter the body what the dangers provides [83].

In addition nanoparticles have more free movement compared to their higher scale materials. This characteristic makes their trace in the body or the environment very difficult. The most important risk which can be mentioned about application of nanotechnology for food packaging is the migration of nanoparticles to food products [84]. This is because of diversity of applications of nanotechnology and the accelerating movements in researches about its utilization for various fields (Figure 10) [18].

Conclusion

Active packaging is a promising method to preserve food products and prolong their shelf life. To date, different kinds of active agents were studied which among them chitosan and its derived materials have shown excellent properties such as biodegradability, antimicrobial activity, antioxidant properties, etc. These unique properties have made chitosan a promising candidate for food packaging applications.

References

[1] Tikhonov, VE; Stepnova, EA; Babak, VG; Yamskov, IA; Palma-Guerrero, J Jansson, H-B; Lopez-Llorca, LV; Salinas, J; Gerasimenko, DV; Avdienko, ID; Varlamov, VP. Bactericidal and antifungal activities of a low molecular weight chitosan and its N-/2(3)-(dodec-2-enyl)succinoyl/-derivatives, *Carbohydrate Polymers*, 64 (2006), 66-72.

[2] Vaghari, H; Jafarizadeh-Malmiri, H; Berenjian, A; Anarjan, N. Recent advances in application of chitosan in fuel cells, *Sustainable Chemical Processes*, 1 (2013), 16.

[3] Malmiri, HJ; Jahanian, MAG; Berenjian, A. Potential Applications of Chitosan Nanoparticles as Novel Support in Enzyme Immobilization, *American Journal of Biochemistry & Biotechnology*, 8 (2012).

[4] Leceta, I; Guerrero, P; Ibarburu, I; Dueñas, MT; de la Caba, K. Characterization and antimicrobial analysis of chitosan-based films, *Journal of Food Engineering*, 116 (2013), 889-899.

[5] Buzek, J; Ask, B. Regulation (EC) No 1223/2009 of the European Parliament and of the Council of 30 November 2009 on cosmetic products, *Official Journal of the European Union L*, 342 (2009).

[6] Cushen, M; Kerry, J; Morris, M; Cruz-Romero, M; Cummins, E. Nanotechnologies in the food industry – Recent developments, risks and regulation, *Trends in Food Science & Technology*, 24 (2012), 30-46.

[7] Neethirajan, S; Jayas, DS. Nanotechnology for the food and bioprocessing industries, *Food Bioprocess Technol*, 4 (2011), 39-47.

[8] Shi, S; Wang, W; Liu, L; Wu, S; Wei, Y; Li, W. Effect of chitosan/nano-silica coating on the physicochemical characteristics of longan fruit under ambient temperature, *Journal of Food Engineering*, 118 (2013), 125-131.

[9] Rhim, J-W; Hong, S-I; Park, H-M Ng, PK. Preparation and characterization of chitosan-based nanocomposite films with antimicrobial activity, *Journal of Agricultural and Food Chemistry*, 54 (2006), 5814-5822.

[10] de Azeredo, HMC. Antimicrobial nanostructures in food packaging, *Trends in Food Science & Technology*, 30 (2013), 56-69.

[11] Cruz-Romero, MC; Murphy, T; Morris, M; Cummins, E; Kerry, JP. Antimicrobial activity of chitosan, organic acids and nano-sized solubilisates for potential use in smart antimicrobially-active packaging for potential food applications, *Food Control*, 34 (2013), 393-397.

[12] Moreira, MdR Pereda, M; Marcovich, NE; Roura, SI. Antimicrobial Effectiveness of Bioactive Packaging Materials from Edible Chitosan and Casein Polymers: Assessment on Carrot, Cheese, and Salami, *Journal of Food Science*, 76 (2011), M54-M63.

[13] Kuorwel, KK; Cran, MJ; Sonneveld, K; Miltz, J; Bigger, SW. Antimicrobial Activity of Biodegradable Polysaccharide and Protein-Based Films Containing Active Agents, *Journal of Food Science*, 76 (2011), R90-R102.

[14] Rhim, J-W; Park, H-M; Ha, C-S. Bio-nanocomposites for food packaging applications, *Progress in Polymer Science*, 38 (2013), 1629-1652

[15] Leceta, I; Guerrero, P; Cabezudo, S; Caba, Kdl. Environmental assessment of chitosan-based films, *Journal of Cleaner Production*, 41 (2013), 312-318.

[16] Dutta, PK; Tripathi, S; Mehrotra, GK; Dutta, J. Perspectives for chitosan based antimicrobial films in food applications, *Food Chemistry*, 114 (2009), 1173-1182.

[17] Joerger, RD. Antimicrobial films for food applications: a quantitative analysis of their effectiveness, *Packaging Technology and Science*, 20 (2007), 231-273.

[18] Restuccia, D; Spizzirri, UG; Parisi, OI; Cirillo, G; Curcio, M; Iemma, F; Puoci, F; Vinci, G; Picci, N. New EU regulation aspects and global market of active and intelligent packaging for food industry applications, *Food Control*, 21 (2010), 1425-1435.

[19] Woranuch, S; Yoksan, R. Eugenol-loaded chitosan nanoparticles: II. Application in bio-based plastics for active packaging, *Carbohydrate Polymers*, 96 (2013), 586-592.

[20] Peng, Y; Wu, Y; Li, Y. Development of tea extracts and chitosan composite films for active packaging materials, *International Journal of Biological Macromolecules*, 59 (2013), 282-289.

[21] Pereira de Abreu, DA; Cruz, JM; Paseiro Losada, P. Active and Intelligent Packaging for the Food Industry, *Food Reviews International*, 28 (2011), 146-187.

[22] De Jong, A; Boumans, H; Slaghek, T; Van Veen, J; Rijk, R; Van Zandvoort, M. Active and intelligent packaging for food: Is it the future?, *Food additives and contaminants*, 22 (2005), 975-979.

[23] Ozdemir, M; Floros, JD. Active Food Packaging Technologies, *Critical Reviews in Food Science and Nutrition*, 44 (2004), 185-193.

[24] Gómez-Estaca, J; López-de-Dicastillo, C; Hernández-Muñoz, P; Catalá, R; Gavara, R. Advances in antioxidant active food packaging, *Trends in Food Science & Technology*, 35 (2014), 42-51.

[25] Siripatrawan, U; Harte, BR. Physical properties and antioxidant activity of an active film from chitosan incorporated with green tea extract, *Food Hydrocolloids*, 24 (2010), 770-775.

[26] Tovar, L; Salafranca, J; Sanchez, C; Nerin, C. Migration Studies To Assess the Safety in Use of a New Antioxidant Active Packaging, *Journal of Agricultural and Food Chemistry*, 53 (2005), 5270-5275.

[27] Torres-Arreola, W; Soto-Valdez, H; Peralta, E; Cárdenas-López, JL; Ezquerra-Brauer, JM. Effect of a Low-Density Polyethylene Film Containing Butylated Hydroxytoluene on Lipid Oxidation and Protein Quality of Sierra Fish (Scomberomorus sierra) Muscle during Frozen Storage, *Journal of Agricultural and Food Chemistry*, 55 (2007), 6140-6146.

[28] Chan, EWC; Lim, YY; Chew, YL. Antioxidant activity of Camellia sinensis leaves and tea from a lowland plantation in Malaysia, *Food Chemistry*, 102 (2007), 1214-1222.

[29] Bolumar, T; Andersen, ML; Orlien, V. Antioxidant active packaging for chicken meat processed by high pressure treatment, *Food Chemistry*, 129 (2011), 1406-1412.

[30] Contini, C; Katsikogianni, MG; O'Neill, FT; O'Sullivan, M; Dowling, DP; Monahan, FJ. Development of active packaging containing natural antioxidants, *Procedia Food Science*, 1 (2011), 224-228.

[31] Contini, C; Álvarez, R; O'Sullivan, M; Dowling, DP; Gargan, SÓ; Monahan, FJ. Effect of an active packaging with citrus extract on lipid oxidation and sensory quality of cooked turkey meat, *Meat Science*, 96 (2014), 1171-1176.

[32] Appendini, P; Hotchkiss, JH. Review of antimicrobial food packaging, *Innovative Food Science & Emerging Technologies*, 3 (2002), 113-126.

[33] Quintavalla, S; Vicini, L. Antimicrobial food packaging in meat industry, *Meat Science*, 62 (2002), 373-380.

[34] Chen, J; Brody, AL. Use of active packaging structures to control the microbial quality of a ready-to-eat meat product, *Food Control*, 30 (2013), 306-310.

[35] Pranoto, Y; Rakshit, SK; Salokhe, VM. Enhancing antimicrobial activity of chitosan films by incorporating garlic oil, potassium sorbate and nisin, *LWT - Food Science and Technology*, 38 (2005), 859-865.

[36] Lee, CH; An, DS; Park, HJ; Lee, DS. Wide-spectrum antimicrobial packaging materials incorporating nisin and chitosan in the coating, *Packaging Technology and Science*, 16 (2003), 99-106.
[37] Kong, M; Chen, XG; Xing, K; Park, HJ. Antimicrobial properties of chitosan and mode of action: A state of the art review, *International Journal of Food Microbiology*, 144 (2010), 51-63.
[38] Suppakul, P; Miltz, J; Sonneveld, K; Bigger, SW. Active Packaging Technologies with an Emphasis on Antimicrobial Packaging and its Applications, *Journal of Food Science*, 68 (2003), 408-420.
[39] Coles, D; Frewer, LJ. Nanotechnology applied to European food production – A review of ethical and regulatory issues, *Trends in Food Science & Technology*, 34 (2013), 32-43.
[40] Neethirajan, S; Jayas, D. Nanotechnology for the Food and Bioprocessing Industries, *Food Bioprocess Technol*, 4 (2011), 39-47.
[41] Sanguansri, P; Augustin, MA. Nanoscale materials development–a food industry perspective, *Trends in Food Science & Technology*, 17 (2006), 547-556.
[42] Degant, O; Schwechten, D. Wheat flour with increased water binding capacity and process and equipment for its manufacture, German Patent DE10107885A1, (2002).
[43] Zhu, K; Huang, S; Peng, W; Qian, H; Zhou, H. Effect of ultrafine grinding on hydration and antioxidant properties of wheat bran dietary fiber, *Food Research International*, 43 (2010), 943-948.
[44] Shibata, T. Method for producing green tea in microfine powder, in, Google Patents, 2002.
[45] Brody, AL; Bugusu, B; Han, JH; Sand, CK; McHugh, TH. Scientific Status Summary, *Journal of Food Science*, 73 (2008), R107-R116.
[46] Roco, MC. Nanotechnology: convergence with modern biology and medicine, *Current Opinion in Biotechnology*, 14 (2003), 337-346.
[47] Sozer, N; Kokini, JL. Nanotechnology and its applications in the food sector, *Trends in Biotechnology*, 27 (2009), 82-89.
[48] Chen, L; Remondetto, GE; Subirade, M. Food protein-based materials as nutraceutical delivery systems, *Trends in Food Science & Technology*, 17 (2006), 272-283.
[49] Shegokar, R; Müller, RH. Nanocrystals: Industrially feasible multifunctional formulation technology for poorly soluble actives, *International Journal of Pharmaceutics*, 399 (2010), 129-139.
[50] Maillard, J-Y., Hartemann, P. Silver as an antimicrobial: facts and gaps in knowledge, *Critical Reviews in Microbiology*, 39 (2013), 373-383.
[51] Duncan, TV. Applications of nanotechnology in food packaging and food safety: Barrier materials, antimicrobials and sensors, *Journal of Colloid and Interface Science*, 363 (2011), 1-24.
[52] Chung, C; Degner, B; McClements, DJ. Controlled biopolymer phase separation in complex food matrices containing fat droplets, starch granules, and hydrocolloids, *Food Research International*, 54 (2013), 829-836.
[53] Rhim, J-W; Kim, Y-T. Chapter 17 - Biopolymer-Based Composite Packaging Materials with Nanoparticles, in: J.H. Han (Ed.) Innovations in Food Packaging (Second Edition), Academic Press, San Diego, 2014, 413-442.

[54] Damm, C; Münstedt, H; Rösch, A. The antimicrobial efficacy of polyamide 6/silver-nano-and microcomposites, *Materials Chemistry and Physics*, 108 (2008), 61-66.

[55] Kumar, R; Münstedt, H. Silver ion release from antimicrobial polyamide/silver composites, *Biomaterials*, 26 (2005), 2081-2088.

[56] Emamifar, A; Kadivar, M; Shahedi, M; Soleimanian-Zad, S. Evaluation of nanocomposite packaging containing Ag and ZnO on shelf life of fresh orange juice, *Innovative food science & emerging technologies*, 11 (2010), 742-748.

[57] Jing, Z; Guo, D; Wang, W; Zhang, S; Qi, W; Ling, B. Comparative study of titania nanoparticles and nanotubes as antibacterial agents, *Solid State Sciences*, 13 (2011), 1797-1803.

[58] Jin, T; Sun, D; Su, J; Zhang, H; Sue, HJ. Antimicrobial efficacy of zinc oxide quantum dots against Listeria monocytogenes, Salmonella enteritidis, and Escherichia coli O157: H7, *Journal of Food Science*, 74 (2009), M46-M52.

[59] Hong, S-I; Rhim, J-W. Antimicrobial activity of organically modified nano-clays, *Journal of nanoscience and nanotechnology*, 8 (2008), 5818-5824.

[60] Woranuch, S; Yoksan, R. Preparation, characterization and antioxidant property of water-soluble ferulic acid grafted chitosan, *Carbohydrate Polymers*, 96 (2013), 495-502.

[61] Harish Prashanth, KV; Tharanathan, RN. Chitin/chitosan: modifications and their unlimited application potential—an overview, *Trends in Food Science & Technology*, 18 (2007), 117-131.

[62] Abdollahi, M; Rezaei, M; Farzi, G. A novel active bionanocomposite film incorporating rosemary essential oil and nanoclay into chitosan, *Journal of Food Engineering*, 111 (2012), 343-350.

[63] Morillon, V; Debeaufort, F; Blond, G; Capelle, M; Voilley, A. Factors Affecting the Moisture Permeability of Lipid-Based Edible Films: A Review, *Critical Reviews in Food Science and Nutrition*, 42 (2002), 67-89.

[64] Arancibia, MY; Alemán, A; Calvo, MM; López-Caballero, ME; Montero, P; Gómez-Guillén, MC. Antimicrobial and antioxidant chitosan solutions enriched with active shrimp (Litopenaeus vannamei) waste materials, *Food Hydrocolloids*, 35 (2014), 710-717.

[65] Fernandes, IP; Amaral, JS; Pinto, V; Ferreira, MJ; Barreiro, MF. Development of chitosan-based antimicrobial leather coatings, *Carbohydrate Polymers*, 98 (2013), 1229-1235.

[66] Bonilla, J; Fortunati, E; Vargas, M; Chiralt, A; Kenny, JM. Effects of chitosan on the physicochemical and antimicrobial properties of PLA films, *Journal of Food Engineering*, 119 (2013), 236-243.

[67] Costa, EM; Silva, S; Pina, C; Tavaria, FK; Pintado, MM. Evaluation and insights into chitosan antimicrobial activity against anaerobic oral pathogens, *Anaerobe*, 18 (2012), 305-309.

[68] Carlson, RP; Taffs, R; Davison, WM; Stewart, PS. Anti-biofilm properties of chitosan-coated surfaces, *Journal of Biomaterials Science, Polymer Edition*, 19 (2008), 1035-1046.

[69] Perdones, Á; Vargas, M; Atarés, L; Chiralt, A. Physical, antioxidant and antimicrobial properties of chitosan–cinnamon leaf oil films as affected by oleic acid, *Food Hydrocolloids*, 36 (2014), 256-264.

[70] Higueras, L; López-Carballo, G; Hernández-Muñoz, P; Gavara, R; Rollini, M. Development of a novel antimicrobial film based on chitosan with LAE (ethyl-Nα-dodecanoyl-l-arginate) and its application to fresh chicken, *International Journal of Food Microbiology*, 165 (2013), 339-345.

[71] Ziani, K; Fernández-Pan, I; Royo, M; Maté, JI. Antifungal activity of films and solutions based on chitosan against typical seed fungi, *Food Hydrocolloids*, 23 (2009), 2309-2314.

[72] Martínez-Camacho, AP; Cortez-Rocha, MO; Ezquerra-Brauer, JM; Graciano-Verdugo, AZ; Rodriguez-Félix, F; Castillo-Ortega, MM; Yépiz-Gómez, MS; Plascencia-Jatomea, M. Chitosan composite films: Thermal, structural, mechanical and antifungal properties, *Carbohydrate Polymers*, 82 (2010), 305-315.

[73] Rao, MS; Chander, R; Sharma, A. Development of Shelf-stable Intermediate-moisture Meat Products Using Active Edible ChitosanCoating and Irradiation, *Journal of Food Science*, 70 (2005), m325-m331.

[74] Xing, K; Chen, XG; Liu, CS; Cha, DS; Park, HJ. Oleoyl-chitosan nanoparticles inhibits Escherichia coli and Staphylococcus aureus by damaging the cell membrane and putative binding to extracellular or intracellular targets, *International Journal of Food Microbiology*, 132 (2009), 127-133.

[75] Zhao, L-M; Shi, L-E; Zhang, Z-L; Chen, J-M; Shi, D-D; Yang, J; Tang, Z-X. Preparation and application of chitosan nanoparticles and nanofibers, *Brazilian Journal of Chemical Engineering,* 28 (2011), 353-362.

[76] Pranoto, Y; Rakshit, S; Salokhe, V. Enhancing antimicrobial activity of chitosan films by incorporating garlic oil, potassium sorbate and nisin, *LWT-Food Science and Technology*, 38 (2005), 859-865.

[77] Chang, PR; Jian, R; Yu, J; Ma, X. Fabrication and characterisation of chitosan nanoparticles/plasticised-starch composites, *Food Chemistry*, 120 (2010), 736-740.

[78] Fernandes, SCM; Freire, CSR; Silvestre, AJD; Pascoal Neto, CM Gandini, A. Novel materials based on chitosan and cellulose, *Polymer International*, 60 (2011), 875-882.

[79] Han, YS; Lee, S-H; Choi, KH; Park, I. Preparation and characterization of chitosan–clay nanocomposites with antimicrobial activity, *Journal of Physics and Chemistry of Solids*, 71 (2010), 464-467.

[80] Ali, SW; Rajendran, S; Joshi, M. Synthesis and characterization of chitosan and silver loaded chitosan nanoparticles for bioactive polyester, *Carbohydrate Polymers*, 83 (2011), 438-446.

[81] Wang, Y; Zhang, Q; Zhang, C-l; Li, P. Characterisation and cooperative antimicrobial properties of chitosan/nano-ZnO composite nanofibrous membranes, *Food Chemistry*, 132 (2012), 419-427.

[82] Reijnders, L. Cleaner nanotechnology and hazard reduction of manufactured nanoparticles, *Journal of Cleaner Production*, 14 (2006), 124-133.

[83] Siegrist, M; Stampfli, N; Kastenholz, H; Keller, C. Perceived risks and perceived benefits of different nanotechnology foods and nanotechnology food packaging, *Appetite*, 51 (2008), 283-290.

[84] Priestly, BG; Bartholomaeus, A; Drew, R. Hazards of Food Contact Material: Nanotechnologies and Nanomaterials, in: Y. Motarjemi (Ed.) *Encyclopedia of Food Safety,* Academic Press, Waltham, 2014, 444-448.

In: Renewable Energy and Sustainable Development
Editors: R. T. Durai Prabhakaran, S. A. Kale et al.
ISBN: 978-1-63463-429-8

Chapter 15

SYMBIOTIC RECYCLING FOR EFFECTIVE WASTE MANAGEMENT

***Gauranga Das*[1*] *and Nimailila Das*[2]**
[1]Director, Govardhan Eco Village, Galtare, Maharashtra, India
[2]Chief Sustainability Officer, Govardhan Eco Village, Galtare, Maharashtra, India

ABSTRACT

Predictions are that waste management is going to be one of the planets biggest concerns in the decades to come. Current waste management techniques like landfills and incineration were devised decades ago when the population density was relatively low. With rising population and steep rise in the volume of waste generated, conventional techniques are proving to be ineffective and ecologically disastrous. In spite of the glaring ecological hazards posed by these techniques they continue to prevail especially in third world nations. Separate segregation and recycling of waste, though a highly effective solution, is not being taken up on a wide-scale due to various technical and financial challenges they pose.

This chapter suggests an immediate solution for urban sewage management and a long term solution in the form of an integrated developmental approach for semi-urban and rural waste management. Both the techniques are based on the concept of "Symbiotic Recycling" - nature's inbuilt mechanism where the waste material from one particular process can be used as a raw material for another process there by overturning the costs of waste management into savings. This chapter presents how by adopting a green technique called as "Soil Biotechnology" a decentralized sewage management system can be effectively implemented without use of any chemicals or energy intensive machinery. Secondly the chapter also discusses the wide-scale applicability of "Symbiotic Recycling" in rural and semi-urban areas by integration of various individual systems comprising of organic farming, cow-barn, biogas plants, composting units and green constructions. Both the techniques have been experimented and adopted in a semi-urban farm community – Govardhan Eco Village (GEV), near Mumbai. At GEV, the philosophy is that nature provides for all our needs and if our lifestyles are in sync with the Nature then nothing in either the production or consumption cycles will be a waste.1. Introduction

* Tel:+91 9930 141 797, E-mail:gaurangadas@ecovillage.org.in.

Keywords: Waste Mangement, Symbiotic Recycling, Soil Biotechnology(SBT), Green Buildings, Organic Farming, Ecological Resilience, Resource-to-trash, Waste-to-resource

Historically it has been observed in various world economies that the manufacturing sector has had a positive influence on the nation's growth rate. Policies governing the manufacturing sector have always aimed towards improving the economy, but have laid little or no emphasis on the impact of such a growth on the ecological infrastructure of the nation. It is a tricky challenge as the systems that are aimed towards improving the economy on the short term, can potentially prove detrimental to the very infrastructure that sustains it in the long term. This dichotomy has created a schism among thinkers who have developed opposing views towards economic sustainability and ecological sustainability, deeming these two paradigms to be incompatible. However for a nation to achieve a balanced growth this dichotomy has to be reconciled. The United Nations in its Millennium development goals [1] set targets for simultaneously improving both the socio-economic conditions of people and ensuring environmental sustainability.

2. Impact of Modern Developmental Systems

The popular model for socio-economic development today is the 2-pronged approach of Industrialization and Urbanization. Industries work on the principle of Demand and Supply. It requires consumer base to prop up the demand while it also needs workforce and natural resources like iron-ore, oil, timber etc., to stroke up the supply. Urbanization solves this problem for industries by providing dedicated workforce and consumer-base clustered around the goods and service industries in the form of cities. This makes a concentrated consumption loop. However the resources needed for running the industries come from distant areas like rural areas, forests, mines, oceans etc. The wastes from these resources are dumped back in these distant areas like landfills, rivers, oceans etc. This can be called the resource loop. The industries extract from the resource loop without contributing any ecological benefit, creating a disconnect between consumption and resource loops. As the need and exploitation of these natural resources is increasing, industries are being forced to explore distant and remote areas for obtaining resources leading to a widening chasm between the 2 loops.

Developmental systems adopted today have failed to acknowledge the aspect of ecological resilience [2] or nature's capacity to absorb disturbance and reorganize while undergoing change so as to still retain essentially the same function, structure, identity, and feedbacks. Our modern developmental systems have been formulated on the premise that nature has abundant resources and have hence adopted a linear model of "resource-to-trash" systems. They take resources from nature, process them into a useful form and at the end of life-cycle dump it back into nature in the form of waste. Wastes are also generated at various stages in this model, whose handling costs are pretty much offloaded into nature. This model has worked for decades and has successfully improved many world economies. However such a model contradicts a basic law of economics, i.e., 'a linear system with definite resources can't continue indefinitely'. Another factor that acts as leakage-hole for this model is that what has been presumed as nature's capacity to produce unlimited inputs and accept unlimited waste, was its mere persistence in the face of changes in ecosystem variables due to anthropogenic causes[3]. Every time we buy a non-biodegradable good, nature heavily subsidizes our purchases. How? Ideally a product lifecycle should span from resources to goods to wastes and finally back to recycled resources. If ‘x’ KJ of energy is needed to convert a set of resources into Goods (which in due course of time degrade into wastes) then, according to the laws of Thermodynamics, the system being irreversible would require more than ‘x’ KJ to convert the wastes back to resources. Now the catch is that we participate in the product life-cycle only in the span from resources to goods to waste and leave the tedious task of converting wastes back to resources on nature. Gradually nature is losing its capacity to continuously subsidize for the increasing wastes.

With more nations of the world adopting this model there is an increasing strain on the planet's ecological resilience resulting in effects like global warming, climate change, loss of biodiversity, food shortage etc. In order to mitigate these effects, a significant amount of financial resources are being spent in the form of environmental protection measures and offering subsidies to the affected populace. Thus the modern developmental systems are straining the economy and proving to be counter productive.

3. Urbanization Impact

A direct consequence of the modern developmental model can be the perceived in the abnormal growth in the urban population. The 2011 census report shows that for the first time in independent India the absolute increase in population is more in urban India than rural India. The trends are no different globally. For the first time in the history of humankind, urbanization has increased at such a massive pace, that more than half the world's population today lives in towns and cities. By 2030 this number is expected to swell to almost 5 billion, with urban growth concentrated in Africa and Asia [4]. This essentially means an undue overload on the urban infrastructure which have not been designed for such inflated numbers. With all modern cities being based on the resource-to-trash systems, one of the key areas that will be badly hit are waste management systems. Predictions are that waste management is going to be one of the planet's biggest concerns in decades to come [5].

With increasing costs of transportation and limitless expansion of cities, the existing models of waste handling are becoming costlier, insufficient and ineffective. For example with water becoming the most in-demand & in-shortage resource, using it indiscriminately for flush based sanitation systems is becoming questionable. In addition clogging and overflow of sewage lines during rains was observed to cause severe problems in all major Indian cities. Current waste management techniques like landfills and incineration were devised decades ago when the population density was relatively low.

With rising population and steep rise in the volume of waste generated the ecological resilience is being tested to its full capacity. Conventional waste management techniques are proving to be ineffective and ecologically disastrous. In spite of the glaring ecological hazards posed by these techniques they continue to prevail especially in third world nations. Separate segregation and recycling of waste, though a highly effective solution, is not being taken up widescale due to the various technical and financial challenges they pose.

4. Symbiotic Recycling Systems

This dichotomy of sustainable economic and ecological development can be harmonized by a close examination of nature, its systems, processes & constituting elements. Nature also has an extensive method of development, evident by the wide variety of life forms it successfully nurtures. The well being of any model can be understood by examining the ease with which the various systems in the model function and how the individual elements constituting each system have a sustained growth.

Nature has a very unique way of waste management. Before we understand that, we need to understand what waste is. Waste can be defined as any object that neither in partial nor in totality contribute to the end goal of a system. Such an object may be produced as a "delayed waste" or as an "immediate waste" in a system. For example, in a manufacturing process, there are wastes generated at various intermediary stages in the process, constituting the "immediate wastes". And the final good produced thereby will also eventually, at the end of its life cycle, be trashed constituting the "delayed waste". The systems comprising the developmental model in nature are in such a way that the wastes from one particular system can be used as raw materials for another system, that way there is no extra energy expended in processing the wastes. In other words, what comes from nature from one system goes back to its proper place in nature in a different system where it is processed and used. Thus the various systems in nature develop a symbiotic dependence on each other and the wastes produced in various stages are recycled within the model. This concept is called Symbiotic recycling.

The modern developmental models are all comprised of disconnect individual systems that follow the resource-to-trash concept. At every stage in this model each system acts like an individual consumer, consuming resources and producing waste. Thus the overall waste produced from this model is very high. If we can formulate models comprising of symbiotically dependent natural systems then we can incorporate an inbuilt mechanism of recycling (a self-sustained feedback loop) and save all hassles of waste management. An in-built reversible system of converting wastes to resources would save the nature of having to pay for ecological Costs. Waste when converted to resources would become an asset & no longer a cost. Such a model then would not be an integration of 'consumers', but a symbiotic integration of 'contributors'.

By developmental models formulated around this concept of Symbiotic recycling we will be able to achieve the desired objective of improving the quality of life and also ensuring environmental sustainability. According to the nature and objectives of any model this concept can be implemented at varying levels of effectiveness. The details of one such successful implementation of Symbiotic recycling in a rural set up is discussed here with.

4. Govardhan Eco Village – A Case Study

Govardhan Eco Village (GEV) is a farm community set up to highlight the importance of living a life in harmony with nature. Unlike the modern day resource-to-trash systems, GEV was modelled to provide a good quality of life by meeting the basic requirements for living like food, shelter, water, waste management etc, by integration of various symbiotically dependent systems. GEV uses specially designed green technologies to ensure comfortable living while maintaining symbiotic harmony with nature. The following table gives few of the techniques employed at GEV:

Sustainable Technologies at GEV

Area	Technology
Alternative Energies	Solar, Bio-Gas, ADPM
Green Building technologies	CSEB, Rammed Earth, Arch Panel & Mangalore Tile Roofing,
Water Resources	Landscaping, ponds, borewells etc. based on Hydrological Survey, Rainwater harvesting, Recycling sewage water in SBT
Organic Farming	Organic Fertilizers & Pesticides, Composting, PRB Farming, Seed Preservation & Perpetuation, Traditional techniques of Rainfall prediction
Waste Management	SBT, Bio-gas Plant, Symbiotic recycling

The following sections describe the organic evolution of different symbiotic systems in GEV in its efforts for attaining sustainable living:

Food

At the heart of this model is organic farming and cow barn, the traditional symbols of sustainable living. The farming system produces enough grass and feed for the animals and the cow barn system produces manure and cow-urine derivatives which provide other necessary ingredients for farming. These symbiotically dependent systems have a built in mechanism of waste management and in return produce grains, vegetables, cloth, natural medicines, milk and other cow products. Thus a self sustainable cycle was established that meets our bare minimum needs.

Housing

Around this basic model we added buildings for human settlements. The challenge was to make buildings that were not only aesthetic and comfortable, but would also create minimal impact on the environment. So after a thorough research, we started making green buildings using an improvised mud brick technology called compressed stabilized mud blocks (CSEB). Unlike the commonly used baked bricks these unfired bricks retain mud's natural property of maintaining optimum temperature. Thus buildings made with these bricks would remaining cool in hot weathers and warm in cool weathers, thus saving a lot of energy in heating and cooling systems. An entire structure made from these bricks have less than 1% the embodied energy as compared to their fired brick counterparts.

Water

So with food and shelter in place our focus was then to ensure a sustainable water source for the village. So we undertook a thorough survey of the ground water flow patterns, called as a hydro-geological survey, and identified ideal locations for extracting water. We also made a facility to recharge the ground water and created various rain water harvesting structures that would serve all the water requirements of our growing community.

Waste Management

With a growing community in place our biggest challenge was waste management. Our intention was to set up symbiotic systems that would harness natures inbuilt concept of recycling within the existing systems for food, housing and water. But at the same time we were also conscious that not any kind of waste could be handled by this natural system, for example highly toxic industrial effluents etc. So we had to simultaneously imbibe in the eco village a culture of using compatible materials. This perhaps is the most important step in the long term sustenance of this model because as much important as it is to create right systems, it is also important to train the people using these systems. Thus the nature of wastes produced in the eco village was mainly fourfold:

1. Animal Waste and other wet wastes like food wastes etc.
2. Dry biodegradable wastes like foliage, paper etc.
3. Dry non-biodegradable wastes like plastics, construction aggregate etc.
4. Human wastes, which constitutes the sewage

To take care of the first category wastes, we added biogas plants that accept wet food wastes and cow dung from the barn system and give out digested slurry, which acts an organic fertilizer input to the farming system. The biogas thus produced is used in cooking thus generating savings on LPG bills. Thus a symbiosis between farming and cowbarn systems was established. We also set up various composting pits that take in other organic wastes like paper, foliage etc., and produce compost for the farming system creating a symbiosis between community living and farming. Construction wastes like broken cement poles and bricks are utilized in making permanent raised beds (PRB) for farming. The PRBs are an innovative way of saving human labor, tractor usage, energy and time before cultivation. The land is prepared by creating these raised beds from cow manure, leaves, compost and soil. These beds are permanent and are more fertile and conducive for growing vegetables and fruits. GEV was recently listed among the top 100 agricultural innovators in India for effective usage of PRBs. The entire boundary of the PRB is made by construction wastes like cement poles and bricks, creating a symbiosis between farming and construction systems. Other construction wastes like quarry dust (an ingredient in making CSEBs) and construction aggregate was used in making cob house construction and in repairing the roads. The broken red bricks are being used in water proofing the roofs in other constructions. Waste card board cartons and cloth are used as mulch in the agriculture field. Mulch is a protective cover placed over the soil to retain moisture, reduce erosion and provide nutrients. Spreading the waste card board cartons and cloth on the land also blocks sunlight, thus preventing growth of any weeds. By using this simple technique one can avoid the labor intensive task of removing weeds or usage of any chemical weedicides. Cement Bags and other plastic bags are utilized to store mud and compost. It is also used to grow plants, esp. grasses like kusha – whose fragrant roots can be easily extracted by cutting the bag open. Wood dust, produced by sawing the wood used for construction, forms an ingredient along with cow dung, in making of chemical free incense sticks. Not only are these incense sticks fragrant, but also have the utility of being a chemical free mosquito repellent. An entire cottage industry is run in GEV which takes as raw ingredients cow dung and cow urine and manufactures various chemical free products like incense sticks, tooth powders, soaps, bathing powder etc.

Need for Safe Sewage Treatment

The final category of waste, namely human waste was a serious issue to be dealt with. Initially we adopted the standard practice of septic tank based sanitation systems. However in due course of time it was found that this process was leading to foul odour, mosquito breeding and stagnation of water due to poor permeability in black cotton soil. Also it was ecologically unethical to dump the contaminated septic tank water into water bodies directly. So we required an alternative technique meeting the following requirements:

1. An ideal sanitation model to run in a self sustaining rural, sub urban and urban community.
2. Saving of ground/surface water resource via recycling of waste water for secondary usage like irrigation of farm
3. Generate photosynthetic biomass for internal use (flower, fruit and excess biomass for green manuring).

5. Soil Biotechnology

Soil Biotechnology [6] (SBT) is an indigenous green technology for sewage processing for recycle and reuse of processed water. SBT has been developed & patented by Chemical Engineering Department, Indian Institute of Technology, Bombay. In this system combined grey water and black water streams are collected and transported via gravity driven underground sewerage network up to the raw water storage tanks. Raw sewage is then pumped and distributed over the SBT bioreactor through a network of pipes. The bioreactor is an impervious containment that incorporates soil, formulated granular filter media, select culture of macro organisms such as earthworms and plants that creates the right ecosytem for the waste processing. Organic solids and liquid wastes of human activity contain both energy and nutrients which, unless harnessed, pollute the environment. Conventional sewage treatment technologies fail to achieve this and as a result the non-utilized nutrients accumulate and disturb the natural balance of soil, water and air through overloading. Salinity, pests, diseases are symptoms of this phenomenon. SBT synergistically engages photosynthesis, respiration and mineral weathering – the three fundamental processes of nature, to process the sewage. This is achieved by soil micro-organisms and regulated by soil macro-organisms viz. geophagus earthworms etc. As a consequence the organic waste constituents of the sewage water are consumed and simultaneously water of desirable quality is produced. The entire process operates in aerobic mode, thus eliminating the possibility of foul odour. SBT removes BOD, COD, ammonia nitrogen, nitrate nitrogen, suspended solids, bacteria, color, odor - all this in a single "all green" system open to atmosphere. The processed water (after multiple passes if necessary) is collected in an intermediate collection tank. The final treated water is recycled by using in the farming and horticulture.

6. SBT Process Validation

The water treated using SBT has the following properties:

1. Always neutral pH
2. Free from any synthetic chemicals
3. Odour free
4. TDS does not build up in the process
5. Complies with the local regulatory treated water standards

The table below gives details of how the SBT process affects individual water quality parameters in samples taken over a period of 2 months.

SBT Data

Parameters	Units	Standard Specifications	Raw Water (Dec '12)	Treated Water (Dec '12)	Raw Water (Jan '13)	Treated Water (Jan'13)
pH	---	5.5 - 9.0	6.28	7.98	6.56	7.48
COD	mg/ltr	250 max	424	42	398	38
BOD 3 days 27°C	mg/ltr	30 max	178	18	168	16
Suspended Solids	mg/ltr	100 max	126	24	118	20
TDS	mg/ltr	2100 max	468	402	482	428
Oil & Grease	mg/ltr	10 max	6	<0.1	4	<0.1

7. Advantages of SBT

Sustainable - Conventional sewage treatment systems are characterized by high mechanization and are highly energy intensive. Consequently the operations & maintenance is very tedious and cost intensive. As a result, in many cases the plants are not in operation, the ground water is polluted and treated water cannot be reused.

SBT stands out as a very sustainable cost-effective solution to the problem of waste water management, specific to the Indian climatic, social and environmental conditions. Besides the process being natural, it depends very little on external energy sources and conserves the same. Ease of installation - An SBT system can be fitted into any garden infrastructure. The civil work involves construction of equalization tank, bar screen chamber, construction of bioreactor containment tanks with suitable partitions to compartmentalize a multi-stage operation and filtrate collection tanks and treated water storage tanks.

The mechanical installations include pumps for distribution of sewage water on bioreactor and then discharge the treated water into storage tank. One pump will be required for each stage of operation. One standby pump per stage can be installed for down time free operation. Recovery Efficiency - The only water loss in SBT systems is due to evapo-transpiration from the plants and soil. Therefore the treated water recovery is very high. Recovery of more than 90% has been observed under all climatic conditions.

Power Saving - Use of SBT system results in substantial saving in electricity cost, as power usage is for pumping alone. The comparative data is as follows for the envisaged capacity of 100 KLD.

Power for SBT system – 36 units per day

Extended Aeration Process – 200 units per day

Total percent saving = (200 – 36) * 100/ 200 = 82% approximately*.

Useful by-products - SBT also facilitates growth of biomass, flowers, biofertiliser apart from the recycled water which is marine compatible.

CONCLUSION

Villages are the backbone of any nation. And if we are looking to take any nation ahead we need to target rural specific developmental models that meet our core requirements of economical and ecological sustainability. The current approach of adapting urban waste management systems to rural and semi urban areas is overkill. Rather one can make use of natures inbuilt systems of symbiotic recycling and thereby add value to the existing processes.

The developmental systems systems showcased in the GEV case study can be expanded upon in a larger scale of operation in rural and semi urban areas and the impact can be studied. However in an urban set up this model will not work due to want of a larger acreage. SBT's functioning, though aligned with nature, is intensive enough to fit in a typical urban scenario that is scarce on space.

An SBT system is quiet & green and forms a perfect solution for decentralized sewage management. Its deceptive look hides the intensive processes that are based on sound principles of terrestrial (soil-plant) ecology. Therefore, the quality of SBT processed products be it treated water or bio-mineral fertilizer, are by far superior to any other systems. Therefore, it qualifies for an ideal zero discharge system even in urban areas.

SBT Comparative chart

Sr. No	PARAMETER	CONVENTIONAL	SOIL BIOTECHNOLOGY
1	pH control	Chemicals (such as Acid/ hydroxides of Ca /Na) required for pH control	No synthetic chemical required. Natural additives are used for simultaneous smell and pH control
2	Smell	Smell due to ammonia and aerosols generated from aqueous phase aeration device	No possibility of aerosol generation. Smell control is achieved via additive addition and high rate ammonia oxidation
3	Green House gases	Green house gas such as carbon dioxide produced is released into atmosphere. Anaerobic process produces other green house gases like methane.	Carbon dioxide produced from organic oxidation is locally trapped in photosynthetic reaction. No Net green house gases are released from the process.
4	Sludge Production	Chemical and biological sludge is produced as waste by product which needs further handling like dewatering and drying for disposal	No sludge is produced. Biomineral fertilizer is produced as useful byproduct.
5	Process loss	15 – 20 percent of water is lost in the process since the water is held within the sludge generated during the process apart from surface loss.	Water loss is only due to evapo-transpiration loss from the filter surface. More than 95 percent recovery is seen.
6	Mechanization	Characterized by high mechanization; high down time Therefore intensive maintenance is required.	Mechanization limited to effluent transfer/distribution pumps only. Practically no process down time.
7	Energy	All conventional aerobic treatment processes are based on aqueous phase reaction and therefore mechanical aeration is highly energy intensive.	Process driven by Natural Aeration in engineered soil ecosystem and therefore no external energy required for aeration; hence energy conservative
8	Byproduct	No useful byproduct is produced.	Harvestable fodder biomass, flowers, biofertiliser apart from fish compatible treated water is produced.
9	Sound Pollution	Due to high mechanization process is characterized by sound pollution	Operates quietly and therefore can be located very close to human habitation.

Sr. No	PARAMETER	CONVENTIONAL	SOIL BIOTECHNOLOGY
10	Tertiary Grade	Tertiary treatment is required subsequent to secondary aerobic treatment to get tertiary grade treated water	Tertiary treatment is achieved in single stage operation. No further treatment required
11	Chemical Intensive	Chemical used for pH control, flocculation etc leads to increase in dissolved solids and sludge.	No synthetic chemical is used in the process. Natural additive is used for process regulation.
12	Operational Issues: manpower	Requires skilled manpower to operation and maintenance	Semi skilled persons can be trained to Operate and maintain.
13	Operational Issues: Start up/Shut down	Start up after prolonged shut down will require heavy expenditure for overhauling	Start up after prolonged shout down is simple and will not require any investment except the green plants.
14	Cost (life cycle, fixed and operating cost	Mechanical components are highly depreciative and therefore cost based on total life cycle is high	Natural system appreciates with time. Due to low on mechanical components the extended life of the system offers low overall lifecycle cost.
15	Aesthetics	Requires secluded place for installation due to smell and other issues	Due to green cover it offers variety of options
16	Green Rating	May not fetch significant rating	Due to item 3, 7, 9, 11 and 14 its green rating for any organization will be substantially high than any other available technology.

REFERENCES

[1] http://www.un.org/millenniumgoals/

[2] Gunderson, L.H.. "Ecological Resilience — In Theory and Application," *Annual Review of Ecology & Systematics* 31: 425, 2000.

[3] Holling, C.S., "Resilience and stability of ecological systems," *Annual Review of Ecology and Systematics* 4: 1–23, 1973.

[4] State of World Population, UNFPA report, 2007.

[5] USEPA report, 2002.

[6] Pattanaik., B.R., "Processing of Wastewater in Soil Filters", Ph.D. Dissertation, *Dept of Chemical Engg.,* IIT Bombay, 2000

In: Renewable Energy and Sustainable Development
Editors: R. T. Durai Prabhakaran, S. A. Kale et al.
ISBN: 978-1-63463-429-8

Chapter 16

LIQUID DESICCANT DEHUMIDIFICATION IN AIR-CONDITIONING SYSTEMS

Abdulrahman Th. Mohammad,[1*] Sohif Bin Mat[2†] and Abduljalil A. Al-abidi[3‡]

[1]Department of Mechanical Engineering, Baqubah Technical Institute, Middle Technical University, Baghdad, Iraq

[2,3]Solar Energy Research Institute, University Kebangsaan Malaysia, Selangor, Malaysia

ABSTRACT

Desiccant dehumidification is the technology for dehumidifying of the air by absorbing the moisture from it. When the moisture absorbs, the air is dehumidified and became most suitable for cooling purpose in humid and hot climates. For this reason, the liquid desiccant dehumidification technology is proposed as an alternative to the traditional vapour compression systems due to its advantage in, removing the air latent load, friendly environment, removing of pollutants from the process air and reduction the electrical energy consumption. This chapter surveys and reviews the liquid desiccant dehumidification concept in air-conditioning system. The chapter first introduces the traditional air-conditioning system. In addition, the principle of liquid desiccants systems is included. This chapter also provides the combination of the liquid desiccant unit with the vapour compression, vapour absorption, direct and indirect evaporative units.

1. INTRODUCTION

The vapor- compression cycle is the most widely used refrigeration cycle in practice. In this cycle, a vapor is compressed, and then condensed into a liquid, following which the pressure is dropped so that fluid can evaporate at a low pressure. The refrigeration cycle

* abd20091976@gmail.com.
† sohif@ukm.my.
‡ abo_anas4@yahoo.com.

requires the addition of external work for its operation. Figure 1 shows the equipment of the refrigeration cycle. The cycle consists of the following processes: (1-2) Adiabatic compression (2-3) Isothermal rejection of heat (3-4) Adiabatic expansion (4-1) Isothermal addition of heat.

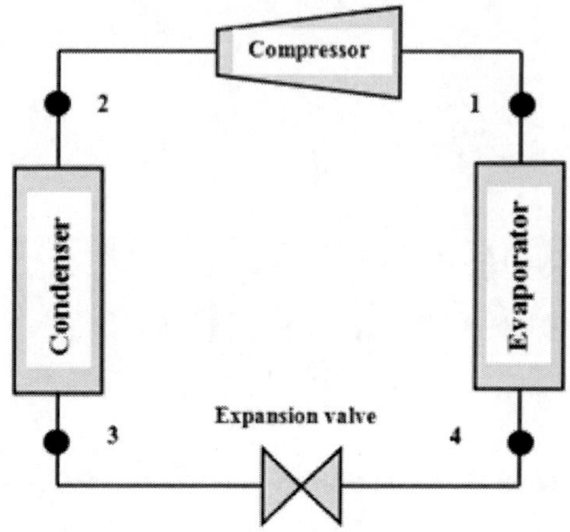

Figure 1. Basic vapor compression.

The refrigeration cycle processes can be represented on pressure-enthalpy diagram as shown in Figure 2:

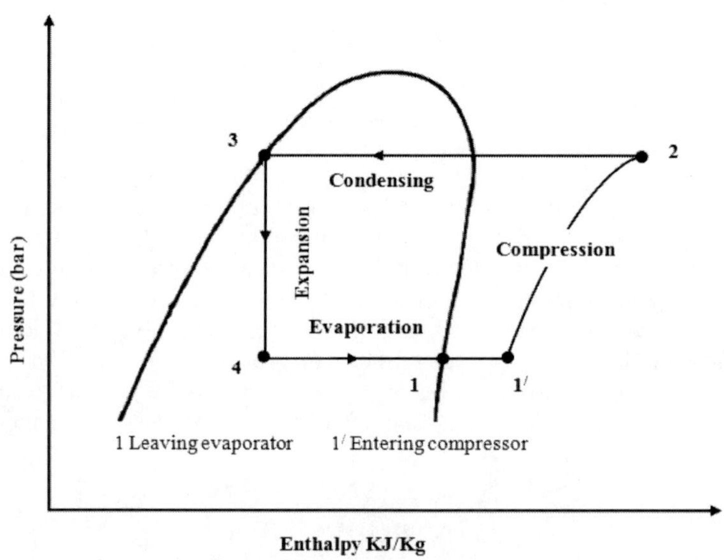

a. The top horizontal line 2-3 represents the condensation process.
b. The vertical line represents 3-4 represents the pressure drop in an expansion device.
c. The bottom horizontal line 4-1 represents the evaporating phase.
d. The horizontal line $1\text{-}1'$ represents the heat gain in the suction line.
e. The diagonal line $1'\text{-}2$ represents the compression of refrigerant gas from suction to the discharge pressure.

Figure 2. Pressure-enthalpy diagram of a refrigeration cycle.

In recent years, many vapor compression systems have been introduced and developed to increase the efficiency of power distribution and to utilize industrial waste heat and renewable clean energy. For reducing the electrical power consumption of the vapor compression system without changing the compressor design; part of the vapor delivered by the compressor is returned back to the compressor. Zubair et al. [1] applied the automatic hot gas bypass technique to reduce the electrical power of refrigeration and air-conditioning systems when operating at part load. Zubair [2] demonstrated the advantages of add-on sub cooler system to improve the performance of refrigeration/air-conditioning systems.

Jianyong and Jianlin [3] obtained a better performance of a new refrigeration cycle for binary non- azeotropic mixed (R22/R134a). The new mixture in a new cycle has a close performance to that is obtainable with R22 and the mixed COP of the system can be improved in a range of 8-9% over that of the conventional refrigeration cycle. Wan et al. [4] performed the thermodynamic and economic analysis of a novel compressed air energy storage refrigeration system; the system is a combination of a gas refrigeration cycle and a vapor compression refrigeration cycle. Liu et al. [5] performed a thermodynamics analysis of an actual air cycle refrigeration. The results show that the pressure ratio, working temperature and isentropic efficiencies of the rotors are factors that effect on the performance of the actual cycle the moast. Toublance and Clausse [6] proposed a novel carnot-type cycle to reach high performances for trans-critical and sub-critical applications. The coefficient of performance of cycle increases by 43-70% when compared to conventional cycle.

2. Liquid Desiccants

The dehumidifier and the regenerator are the main components in liquid desiccant systems. The most common technology today is the packed bed for the dehumidifier/regenerator, however; packed beds must work with high desiccant flow rates in order to achieve good dehumidification levels without internal cooling [7]. The main purpose of the desiccant is to attract the vapor water from the air, and can be classified it into a solid and liquid desiccant. There are several types of solid materials that have the ability to hold water vapor such as: (silicas, polymers, zeolites, aluminas, hydratable salts and mixture). Other available liquid desiccants are lithium chloride (LiCl), lithium bromide (LiBr), calcium chloride ($CaCl_2$), tri-ethylene glycol (TEG), and a mixture of 50% $CaCl_2$ and 50% LiBr. These liquid desiccants have common general requirements including: low vapor pressure, low crystallization point, high density, low viscosity, low regeneration temperature, and low cost.

The vapor pressure of the desiccants increases with an increase of temperature and moisture contents. Figure 3 shows the vapor pressure of lithium chloride as function of solution concentration and temperature. If the desiccant vapor pressure is lower than the vapor pressure of the process air, the moisture content is transferred from the air to the solution.

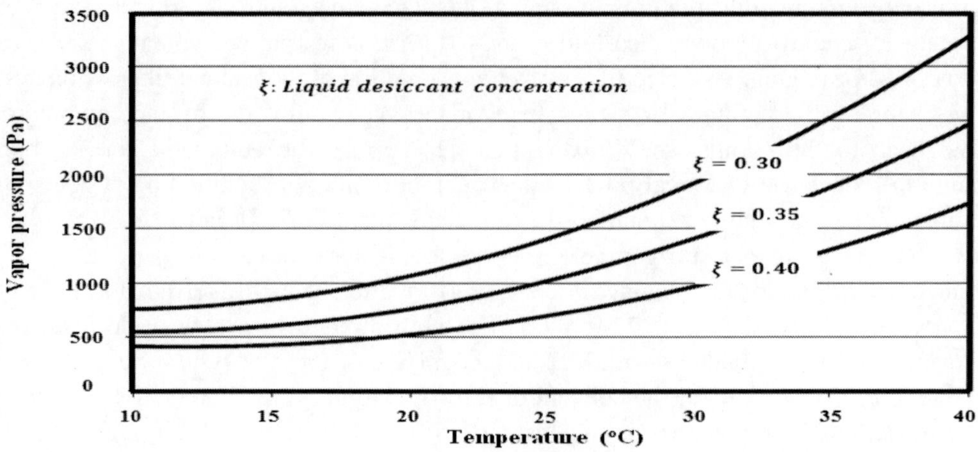

Figure 3. Vapor pressure of LiCl as a function of solution concentration and temperature [8].

The liquid desiccants can be regenerated at a low temperature of about 50°C to 80°C. For this reason, the regeneration process could be driven by solar energy, waste heat or geothermal power [9]. Several works were done to investigate the characteristic of single and mixture of two liquid desiccants, these characteristics can be represented in Table 1.

Table 1. Main characteristics of liquid desiccants

Characteristic	TEG	LiCL and Li Br	CaCl2
Suitability of equilibrium Characteristics	Moderate dehumidification when regenerated at 65-80°C	Good dehumidification with regeneration temperatures above 80°C	Poor dehumidification
	Requires high circulation rate between dehumidifier and regenerator	Does not require high circulation rate between dehumidifier and regenerator	Regeneration possible at temperatures of about 60°C
			Does not require high circulation rate between dehumidifier and regenerator
Loss desiccant through Evaporation	Present during regeneration	No loss	No loss
	Minimal during dehumidification		
Possibility of crystallization	Does not crystallize	Present	Present--more likely than for lithium salts
Corrosion hazard	Moderate, requires inhibitors and hydrogen ion concentration (PH) control	High, requires inhibitors and hydrogen ion concentration (PH) control	Moderate, requires inhibitors and hydrogen ion concentration (PH) control
Toxicity of vapors	Non toxic	Does not evaporate	Does not evaporate
Cost	High	High	Low
Notes: TEG: Tri-ethylene glycol	LiCL: Lithium chloride	LiBr: Lithium bromide	$CaCl_2$: Calcium chloride

3. Heat and Mass Transfer Process

After the air is dehumidified by being brought into contact with strong liquid or solid desiccant, vapor compression, vapor absorption, direct or indirect evaporation cooler are used to provide sensible cooling to dehumidification air. When the solution is weakened by absorption of moisture, it is send direct to regeneration process to release the moisture by using external heat resource. This is called "reactivating"the desiccant [10]. The typical cycle of the desiccant is made up by three processes as shown in Figure 4.

The combination of heat and mass transfer process between the liquid desiccant and the air is very complex and has significant effect on the dehumidifier performance. The dehumidifier performance is measured in terms of the amount of water transfer from the air to the desiccant solution (moisture removal rate). Models for heat and mass transfer in the liquid desiccant dehumidification process are described below according to the flow configuration.

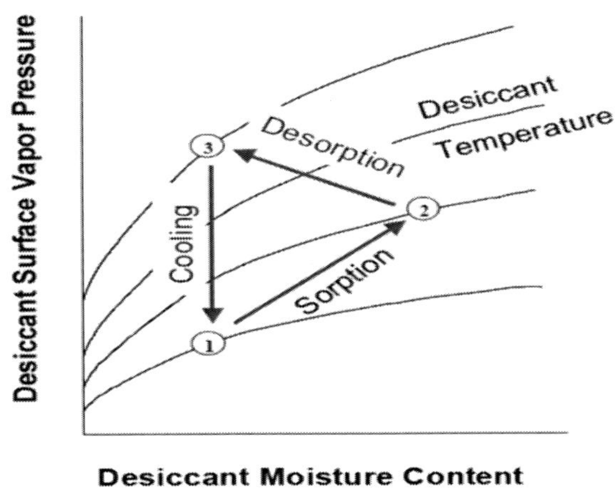

Figure 4. Desiccant Dehumidification and Regeneration Processes [11].

4. A Packed Tower Configration

The liquid desiccant film flows downward over the packing material while the air is passed in different directions through the packing material; there are four basic configurations of a packed tower.

4.1. A Counter-Flow Configuration

The strong liquid desiccant is sprayed at the top of the absorber while the humid air enters the absorber at the bottom in opposite direction. The schematic of the counter flow absorber is shown in Figure 5. The performance of a counter flow configuration tends to be better than the parallel and a cross flow configuration. Therefore a counter flow configurations are drawing more attention in recent years. The ordinary differential equations using a self adaptive integration technique was used by Klimanek & Biatecki [12] to evaluate the distribution of heat and mass transfer sources in a counter flow configuration. Ren [13] developed analytical model to the simple effectiveness-NTU method, the model takes into consideration the effect of nonlinearities of the enthalpy and humidity ratio of air in equilibrium with the solution and water. The effect of absorber packing height was reported in Ani et al. [14] study to determine the optimal performance of a hybrid liquid desiccant system. Fumo & Goswami [15] studied the performance of a packed tower absorber and regenerator for an aqueous lithium chloride desiccant dehumidification system. The rates of dehumidification and regeneration, as well as the effectiveness of the dehumidification and regeneration processes were assessed under the effects of variables such as air and desiccant flow rates, air temperature and humidity, and desiccant temperature and concentration. The packed bed absorption tower was constructed from a 25.4 cm (24 cm I.D.) diameter acrylic tube to allow for flow visualization. The height of the tower is constant and equal to 60 cm. The packing used were 2.54 cm polypropylene Rauschert Hiflow rings with specific surface area of 210 m^2/m^3. Fresh, unused lithium chloride was stored in a tank, and its temperature was adjusted by circulating cold or warm water through a submerged stainless steel coil. Air was blown past an air heater or a cooling.

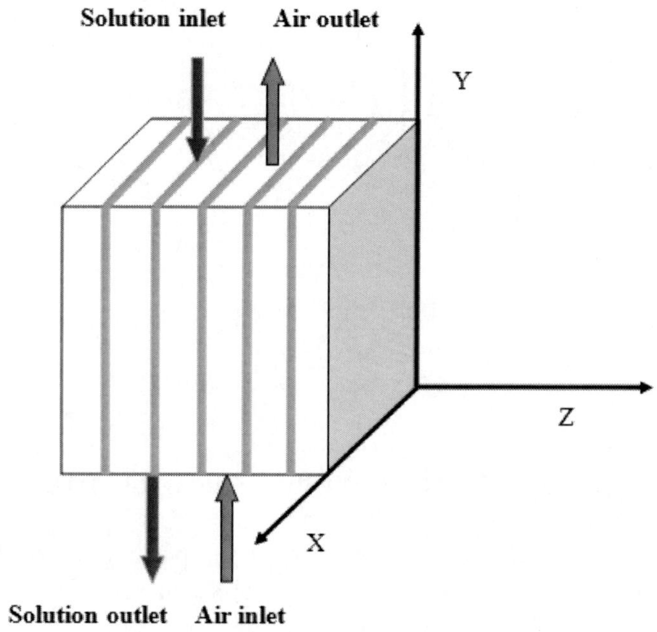

Figure 5. Schematic of a counter–flow liquid desiccant dehumidifier.

Gandhidasan [16] described a simple model to define a suitable dimensionless vapor pressure and temperature difference ratio for liquid desiccant dehumidification, the model predicts the rate of moisture removal from the humid air to the solution. Ren et al. [17] developed a controlling equation to study on the process analysis of adiabatic liquid desiccant dehumidification/regeneration processes. Ren [18] developed one-dimensional model to describe the heat and mass transfer process in both parallel and counter current flow configurations. The effects of Lewis number, changes in solution mass flow rate and concentration, liquid phase heat and mass transfer resistances, and unequal effective heat and mass transfer area are considered. Numerical analysis was used by Shoushi et al. [19] to investigate the heat and mass transfer in aqueous lithium bromide outside the vertical tube using CFD software package. The results show that the maximum heat and mass flux occur at the inlet region.

Liu & Jiang [20] proposed a zonal method to give the heat and mass transfer characteristics of various liquid desiccants in parallel and counter current flow dehumidifier /regenerator. Liu et al. [21] provided and developed analytical solutions to analyze the parameters that effect on the heat and mass transfer performance inside the counter flow, parallel flow and cross flow dehumidifier /regenerator.

4.2. A Parallel-Flow Configuration

As shown in Figure 6, the air flow configuration in the parallel flow dehumidifier is at the same direction of desiccant solution flow.

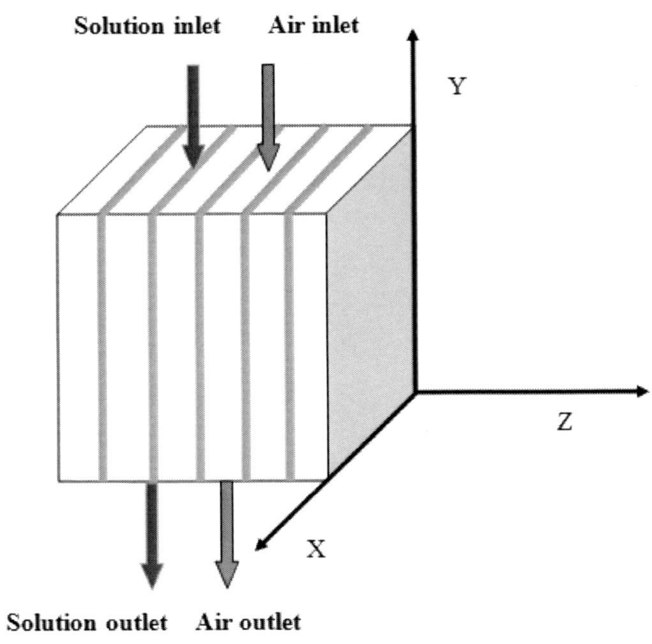

Figure 6. Schematic of a parallel–flow liquid desiccant dehumidifier.

The study of heat and mass transfer processes in a parallel flow dehumidifier/regenerator has revived a little attention compared to study in a counter flow configuration. Rahmah et al. [22] used the control volume approach to predict the outlet parameters of air and liquid desiccant in a parallel flow heat exchanger. Peng & Zhang [23] simulated the heat and mass transfer process in a new solar collector/generator in order to conquer the adverse influence of high humidity ratio of air on regenerator performance. Liu et al. [24] analyzed and compared the heat and mass transfer performances of the two regenerator modes (hot air mode and hot solution mode). In the hot air driven regenerator, the parallel flow configuration has better mass transfer performance than the counter flow configuration while in the hot solution driven regenerator, the counter flow configuration has better mass transfer performance than a parallel flow.

4.3. A Cross-Flow Configuration

The schematic of a cross–flow liquid desiccant dehumidifier/regenerator is shown in Figure 7.

Although the performance of cross-flow liquid desiccant type may be lower than that of the counter-flow type, many studies have performed experimental test on this type also. Moon et al. [26] developed a new empirical correlation method to investigate the performance of a cross-flow liquid desiccant dehumidification system. A cross flow dehumidification with calcium chloride as the liquid desiccant was used. The humid air is drawn to the structured packing tower. Packed tower was used a cross corrugated cellulose structured packing with a flute angle of 45° and height of 4.3 mm, which provides a surface area per unit volume of 608 m^2m^{-3}. The cellulose structured packing pad is wet with liquid desiccant that flows from top to bottom at set flow rates between 0.4-25 m/s. The concentrated liquid desiccant stored in the strong solution tank (400 l) was adjusted to the required concentration. The strong solution was supplied with a solution pump (88 W). The data was corresponded to m_s/m_a ratios of 0.42-2.57.

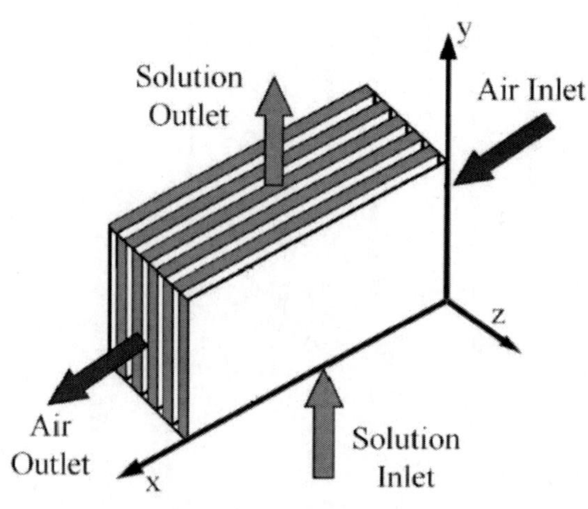

Figure 7. Schematic of a cross – flow liquid desiccant dehumidifier [25].

A theoretical model was used by Liu et al. [27] to predict the air and the solution parameters inside a cross-flow dehumidifier as well as the outlet parameters. Pieneda and Diaz [28] analyzed the performance of a small-scale prototype a cross-flow liquid desiccant dehumidifier installed at a commercial refrigerated warehouse. Liu et al. [29, 30] investigated experimentally the performance of a cross- flow configuration regenerator and dehumidifier in terms of the effectiveness and the moisture removal rate. Celdek structured packing were used in the dehumidifier/regenerator and a LiBr aqueous solution was used as the liquid desiccant. The effects of the inlet parameters, including air and desiccant flow rates, air inlet temperature and humidity ratio and desiccant inlet temperature and concentration, on the two indices were investigated. The packed module was the core device of the setup, in which the cross flowing air and desiccant exchanged heat and moisture. The module used Celdek structured packing with the specific surface area of 396 m^2/m^3. Its height, length and width were 0.55 m, 0.40 m and 0.35 m, respectively. The air handling system, liquid desiccant system and cooling water system provided the air and desiccant to the packed module with certain parameters. The air handling system consisted of the cooling coil, the heater, the humidifier and the fan. The liquid desiccant system consisted of two solution tanks, the solution pump, solution valve and heat exchanger. The cooling water system supplied cooling water to the heat exchanger to adjust the desiccant temperature coming into the packed module.

Liu et al.[31] used analytical solution to investigate the heat and mass transfer processes between the air and desiccant solution in a cross flow packed bed dehumidifier. Wang et al. [32] developed and analyzed a new method called h_d-L_e evaluation method for determining the performance of coupled heat and mass transfer in the liquid desiccant dehumidifier. As reported in Dia and Zhang [33] study, a mathematical model was investigated to determine the performance of honeycomb papers-liquid desiccant dehumidifier. Ali et al. [34] investigated the heat and mass transfer processes between the air and desiccant solution of a cross flow configuration liquid desiccant dehumidifier as well as examined the effects of Cu-ultrafine particles in enhancing heat and mass transfer. Zhang et al. [35] tested experimentally the mass transfer characteristics of liquid desiccant dehumidifier/regenerator with a lithium chloride as a solution. Experimental and theoretical studies was investigated by Saman and Alizadah [36] to describe the performance of a cross flow-heat exchangers as a dehumidifier with indirect evaporative cooler. Liu et al. [37] proposed a simplified approach to estimate the hour-by-hour performance of dehumidifier. They described the heat and mass transfer performance of dehumidifier in terms of enthalpy and moisture effectiveness. Liu and Jaing [38] investigated the performance of a cross flow dehumidifier/ regenerator using calcium chloride solution as a desiccant, as well as analyzed and designed the effects of some operating parameters on the performance of the system.

4.4. A Sprayed Coil Arrangement Configuration

A sprayed coil dehumidifier consists of dehumidification process and sensible cooling process in a single piece of equipment. The water supplied from the cooling tower inside the finned coil while the liquid desiccant spray over the coil. Figure 8 represents Parallel-plate, internally cooled absorber. Khan [39] solved numerically the dehumidification process of an internally-cooling desiccant absorber. Kumar et al. [40] carried out the modeling and simulations of dehumidifier/regenerator performance of liquid desiccant with heat exchanger by using solar cooling technology. The non adiabatic structured packed bed with calcium chloride as a desiccant used by Bansal et al. [41] to improve the quality of indoor air and achieved lower output humidity ratio.

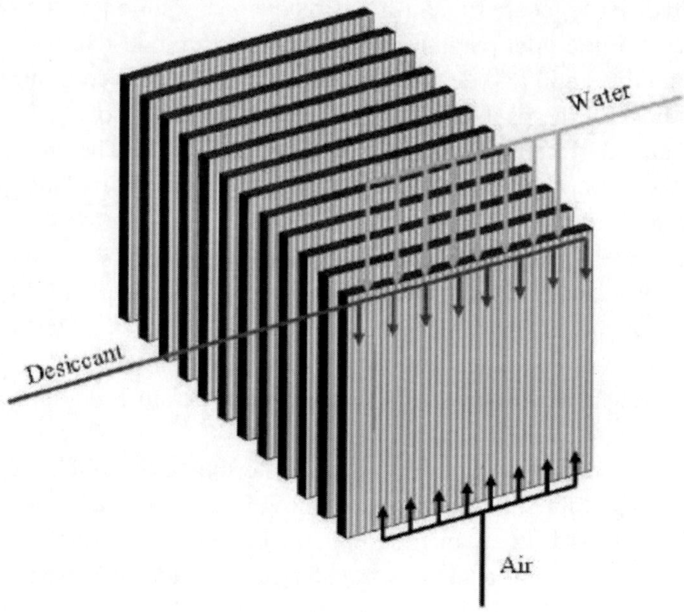

Figure 8. Parallel-plate, internally cooled absorber [42].

5. LIQUID DESICCANT SYSTEMS

The Kathabar-air conditioning system was the first system that used liquid desiccant technology in the United States, in 1910, and the other is the Hygrol system manufactured by Niagara Blower Company, in the 1930s, that consists of two components, the absorber component and regenerator component and used the TEG as a solution to remove frost from evaporator coil in refrigeration systems. Among the current development technology of the liquid desiccant systems, there are different kinds of liquid desiccant air conditioning systems with various types of desiccants have been used in industrial applications. This section briefly surveys the recent researches and development activities in the liquid desiccant systems.

5.1. Pure Liquid Desiccant Systems

The simplest liquid desiccant system configuration is shown in Figure 9, consists of the conditioner, regenerator, heat exchanger for pre-heating the solution and another one for pre-cooling the solution. The solution attracts moisture from the air in the conditioner, which makes the solution weaker. The weak solution is sent through heat exchanger to heated and spray into the regenerator. In the regenerator, the air carries away water vapor from the solution. After this process the solution is sent to the conditioner across heat exchanger for cooling the solution.

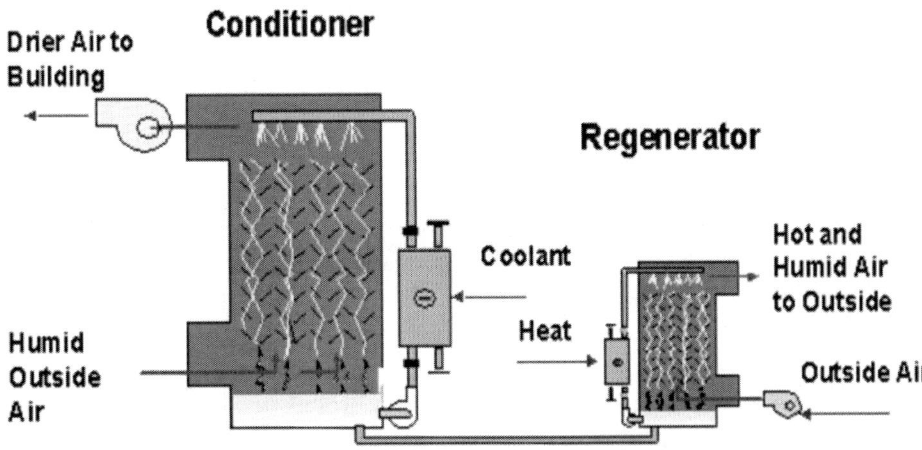

Figure 9. A simple liquid desiccant systems.

Many research works can be found on pure liquid desiccant system. Scalabrin and Scaltriti [43] used internally cooled dehumidifier and heated regenerator to simulate an open process of summer air conditioning. Ameel et al. [44] used internally cooled absorber to investigate an open-cycle-absorption refrigeration system. Jain et al. [45] investigated experimentally study on a liquid desiccant system that had a falling film tubular absorber and a falling film plate regenerator. Some advantages of the liquid desiccant air conditioning were listed by Factor & Gershon [46] as follows: low pressure drop in the contactors with air, requiring lower temperature for regeneration, and possibility of filtering to remove dirt taken from the air. Hwang et al. [47] reported that the absorption cycle is preferred to the adsorption cycle, and the liquid desiccant system is preferred to the solid desiccant system.

Yutong & Yang [48] attempted to obtain the best configurations of an open cycle liquid desiccant dehumidification system. The system uses the counter flow dehumidifier type and solar collector to regenerate the dilute solution. Two loops in system: liquid desiccant dehumidification loop and air dehumidification loop, also the system had two solution tanks connect to the liquid desiccant loop, the first for strong solution and the other for weak solution. Oliveira [49] modeled a new liquid desiccant system uses needle impeller rotors, evaporator and absorber. The cycle without indoor air circulation, air move though needle impeller fan mounted on a shaft, the flexible impeller is constructed by modeling a chimney sweep brush consist of 6 rings each on have 2800, 1 mm thick nylon fibers. The objective of the system were applied several operating conditioning and had the potential to complete with conventional system. Ranjneesh [50] proposed new cycles to improve the COP and carried out experiments to study the impact of various factors on the performance of regenerator and dehumidifier.

Liu et al. [51] designed a new liquid desiccant system using 40% LiCl solution as a desiccant and evaporative cooling capable of treating ambient air at 35 °C to supply air at 18 °C with specific humidity of about 12-13 g/kg, The results show that the COP and cooling capacity was 0.8 at the heat source temperature 70 °C and 40 KW respective. Xiong et al. [52] investigated that the COP of the system increased from 0.24 to 0.73 when used a novel two-stage liquid desiccant dehumidification. Many researchers used chilled water to cool the process air in liquid desiccant dehumidifier, such as [53-57], so the reheat of the process air may be needed for the precise control of room air conditions. Jones [58] installed and tested low-flow liquid desiccant air handling system; a natural gas boiler uses to supply the heat and a cooling tower for heat rejection. COP of the system varied between 0.58 and 4.48. The main components of the system (1) conditioner/regeneration unit (2) cooling tower (3) solar thermal collector (4) gas boiler. In Taiwan, Yang and Wang [59, 60] conducted a parametric study on a 9 m long solar collector and investigated that the double-glazed forced convection suitable in humid climates. These systems have several advantages over solid desiccants, including lower pressure drop of air across the desiccant material, suitability for dust removal by filtration, ease of manipulation and greater mobility. Kumar et al. [61] simulated, analyzed and modified the performance of standalone liquid desiccant system (Jain cycle) [62]. The two major components of the system were absorber and regenerator type falling film, liquid desiccant falls under gravity in the form of a thin film on the side surface of the tubes from the top of absorber /regenerator. Two ways to modify the cycle, the first, by adding one absorber in parallel with the old absorber to reduce the load on the absorber, and the COP increases from 0.24 to 0.4 and another is by adding a third absorber with COP increase to 0.52.

A new regeneration method was proposed by Li & Zhang [63] to regenerate the solution in liquid desiccant system. The regeneration system is named photovoltaic-electro dialysis (PV-ED); a membrane is employed to regenerate the liquid desiccant in an electro dialysis, while solar photovoltaic generator is adopted to supply electric power for this process. The new regeneration method achieves good stability with the immunity against the adverse impact from the outside high humidity; its performance is much higher than that of the thermal regeneration manner while putting aside the low efficiency of the photovoltaic system. Audah et al. [64] studied the feasibility of using a solar–powered liquid desiccant system used calcium chloride as liquid desiccant, parabolic solar concentrators as a heat source for regenerating the liquid desiccant. The liquid desiccant model predicted the amount of condensate obtained from the humid air leaving the regenerator bed when directed through a coil submerged in cold sea water. The optimal regeneration temperature increases with decreased heat sink temperature with values of 50.5 ^{o}C and 52 ^{o}C corresponding to sink temperatures of 19 ^{o}C and 16 ^{o}C. Jain et al. [65] used a liquid desiccant system had a double channeled exchanger for air to liquid desiccant heat and mass transfer; this way provides a large area of mass transfer between the solution and air. The performance of the system was presented in terms of moisture removal rates and efficiency of dehumidifier /regenerator. Most of the physical parameters are measured against a suitable reference point directly. Grossman [66] developed an open-cycle absorption chiller and desiccant system for use with low temperature heat sources such as flat plate collector. The system consists of numbers of major components, an indirect contact evaporative cooler, an air dehumidifier, two air to air heat exchangers and solution to solution heat exchanger. Bakhitar et al. [67] proposed a novel method to evaluate the performance of liquid desiccant air dehumidifier system.

5.2. Hybrid Liquid Desiccant Air Conditioning Systems

Combining the ability of the liquid desiccant dehumidification process for removing the latent load and the ability of the cooling unit to meet a sensible cooling load is defined as a hybrid liquid desiccant air-conditioning system. The dehumidifier subsystem in the hybrid system needs to remove only enough moisture to satisfy the building's needs by reducing the size of the dehumidifier and lowering the required regeneration temperature needed by the desiccant cooling system. When the liquid desiccant system is integrated into a conventional cooling system to remove latent heat, it enables the conventional system to handle the sensible load only. This removal of latent load will reduce the size of compressors, fans, ductworks and thus will reduce the initial equipment cost. Two important advantages of a hybrid liquid desiccant systems are that can reduce the chlorofluorocarbon (CFC) and reduce the electricity peak load.

6. Classification of Hybrid Liquid Desiccant Systems

The main classification of the hybrid liquid desiccant cooling system is based on the cooling units used in the reduction of the dehumidified air temperature. As shown in Figure 10, there are several kinds of cooling units: vapor compression, vapor absorption, and direct or indirect evaporative cooler units. The application of these units depends on the operating conditions such as air temperature, air humidity ratio and the unit cost.

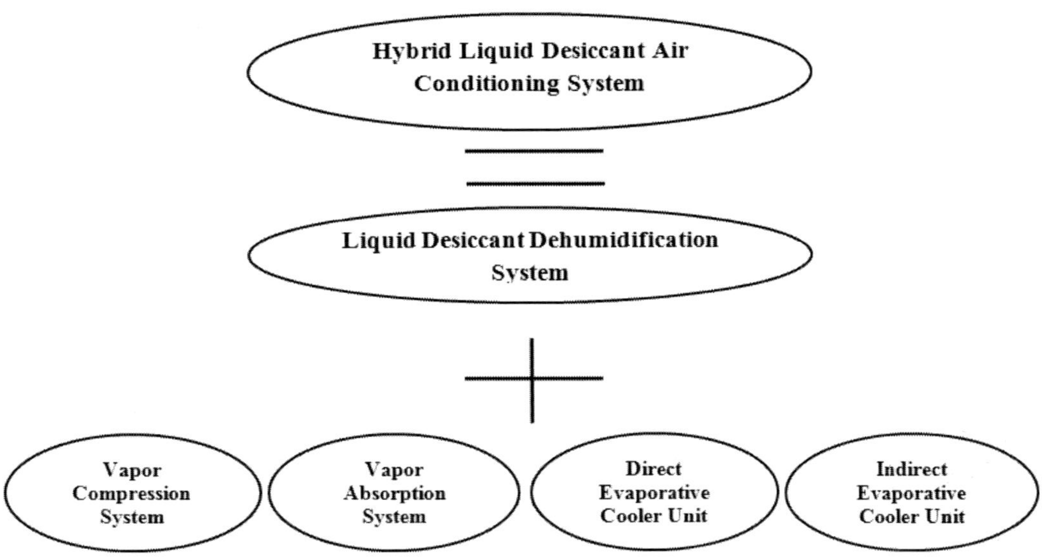

Figure 10. Hybrid liquid Desiccant air conditioning systems classifications.

Conclusion

A liquid desiccant dehumidification removes the latent heat of the processed air and puts it into the liquid desiccant. The review demonstrates that the liquid desiccant dehumidification is a simple technology that can be improved by combining the conventional vapor compression, vapor absorption, and evaporative cooling technologies. One of the important advantages of a liquid desiccant is that it can be removed the latent heat of the process air and regenerated with low temperature by free energy, such as solar energy and waste energy.

In hot and humid regions, Application of the liquid desiccant systems combined with the conventional compression unit minimizes the electric power consumption of VCS and its size. The reason lies to the dehumidification unit that removes of the air moisture content. From the extensive review of the recent literature, it has been seen that the COP of hybrid liquid desiccant cooling has more than the vapour compression system alone by 23.1-73.8% and the energy saving about 26-80%, especially.

REFERENCES

[1] Zubair, S.M., Yaqub, M. & Khan, S.H. 1995. Second-law-based rhermodynamic analysis of hot gas, by pass, capacity-control schemes for refrigeration and air-conditioning systems. *Energy* 20: 483-493.

[2] Zubair, S.M. 1999. Improvement of refrigeration/air conditioning performance with mechanical sub- cooling. *Energy* 15: 427-433.

[3] Jianyong, C. & Jianlin, Y. 2008. Performance of a new refrigeration cycle using refrigerant mixture R32/R134a for residential air-conditioner applications. *Energy and Building* 40: 2022-2027.

[4] Wang, S., Chen, G., Ming, F. & Wang, Q. 2006. A new compressed air energy storage refrigeration system. *Energy Conversation and Management* 47: 3408–3416.

[5] Liu, S., Zhang, Z. & Tian, L. 2011. Thermodynamic analysis of the actual air cycle refrigeration system. *Systems Engineering Procedia* 1: 112–116.

[6] Toublanc, C. & Clausse, M. 2008. Analysis of a novel refrigeration Carnot-type cycle based on isothermal vapour compression. *International Journal of Refrigeration* 31: 1190-1197.

[7] Mesquite, L.C.S., Harrison, S.J. & Thomey, D. 2006. Modeling of heat and mass transfer in parallel plate liquid desiccant dehumidifiers. *Solar Energy,* 80: 1475-1482.

[8] Hueffed, A. K.2007. Asimplified model of heat and mass transfer between air and falling -film desiccant in a parallel-plate dehumidifier. *Master of Science, Mechanical Engineering,* Mississippi State University.

[9] Zhang, X. S., Fei, X. F., Shi, M. H., & Cao, Y. R. 2003. Study on the dehumidifier of the energy storage liquid desiccant cooling systems. *Journal of Southheast University* 33: 1-4.

[10] Mei, V.C., Chen, F.C., Lanan, Z., Collier, R.K., Jr & Meckler, G. 1992. An assessment of desiccant cooling and dehumidification technology, Oak Ridge National Liboratory.

[11] Nayak, S.M. 2005. Experimental and theoretical investigation of integrated engine generator-liquid desiccant system, PhD thesis Faculty of the Graduate School of the University of Maryland, College Park.

[12] Klimanek, A. & Biatecki, R.A. 2009. Solution of heat and mass transfer in counterflow wet-cooling tower fills. *International Communications in Heat and Mass Transfer* 36: 547-553.

[13] Ren, C-Q. 2008. Corrections to the simple effectiveness-NTU method for counterflow cooling towers and packed bed liquid desiccant–air contact systems. *International Journal of Heat and Mass Transfer* 51: 237-245.

[14] Ani, F.N., Badawi, E.M., & Kannan, K.S. 2005. The effect of absorber packing height on the performance of a hybrid liquid desiccant system. *Renewable Energy* 30: 2247-2256.

[15] Fumo, N. & Goswami, D.Y. 2002. Study of an aqueous lithium chloride desiccant system:Air dehumidification and desiccant. *Solar Energy* 72: 351-361.

[16] Gandhidasan, P. 2004. A simplified model for air dehumidification with liquid desiccant. *Solar Energy* 76: 409-416.

[17] Ren, C-Q., Yi J., Guangfa, T. & Yianpin, Z. 2005. A characteristic study of liquid desiccant dehumidification/regeneration processes. *Solar Energy* 79: 483-494.

[18] Ren, C-Q. 2008. Effectiveness–NTU relation for packed bed liquid desiccant–air contact systems with a double film model for heat and mass transfer. *International Journal of Heat and Mass Transfer* 51: 1793-1803.

[19] Shoushi, Bo, Ma, X., Lan, Z., Chen, J. & Chen, H. 2010. Numerical simulation on the falling film absorption process in a counter-flow absorber. *Chemical Engineering Journal* 156: 607-612.

[20] Liu, X.-H. & Jiang, Y. 2009. Handling zone dividing method in packed bed liquid desiccant dehumidification/regeneration process. *Energy Conversion and Management* 50: 3024-3034.

[21] Liu, X.-H., Jian, Y., Xia, J. & Chang, X. 2007. Analytical solutions of coupled heat and mass transfer processes in liquid desiccant air dehumidifier/regenerator. *Energy Conversion and Management* 48: 2221-2232.

[22] Rahamah, A., Elsayed, M.M. & AL-Najem, N. M. 1998. A numerical solution for cooling and dehumidification of air by a falling desiccant film in parallel flow. *Renewable Enery* 13: 305-322.

[23] Peng, D. & Zhang, X. 2009. Modeling and performance analysis of solar air pretreatment collector/regenerator using liquid desiccant. *Renewable Energy* 34: 699-705.

[24] Liu, X.-H., Jing, Y. & Yi, X.Q. 2009. Effect of regeneration mode on the performance of liquid desiccant packed bed regenerator. *Renewable Energy* 34: 209-216.

[25] Afshin, M. 2010. Selection of the Liquid Desiccant in a Run-Around Membrane Energy Exchanger. *Master of Science, Mechanical Engineering,* University of Saskatchewan, Saskatoon.

[26] Moon, C.G, Bansai, P.K. & Jain, S. 2009. New mass transfer performance data of a cross-flow liquid desiccant dehumidification system. *International Journal of Refrigeration* 32: 524-533.

[27] Liu, X.-H., Jiang, Y. & Qu, K.Y. 2007. Heat and mass transfer model of cross flow liquid desiccant air dehumidifier/regenerator. *Energy Conversion and Management* 48: 546-554.

[28] Pineda, S.M. & Diaz, G. 2011. Performance of an adiabatic cross-flow liquid-desiccant absorber inside a refrigerated warehouse *International Journal of Refrigeration* 34: 138-147.

[29] Liu, X.-H., Qu, K.Y. & Jiang, Y. 2006. Empirical correlations to predict the performance of the dehumidifier using liquid desiccant in heat and mass transfer. *Renewable Energy* 31: 1627-1639.

[30] Liu, X.-H., Jian, Y., Chang, X.M. & Yi, X.Q. 2007. Experimental investigation of the heat and mass transfer between air and liquid desiccant in a cross-flow regenerator. *Renewable Energy* 32: 1623-1636.

[31] Liu, X.-H., Jiang, Y. & Qu, K.Y. 2008. Analytical solution of combined heat and mass transfer performance in a cross-flow packed bed liquid desiccant air dehumidifier *International Journal of Heat and Mass Transfer* 51: 4563–4572.

[32] Wang, J., Tang, C., An, S. & Liu, S. 2009. The Theoretical Study and Experimental Analysis for the Regeneration Performance of the Desiccant Solution in the Padding Tower, in International Conference on Energy and Environment Technology 1: 462-465.

[33] Dai, Y.J. & Zhang, H.F. 2004. Numerical simulation and theoretical analysis of heat and mass transfer in a cross flow liquid desiccant air dehumidifier packed with honeycomb paper. *Energy Conversion and Management* 45: 1343-1356.

[34] Ali, A., Vafai K. & Khaled, A.-R.A. 2004. Analysis of heat and mass transfer between air and falling film in a cross flow configuration. *International Journal of Heat and Mass Transfer* 47: 743-755.

[35] Zhang, L., Dang, C. & Hihara, E. 2010. Performance analysis of a no-frost hybrid air conditioning system with integrated liquid desiccant dehumidification. *International Journal of Refrigeration* 33: 116-124.

[36] Saman, W.Y. & Alizadah, S. 2001. Modelling and performance analysis of a cross-flow type plate heat exchanger for dehumidification/cooling. *Solar Energy* 7: 361-372.

[37] Liu, X.-H., Zhang, Y., Qu, K.Y. & Jiang, Y. 2006. Experimental study on mass transfer performances of cross flow dehumidifier using liquid desiccant. *Energy Conversion and Management* 47: 2682-2692.

[38] Liu, X-H. & Jaing, Y. 2009. Handiling zone dividing in packed bed desiccant dehumidification/regeneration process. *Energy Conversion and Mngement* 50: 3024-3034.

[39] Khan, A.Y. 1998. Cooling and dehumidification performance analysis of internally cooled liquid desiccant absorber. *Applied Thermal Engineering* 18:265-281.

[40] Kumar, N., Srivastava, G. & Gangil, B. 2010. Heat and Mass Transfer Study of Liquid Desiccant with Heat Exchanger by Solar Cooling Technology. *International Journal of Engineering Studies* 2: 105-118.

[41] Bansal, B., Jain, S. & Moon, C. 2011. Performance comparison of an adiabatic and an internally cooled structured packed-bed dehumidifier. *Applied Thermal Engineering* 31: 14-19.

[42] Luccio, C.D. 2007. Analysis of a flat-plate, liquid desiccant, dehumidifier and regenerator. Degree of Doctor of Philosophy, Department of Mechanical and Materials, Queen's University.

[43] Scalabrin, G. & Scaltriti, G. 1990. A liquid sorption–desorption system for air conditioning with heat at lower temperature. *Journal of Solar Energy Engineering* 112: 70-75.

[44] Ameel, T.A., K. G. G. & Wood, B.D. 1995. Performance predictions of alternative low cost absorbents for open-cycle absorption solar cooling. *Sol Energy* 54: 65-73.

[45] Jain, S., Dhar, P.L.& Kaushik, S.C. 2000. Experimental studies on the dehumidifier and regenerator of a liquid desiccant cooling system. *Applied Thermal Engineering* 20: 253-267.

[46] Factor, H.M. & Gershon, G. 1980. A packed bed dehumidifier/regenerator for solar airconditioning with liquid desiccant. *Solar Energy* 24: 541-550.

[47] Hwang, Y., Radermacher, R., Al Alili, A. & Isoroku, K. 2008. Review of solar cooling technologies. *HVAC&R Research* 14: 507-528.

[48] Yutong, L. & Yang, H. 2008. Investigation on solar desiccant dehumidification process for energy conservation of central air-conditioning systems. *Applied Thermal Engineering* 28: 1118-1126.

[49] Oliveira, A.C., Afonso, C.F., Riffat, S.B. & Doherty, P.S. 2000. Thermal performance of a novel air conditioning system using liquid desiccant. *Applied Thermal Engineering* 20: 1213-1223.

[50] Ranjneesh, K. 2008. Studies on standalone liquid desiccant based air conditioning systems. PhD thesis. Mechanical Engineering Department, IIT Delhi, India.

[51] Li, C.N., Liu, J.H., Liu, S.C. & Chen, J. 2010. A new air-conditioning system of liquid desiccant and evaporation cooling, Leading Academic Discipline Project of Shanghai Municipal Education Commission, Project No. J50502, China.

[52] Xiong, Z.Q., Dai, Y.J. & Wang, R.Z. 2010. Development of a novel two-stage liquid desiccant dehumidification system assisted by $CaCl_2$ solution using exergy analysis method. *Appl Energy* 87: 1495-1504.

[53] Hellmann, H-M. & Grossman, G. 1995. Simulation and analysis of an open-cycle dehumidifier-evaporator-regenerator (DER) absorption chiller for low-grade heat utilization. *International Journal of Refrigeration* 18: 1177-1189.

[54] Pohi, J.P., Hellmann, H-M. & Grossman, G. 1998. Investigation and comparison of two configurations of a novel open-cycle absorption chiller. *International Journal Refrigeration* 21: 142-149.

[55] Gommed, K., Grossman, G. & Ziegler, F. 2004. Experimental investigation of a LiCl-water open absorption system for cooling and dehumidification. *Journal of Solar Energy -T ASME* 126: 710-715.

[56] Gommed, K. & Grossman, G. 2004. A liquid desiccant system for solar cooling and dehumidification. *Journal of Solar Energy-T ASME* 126: 879-885.

[57] Gommed, K. & Grossman, G. 2007. Experimental investigation of a liquid desiccant system for solar cooling and dehumidification. *Solar Energy* 81: 131-138.

[58] Jones, B. M. 2008. First results of a solar-thermal liquid desiccant air conditioning concept, 1st International Congress on Heating, Cooling and Building, 7th to 10th October 2008.

[59] Yang, R. & Wang, P.-L. 1994. Experimental study of a forced convection solar collector/regenerator for open cycle absorption cooling. *Solar Energy Engineering* 116: 194-199.

[60] Yang, R. & Wang, P-L. 2001. A simulation study of performance evaluation of single-glazed and double-glazed collectors/regenerators for an open cycle absorption solar cooling system. *Solar Energy* 71: 263-268.

[61] Katejanekarn, T., Chirarattananon, S. & Kumar, S. 2009. An experimental study of a solar-regenerated liquid desiccant ventilation pre-conditioning system. *Solar Energy* 83: 920-933.

[62] Jain, S., Dhar, P.L. & Kaushik, S.C. 1994. Evaluation of liquid desiccant based evaporative cooling cycles for typical hot and humid climates. *Heat Recovery System and CHP* 14: 621-632.

[63] Li, X.-W. & Zhang, X.-S. 2009. Photovoltaic-electrodialysis regeneration method for liquid desiccant cooling system. *Solar Energy* 83: 2195-2204.

[64] Audah, N., Ghaddar, N. & Ghali, K. 2011. Optimized solar-powered liquid desiccant system to supply building fresh water and cooling needs. *Applied Energy* 88: 3726-3736.

[65] Jain, S., Tripathi, S. & Das, R.S. 2011. Experimental performance of a liquid desiccant dehumidification system under tropical climates. *Energy Conversion and Management* 52: 2461-2466.

[66] Grossman, G. 2002. Solar-powerd system for cooling, dehumidification and airconditioning. *Solar Energy* 72: 53-62.

[67] Bakhitar, A., Rokhman, F. & Choi, K.H. 2010. A Novel Method to Evaluate the Performance of Liquid Desiccant Air Dehumidifier System," Energy and Buildings, doi:10.1016/j.enbuild.2011.10.011.

About the Editors

R. T. Durai Prabhakaran: R. T. Durai Prabhakaran prior to joining at Advanced Composites Group - University of Strathclyde, UK, worked for five years as a postdoctoral researcher at "Section of Composites and Materials Mechanics", Department of Wind Energy, The Technical University of Denmark (DTU). From 2007-2008, he worked as an Assistant Professor in the Department of Mechanical Engineering at Birla Institute of Technology and Science (BITS – Pilani), Goa campus, India for a period of one and half years. He received his Ph.D. from the Department of Mechanical Engineering, BITS - Pilani in 2007 and obtained Master's degree (M. Tech) from Indian Institute of Technology, New Delhi in 2000. His research field covers experimental investigations and modelling of fibre/matrix interfaces, mechanical characterization of hybrid and natural fibre composites, and materials for wind energy.

Sandip A. Kale: Sandip A. Kale is working as Associate Professor at Mechanical Engineering Department, Trinity College of Engineering and Research, Pune affiliated to University of Pune, India from 2010. He received his Master's Degree and pursuing Ph.D. from University of Pune. He handled various responsibilities in engineering educational institutes and industries from 2000 to 2010. His current research area covers experimental and numerical investigations in wind energy, renewable energy and composite materials.

Kandasamy Prabakar: Kandasamy Prabakar received his M.Sc and Ph.D from Bharathiar University, Coimbatore, Tamil Nadu, India. He was awarded JSPS fellowship during 2004-2006, University of Toyama, Japan. He is working at Pusan National University, School of Electrical Engineering from 2008. His past and current research interest are on structural, optical, interface and electrical properties of semiconducting thin films, pervoskite solar cells, synthesis of quantum dots and wires applied to solar cells and nanophotonic devices.

INDEX

A

B

C

D

E

F

G

H

I

N

O

P

T

U

V

W

Y

Z